Introduction to Microprocessors

Introduction to Microprocessors

Editor:

Nelson Bolívar, Ph.D

ArclerPress

www.arclerpress.com

Introduction to Microprocessors

Editor: Nelson Bolívar, Ph.D

708 3rd Avenue, 6th Floor
New York
NY 10017
United States of America
www.arclerpress.com

ISBN: 978-1-68094-357-3

Library of Congress Control Number: 2016951044

Arcler Press LLC publishes wide variety of books and eBooks. For more information about Arcler Press and its products, visit our website at www.arclerpress.com

About the Editor

Nelson Bolívar, Ph.D

Nelson Bolivar has a PhD. in physics from the University of Lorraine in France finished in 2014. His expertise is in quantum systems and condensed matter. His interest includes spintronic devices and correspondences between general relativity and condensed matter. He is currently an associate professor at the Central University of Venezuela

Contents

List of Contributors ... *xi*

Preface ... *xv*

Introduction ... *xvii*

Chapter 1 Microprocessor Power Modeling Technique ... 1

1. Introduction ... 1
2. The Sesame Environment ... 2
3. Event Signatures ... 2
4. Microprocessor Power Model ... 4
5. Experiments ... 5
6. Related Work ... 7

Conclusions ... 8

Refererences ... 8

Chapter 2 Time-Predictable Computer Architecture ... 11

1. Introduction ... 11
2. Related Work ... 12
3. Wcet Analysis Issues ... 13
4. Time-Predictable Architecture ... 16
5. Evaluation ... 21

Conclusion ... 25

References ... 25

Chapter 3 "Power and Energy Consumption of Nonvolatile Microcontrollers" ... 29

1. Introduction ... 29
2. Features of Our Research ... 29
3. Architecture ... 30
4. Evaluation ... 32

Conclusions ... 38

References ... 38

Chapter 4 Jitter Reduced Self Biased Plls ... 41

1. Introduction ... 41
2. Design of Reduced Jitter Self Biased Pll ... 42
3. Results and Discussions ... 44

Conclusion ... 47

References ... 47

Chapter 5 Generation of Silicon-Based Hardware PLC by Direct Conversion of the Ladder Diagram to Circuit Design Language ... 49

1. Introduction ... 49
2. FPGA-Based PLC Design ... 50
3. Optimization of Ladder Diagram Program with Sequential and Parallel Structure ... 51
4. Implementation of Sequential and Parallel Operations in FPGA ... 55
5. Implementation of the Sequential Operations in VHDL ... 55
6. Case Study ... 57

Conclusions ... 58

Acknowledgment 58
References 58

Chapter 6 **Future Chip Multiprocessors** **59**
1. Introduction 59
2. Related Work 60
3. Cmp Thermal Issues 60
4. Experimental Setup 64
5. Results 66
References 70

Chapter 7 **Design and Architecture of Mixed-Signal Embedded Systems** **73**
1. Introduction 73

Chapter 8 **Design and Implementation of Space Vector PWM Inverter Based on a Low Cost Microcontroller** **75**
1. Introduction 75
2. Experimental Setup 82
Conclusions 84
References 84

Chapter 9 **A Comparison of Fuzzy Logic and Pid Controller for a Single-Axis Solar Tracking System** **87**
1. Background 87
2. Results and Discussion 92
Conclusion 97
Declarations 97
References 97

Chapter 10 **Virtual Prototyping and Performance Analysis of Two Memory Architectures** **99**
1. Introduction 99
2. Overview of Processors Memory Architecture 100
3. Results and Analysis 104
Conclusions 109
References 109

Chapter 11 **A Quantitative Evaluation of a Network on Chip Design Flow for Multi-Core Consumer Multimedia Applications** **111**
1. Introduction 111
2. Related Work 112
3. Instantiation 122
4. Verification 123
5. Digital Tv 123
6. Experimental Results 126
References 132

Chapter 12 **An Ieee 1149.X Embedded Test Coprocessor** **137**
1. Introduction 137
2. Leee 1149.X Standards and Test Command Set 137
3. A Micro-Programmed Architecture for an Ieee 1449.7 Coprocessor 138
4. Test Command Design 142
5. Implementation Cost and Performance 143
Conclusions 145
References 146

Chapter 13	**Modelling Field Bus Communications in Mixed-Signal Embedded Systems**	**147**
1.	Introduction	147
2.	Digital Models	148
3.	Analog Modelling	150
4.	Platform Simulation	155
	Conclusion	158
	References	160
Chapter 14	**Parallel Backprojection: A Case Study in High-Performance Reconfigurable Computing**	**161**
1.	Introduction	161
2.	Background	162
3.	Related Work	163
4.	Experimental Design	163
5.	Experimental Results	169
6.	Discussion	172
	Conclusions	173
	Declarations	173
	References	173
Chapter 15	**Design of Processors with Reconfigurable Microarchitecture**	**175**
1.	Introduction	175
2.	Reconfigurability and Power Proportionality	176
3.	Conditional Partial Order Graphs	176
4.	Design of Reconfigurable Microarchitectures	178
5.	Reconfigurable 8051 Microcontroller	180
	Conclusions	183
	Acknowledgements	183
	References	183
Chapter 16	**Microprocessor Protection Devices**	**185**
1.	Introduction	185
2.	Modern Lines In Designing Microprocessor Protection Devices	185
3.	Offered Solutions	189
	Conclusion	191
	References	192
Chapter 17	**Reliability of Microprocessor-Based Relay Protection Devices**	**193**
1.	Introduction	193
2.	Reliability Myths	193
3.	One More Class of Problems Which MPD Manufacturers Prefer Not to Mention	201
	References	201
Chapter 10	**Small Private Key MQPKS on an Embedded Microprocessor**	**203**
1.	Introduction	203
2.	Related Work	203
3.	Implementation of Small Private Key PKS	207
4.	Results	211
	Conclusions	212
	Acknowledgments	212

Appendix 212
References 213

Chapter 19 Performance of Multi-Core Processors 215
1. Introduction 215
2. Literature Review 216
3. Simulation Methodology and Application 218
4. Results 218
Conclusions 221
References 221

Index 223

List of Contributors

Peter van Stralen
Computer Systems Architecture group, Informatics Institute, University of Amsterdam, Kruislaan 403, 1098 SJ, Amsterdam, The Netherlands

Andy D. Pimentel
Computer Systems Architecture group, Informatics Institute, University of Amsterdam, Kruislaan 403, 1098 SJ, Amsterdam, The Netherlands

Martin Schoeberl
Institute of Computer Engineering, Vienna University of Technology

Dahoo Kim
Graduate School of Information Science and Technology, Hokkaido University, Sapporo, Japan

Itaru Hida
Graduate School of Information Science and Technology, Hokkaido University, Sapporo, Japan

Eric Shun Fukuda
Graduate School of Information Science and Technology, Hokkaido University, Sapporo, Japan

Tetsuya Asai
Graduate School of Information Science and Technology, Hokkaido University, Sapporo, Japan

Masato Motomura
Graduate School of Information Science and Technology, Hokkaido University, Sapporo, Japan

J. Dhurga Devi
Department of ECE, College of Engineering, Anna University, Chennai, India

Daoshan Du
IMS-Mechatronics Laboratory, Department of Mechanical and Aeronautic Engineering*University of California*

Xiaodong Xu
IMS-Mechatronics Laboratory, Department of Mechanical and Aeronautic Engineering*University of California*

Kazuo Yamazaki
IMS-Mechatronics Laboratory, Department of Mechanical and Aeronautic Engineering*University of California*

Kyriakos Stavrou
Department of Computer Science, University of Cyprus, 75 Kallipoleos Street, 1678 Nicosia, Cyprus

Pedro Trancoso
Department of Computer Science, University of Cyprus, 75 Kallipoleos Street, 1678 Nicosia, Cyprus

Sergio Saponara
Dipartimento Ingegneria dell' Informazione, Universita di Pisa, Via G. Caruso 16, 56122 Pisa, Italy `

Pierluigi Nuzzo
EECS Department, University of California at Berkeley, 205 Cory Hall, Berkeley, CA 94720, USA

Paolo D'Abramo
Austriamicrosystems AG, Tobelbaderstrasse 30, Schloss Premstaetten, 8141 Unterpremstaetten, Austria

Luca Fanucci
Dipartimento Ingegneria dell' Informazione, Universita di Pisa, Via G. Caruso 16, 56122 Pisa, Italy `

Mahmoud M. Gaballah
Department of Industrial Electronics and Control Engineering, Faculty of Electronic Engineering, Minufiya University, Menufia 32852, Egypt
Research and development, Sigma Electronic Industries, 9 Obour buildings, Salah Salem St, Cairo, Egypt

Emre Kiyak
Avionics Department, Faculty of Aeronautics and Astronautics, Anadolu University, 26470 Eskisehir, Turkey

Gokhan Gol
Avionics Department, Faculty of Aeronautics and Astronautics, Anadolu University, 26470 Eskisehir, Turkey

Huda S. Muhammad
Schlumberger Corp., Dubai Internet City, Bldg. 14, Dubai, UAE

Assim Sagahyroon
Department of Computer Science and Engineering, American University of Sharjah, Sharjah, UAE

Andreas Hansson
University of Twente, Enschede, The Netherlands

Kees Goossens
Eindhoven University of Technology, Eindhoven, The Netherlands

Ukbagiorgis Iyasu Gebremeskel
Faculty of Technology and Maritime Sciences, Buskerud and Vestfold University College, Kongsberg, Norway

José Manuel Martins Ferreira
Faculty of Technology and Maritime Sciences, Buskerud and Vestfold University College, Kongsberg, Norway

Mohamad Alassir
Universit´e Pierre et Marie Curie – Paris 6, EA 2385, BC 252-4 Place Jussieu, 75252 Paris Cedex 05, France

Julien Denoulet
Universit´e Pierre et Marie Curie – Paris 6, EA 2385, BC 252-4 Place Jussieu, 75252 Paris Cedex 05, France

Olivier Romain
Universit´e Pierre et Marie Curie – Paris 6, EA 2385, BC 252-4 Place Jussieu, 75252 Paris Cedex 05, France

Abraham Suissa
Universit´e Pierre et Marie Curie – Paris 6, EA 2385, BC 252-4 Place Jussieu, 75252 Paris Cedex 05, France

Patrick Garda
Universit´e Pierre et Marie Curie – Paris 6, EA 2385, BC 252-4 Place Jussieu, 75252 Paris Cedex 05, France

Ben Cordes
Department of Electrical and Computer Engineering, Northeastern University, Boston, MA 02115, USA

Miriam Leeser
Department of Electrical and Computer Engineering, Northeastern University, Boston, MA 02115, USA

Andrey Mokhov
School of Electrical and Electronic Engineering, Newcastle University, Newcastle upon Tyne NE1 7RU, UK

Maxim Rykunov
School of Electrical and Electronic Engineering, Newcastle University, Newcastle upon Tyne NE1 7RU, UK

Danil Sokolov
School of Electrical and Electronic Engineering, Newcastle University, Newcastle upon Tyne NE1 7RU, UK

Alex Yakovlev
School of Electrical and Electronic Engineering, Newcastle University, Newcastle upon Tyne NE1 7RU, UK

Vladimir Gurevich
Central Electric Laboratory, Israel Electric Corp., Haifa 31000, Israel

Vladimir Gurevich
Central Electric Laboratory, Israel Electric Corp, Haifa 3100, Israel

Hwajeong Seo
Computer Engineering, Pusan National University, Pusan 609-735, Korea

Jihyun Kim
Computer Engineering, Pusan National University, Pusan 609-735, Korea

Jongseok Choi
Computer Engineering, Pusan National University, Pusan 609-735, Korea

Taehwan Park
Computer Engineering, Pusan National University, Pusan 609-735, Korea

Zhe Liu
Laboratory of Algorithmics, Cryptology and Security, University of Luxembourg, 6 Rue Richard Coudenhove-Kalergi, Luxembourg L–1359, Luxembourg

Howon Kim
Computer Engineering, Pusan National University, Pusan 609-735, Korea

Nitin Chaturvedi
Department of Electrical Electronics Engineering, BITS, Pilani, India

S Gurunarayanan
Department of Electrical Electronics Engineering, BITS, Pilani, India

Preface

Microprocessors are electronics chips that functions as the central processing unit (CPU) of a computer. A processor uses the basic concept of stored program execution. Every microprocessor has its own associated set of instructions that adjust and contains ways of computing and communicating with other hardware. A minimal hypothetical microprocessor contains arithmetic and logic unit (ALU), Instruction decode and control unit, Instruction register, Program counter (PC), clock circuit (internal or external), reset circuit (internal or external) and registers. As integrated circuit technology advanced, it has been feasible to manufacture more and more complex processors on a single chip, allowing more transistors on a single chip and then increasing allowed word sizes from 4- and 8-bit words up to today's 64-bit words. Additional features were added to the processor architecture; more on-chip registers speed up programs, and complex instructions could be used to make more compact programs.

This book, Introduction to Microprocessors, provides a solid foundation in the technology of microprocessors and microcontrollers, their principles and applications. It introduces the practicing engineer to microprocessors and covers topics ranging from components for information processing to hardware structures and addressing modes, along with support software and structured programming. General principles are illustrated with examples from commercial microprocessors. In today's life, we uses microprocessor in exactly all devices. Our computer, calculators and Mobiles are some examples. It is very useful in the field of instrumentation. It is also used in medical instrumentation. Microprocessor based controllers are available in home appliances, such as microwave oven, washing machine etc., microprocessors are being used in controlling various parameters like speed, pressure, temperature etc. These are used with the help of suitable transduction. Microprocessors are being used in a wide range of communication equipments. In telephone industry, these are used in digital telephone sets. Telephone exchanges and modem etc. The use of microprocessor in television, satellite communication have made teleconferencing possible. Railway reservation and air reservation system also uses this technology. LAN and WAN for communication of vertical information through computer network. Microprocessor based microcomputer with software packages has changed the office environment. Microprocessors based systems are being used for word processing, spread sheet operations, storage etc. The microprocessor has transform the publication technology. The usage of microprocessor in toys, entertainment equipment and home solicitations is making them more entertaining and full of features. The use of microprocessors is more widespread and popular. This is an invaluable guide for both students and practicing engineers, as well as researchers. It will lead the reader even beyond the simplest cases to comparative studies and so a better understanding of its functioning. The text is comprised of nineteen chapters.

Editor
Nelson Bolívar, Ph.D

INTRODUCTION

A microprocessors is simply a computer processor that has been configured into the design and function of a microchip. Sometimes referred to as a logic chip, this small component functions as the means of executing the command to start booting up a computer. As part of the process, the microprocessor initiates the activation of all the components necessary to allow the computer to be used, such as waking the operating system.

The basic functionality of this processor is all based on its inherent ability to respond to and generate mathematical and logical operations. This is made possible with the use of registers, within which resides all the data required to execute such basic functions as addition and subtraction. The configuration of the registers also allow the microprocessor to handle tasks like comparing two different numbers and retrieving numbers from various areas and redistributing them.

At the core of the design is a series of logical instructions that regulate the order that tasks are received and executed. The presence of the instructions assure that tasks are carried out in a sequence that will flow from the completion of one task and into the logical initiation of the next task in the sequence. For example, when an end user boots up a computer, the first instruction that will be executed is the activation of the BIOS, or basic input/output system. From that point, the BIOS will load the operating system into the active memory and move on to the next task in the starting process. Each successive task will be achieved once the previous task is complete.

Generally, a microprocessor is a component that remains efficient and productive for the life of the computer. It is possible for the functionality of the chip to become damaged in some manner, however, and when this happens, it is usually an easy process for a computer technician to replace the corrupted chip with a fully functioning one.

Structure

The internal arrangement of a microprocessor varies depending on the age of the design and the intended purposes of the microprocessor. The complexity of an integrated circuit (IC) is bounded by physical limitations of the number of transistors that can be put onto one chip, the number of package terminations that can connect the processor to other parts of the system, the number of interconnections it is possible to make on the chip, and the heat that the chip can dissipate. Advancing technology makes more complex and powerful chips feasible to manufacture.

A minimal hypothetical microprocessor might only include an arithmetic logic unit (ALU) and a control logic section. The ALU performs operations such as addition, subtraction, and operations such as AND or OR. Each operation of the ALU sets one or more flags in a status register, which indicate the results of the last operation (zero value, negative number, overflow, or others). The control logic retrieves instruction codes from memory and initiates the sequence of operations required for the ALU to carry out the instruction. A single operation code might affect many individual data paths, registers, and other elements of the processor.

As integrated circuit technology advanced, it was feasible to manufacture more and more complex processors on a single chip. The size of data objects became larger; allowing more transistors on a chip allowed word sizes to increase from 4- and 8-bit words up to today's 64-bit words. Additional features were added to the processor architecture; more on-chip registers sped up programs, and complex instructions could be used to make more compact programs. Floating-point arithmetic, for example, was often not available on 8-bit microprocessors, but had to be carried out in software. Integration of the floating point unit first as a separate integrated circuit and then as part of the same microprocessor chip, sped up floating point calculations.

Occasionally, physical limitations of integrated circuits made such practices as a bit slice approach necessary. Instead of processing all of a long word on one integrated circuit, multiple circuits in parallel processed subsets of each data word. While this required extra logic to handle, for example, carry and overflow within each slice, the result was a system that could handle, for example, 32-bit words using integrated circuits with a capacity for only four bits each.

With the ability to put large numbers of transistors on one chip, it becomes feasible to integrate memory on the same die as the processor. This CPU cache has the advantage of faster access than off-chip memory, and increases the processing speed of the system for many applications. Processor clock frequency has increased more rapidly than external memory speed, except in the recent past, so cache memory is necessary if the processor is not delayed by slower external memory.

What is a Boot Sector?

A boot sector is the portion of a hard disk or floppy disk that has the code stored on it to boot special programs, and to reference other key features to keep the disk working. There are many types, but there are two main ones: the master boot record and the volume boot record. A master boot record exists on a drive that has been partitioned, and it usually finds the active partition and run its own volume boot record. The volume boot record, in turn, often contains code to run the operating system on the computer.

Hard disk drives contain a master boot record as their first boot sector, while floppy disks or USB disks usually only contain a volume boot record as their first boot sector, as they cannot be partitioned. The BIOS of a computer, the part that runs before anything else, immediately looks to this sector of a drive, whether it is a master or a volume, for instructions on what to do next. The boot sector can actually include instructions to do fairly complex things, which comes in use for things like giving a user the option to run one of many operating systems, but also means there is the potential for abuse in the form of viruses.

To be a boot sector, the sector need meet only one criteria, which is to have a signature of 0xAA55 as its final two bytes. A failure to have this signature can result in an error, and the computer may not complete booting. This can happen for a number of reasons, including a virus, or simply a corrupted sector from a physical error on the drive itself.

This type of virus is simply one that replaces the normal boot sector code with code of its own choosing. Because the boot sector is loaded whenever a computer starts up, such viruses can be incredibly destructive, and in some cases may be quite difficult to properly remove. Since the virus is loaded into memory as soon as the computer starts up, it can also be spread quite easily to every drive or disk that the infected computer comes into contact with.

The most common way that a boot sector virus is spread is by leaving an infected disk in the disk drive of a computer. When it next starts up, the BIOS reads the volume boot record of that disk, receives the virus, and passes it into memory. From there it can spread to other drives, and to other disks inserted. A virus can also be passed over a network, however, if it is not properly protected, and may even be transmitted as an attachment to an email.

Removing a boot sector virus requires a good anti-virus program. Many encrypt the boot sector when they infect it, so that it is not a straight-forward matter to remove the virus. For this reason, it's important to use a good antivirus program that has a registry of many such viruses, and so can carefully remove the virus without causing damage to your computer.

What is a Microchip?

A microchip is a small semiconductor used to relay information via specific electrical characteristics. In some cases, the term can be used interchangeably with integrated circuit. The microchip is at the heart of many electronics, including computers, cell phones and even microwave ovens.

The first microchip is credited jointly to Robert Noyce and Jack Kilby in 1958. Though both were working for different companies and coming at the invention from slightly different angles, the two companies decided both had part of the overall answer and decided to cross license their inventions to come up with one unified piece of technology. After being demonstrated in 1958, it was first available commercially in 1961.

The technology was basic compared to modern standards. The first chip held one transistor, three resistors and one capacitor; modern ones commonly hold millions of transistors in a space smaller than a U.S. penny. The increase in smaller and smaller semiconductor chips has led to numerous other benefits. Beyond being used in electronic gadgets, they can be inserted into biological organisms as well.

The microchip has even been credited as an invention which is used to save lives. Pacemakers use them in order to keep their timing so that they can run hearts efficiently. Mechanical hearts, themselves, use pacemakers to fully take over the function of a biological heart.

As the cost of biologically-centric microchip applications has come down, the number of uses has increased. For example, many pet owners now get them inserted in their animals; if the pet is lost and taken to an animal shelter or veterinary clinic, one of the first things workers will usually do is scan the animal to see if it has a microchip. The chip will reveal the owner's name and contact information.

Likewise, some people have had microchips inserted into themselves. If they are ever incapacitated and need to be taken to the hospital, they are already carrying their full medical history inside their body. These can be easily scanned and the information downloaded for medical professionals to make the most appropriate decisions possible. In the future, it is envisioned people may have these devices implanted that will take care of many of the everyday tasks. Cars may start as the owner approaches, or doors to homes may unlock only for those who have chips programmed with the key.

How do Microprocessors Work?

Microprocessors use a number of different processes to function. Their main purpose is to process a series of numbers placed into sequences which make up a program. Each of these sequences gives some sort of instruction to the microprocessor which, in turn, relates information to other parts of the computer. This facilitates the actions necessary for the program to function. Microprocessors are types of central processing units (CPUs), essentially the central brain of a computer. A microprocessor takes the form of a computer chip that is placed in a motherboard, which operates as the relay center for all the higher functions processed from the CPU.

When a microprocessor is activated, it performs a series of actions, each one defining an exact point of communication. This communication gives instructions in the form of binary code, a series of ones and zeros. The CPU then responds to the instructions by processing the code, taking the necessary actions requested by the code, and relaying to the responsible input section that the action has successfully taken place.

The first step in this process is known as the fetch action. A program will elicit a series of ones and zeroes that define an exact action. Part of the sequence is responsible for informing microprocessors of the location of the necessary code within the program. This is the portion in which random access memory (RAM) is used. The RAM provides the memory for the CPU to be able to hold the instructions long enough for them to be used. When there is not enough RAM in a computer, the computer slows down.

The next step involving the workload of a microprocessor is known as the decoding action. Each set of numbers within the sequence is responsible for a certain action. In order for the CPU to order the correct components to do their jobs, each part of the sequence of numbers must be identified and given the correct operational parameters. For example, if a user is burning a DVD, the CPU needs to communicate certain numerical values to the DVD unit that burns the disk, the hard drive which supplies the information and the video card for display of the status for the user.

Execution is the next step in the function of microprocessors. Essentially, the CPU tells the computer components to do their jobs. During the execution phase, the microprocessor stays in constant contact with the components, making sure each portion of the activity is successfully completed according to the instructions gathered and sent during the previous two steps.

The final action for microprocessors involves the writeback function. This is simply the CPU making a copy of the actions and their results onto the computer's main memory, usually found in the hard drive. The writeback step is essential to determining problematic issues when something goes wrong. For example, if the DVD did not burn correctly, a user can access the writeback files and find out which step occurred without success. These files are placed in a section of the memory known as the registry, which often suffers from increased levels of corruption as redundant actions are completed regularly.

Difference between Vector and Scalar Processors

The biggest difference between vector and scalar processors is how many data items each is handling at once. Computer processing is often a pretty complex science, and understanding how it works on a technical level frequently requires a lot of knowledge and expertise. When it comes to the basic processing types, though, it's often easier to see things more simply. In essence, a vector processor aggregates multiple data points, processing each in turn. It's often really good for complicated tasks that can be broken down into smaller jobs that will respond to a similar instruction. Vector processors are efficient when it comes to getting things done, but this efficiency can cause other parts of the computer system to go slowly. Scalar processors, on the other hand, typically handle just one job at a time, and work on what's basically a point-to-point basis. This type of processor doesn't usually impact the speed of the machine as a whole, but can be slower when it comes to finishing more complicated jobs. Both are important to many sectors, and some computers and devices actually use both simultaneously to maximize efficiency.

Broad Importance of Computer Processing

The part of a computer that allows it to function, at least on a very broad level, is generally known as the central processing unit (CPU). This unit carries out the instructions of various programs; it receives a program's instructions, decodes those instructions, and breaks them into individual parts. It then executes those instructions and reports the results, writing them back into the device's temporary or permanent memory. Processors are usually formatted from the beginning as either vector or scalar.

Scalar Basics

Scalar processors are the most basic type of processor. These usually process just one item at a time, typically integers or floating point numbers. Floating point numbers are numbers that are either too large or small to be represented by integers. According to the scalar system of ordering information, each instruction is handled sequentially. As a result, scalar processing can take up some time.

How Vector Processors Work

In contrast, vector processors typically operate on an array of data points. This means that rather than handling each item individually, multiple items that all have the same instruction can be completed at once. This can save time over scalar processing, but also adds complexity to a system; this can and often does slow other functions. Vector processing usually works best when there is a large amount of data to be processed. In these instances, groups of data and individual data sets can be handled by one instruction.

Startup Times

Vector and scalar processors also differ in their startup times. A vector processor often requires a prolonged startup of the computer because of the multiple tasks being performed. Scalar processors, on the other hand, tend to start a computer in a much shorter amount of time since only single tasks are being executed.

Using the Two Together

Not all computer systems have to use one over the other, and in certain settings the two actually work in tandem. The superscalar processor is one example. This system takes elements of each type and combines them for even faster processing. Using instruction-level parallelism, superscalar processing can perform multiple operations at the same time. This allows for the CPU to perform at a much faster level than a basic scalar processor, without the additional complexity and other limitations of the vector system.

There can be problems with this type of processor, however, as it must determine which tasks can be performed in parallel and which are dependent on other tasks being completed first. Errors in data assignment often lead to crashes and other malfunctions.

CHAPTER 1 MICROPROCESSOR POWER MODELING TECHNIQUE

Peter van Stralen, Andy D. Pimentel

Keywords: High-level power modeling, Profiling, Event signatures

INTRODUCTION

The increasing complexity of modern embedded systems, which are more and more based on heterogeneous multiprocessor-system-on-chip (MP-SoC) architectures, has led to the emergence of system-level design. A key ingredient of system-level design is the notion of high-level modeling and simulation in which the models allow for capturing the behavior of system components and their interactions at a high level of abstraction. As these high-level models minimize the modeling effort and are optimized for execution speed, they can be applied at the early design stages to perform, for example, architectural design space exploration (DSE). Such early DSE is of eminent importance as early design choices heavily influence the success or failure of the final product.

The Sesame modeling and simulation framework [7, 16] facilitates efficient system-level DSE of embedded multimedia systems, allowing rapid performance evaluation of different architecture designs, application to architecture mappings, and hardware/software partitionings. Key to this flexibility is the separation of application and architecture models, together with an explicit mapping step to map an application model onto an architecture model.

Until now, the Sesame modeling and simulation framework has purely focused on the performance analysis of multimedia MP-SoC architectures. Evidently, to make good design trade-offs, also power consumption needs to be taken into account during the process of DSE. Therefore, this paper presents the first step towards including system-level power models in Sesame. More specifically, we elaborate on the concept of event signatures, which was recently introduced in [24], that allows for high-level power modeling of microprocessors (and their local memory hierarchy). This signature-based power modeling operates at a higher level of abstraction than commonly-used instruction-set simulator (ISS) based power models and should thus be capable of achieving good evaluation performance. This is important since ISS-based power estimation generally is not suited for early DSE as it is too slow for evaluating a large design space: the evaluation of a single design point via ISS-based simulation with a realistic benchmark program may take in the order of seconds to hundreds of seconds [12]. Moreover, unlike many other high-level power estimation techniques, our signature-based power modeling technique still incorporates an explicit micro-architecture model of the processor, and thus is able to perform micro-architectural DSE as well.

Using several experiments, we compare the results from our signature-based power modeling with those from Wattch [6] and Sim-Panalyzer [1], which are widely-used ISS-based power analysis tools. Here, we show that with a good underlying power model, the signature-based power modeling technique can yield accurate estimations, while being at least an order of magnitude faster than Wattch or Sim-Panalyzer. In order to perform system-level power modeling of an entire MP-SoC, the next step (not addressed in this paper) will be to extend the power modeling framework with models for the interconnect and possible dedicated components in the MP-SoC.

In the next section, we briefly describe the Sesame framework. Section 3 introduces the concept of event signatures and explains how they can be used for high-level

Citation: Peter van Stralen, Andy D. Pimentel, Microprocessor Power Modeling Technique, doi:10.1007/s11265-008-0301-8.

power modeling of microprocessors. In Section 4, we describe the power models used for modeling different aspects of microprocessors. Section 5 presents a number of experiments in which we compare the results from our models against those from Wattch and Sim-Panalyzer. In Section 6, we describe related work, after which Section 7 concludes the paper.

THE SESAME ENVIRONMENT

To facilitate flexible performance analysis of embedded (media) systems architectures, the Sesame modeling and simulation environment [7, 16] uses separate application and architecture models. An application model describes the functional behavior of an application while the architecture model defines architecture resources and captures their performance constraints. After explicitly mapping an application model onto an architecture model, they are co-simulated via trace-driven simulation. This allows for evaluation of the system performance of a particular application, mapping, and underlying architecture. Essential in this methodology is that an application model is independent from architectural specifics and assumptions on hardware/software partitioning. As a result, a single application model can be used to exercise different hardware/software partitionings and can be mapped onto a range of architecture models, possibly representing different architecture designs or modeling the same architecture design at various levels of abstraction. In Fig. 1, the layered infrastructure of Sesame is illustrated using an example in which a Motion-JPEG encoder application model is mapped onto a bus-based MP-SoC architecture model.

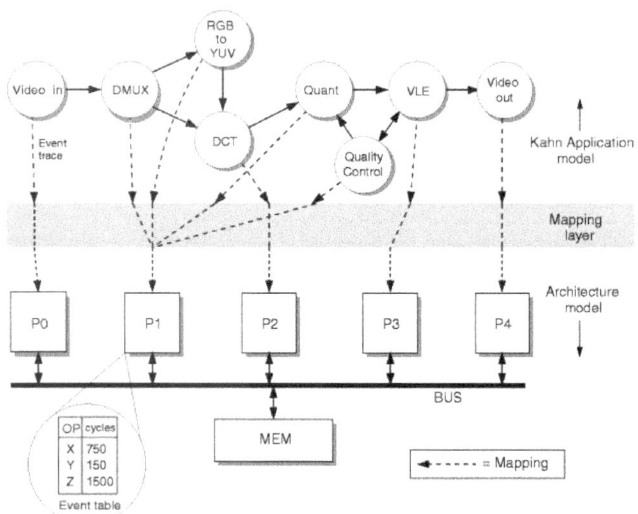

Figure 1 Modeling an Motion-JPEG application on a bus-based MP-SoC architecture in Sesame.

For application modeling, Sesame uses the Kahn Process Network (KPN) model of computation [10], which fits well to the multimedia application domain. In a KPN, parallel processes communicate with each other via unbounded FIFO channels, where reading from these channels is blocking and writing is non-blocking. The computational behavior of an application is captured by instrumenting the code of each Kahn process with annotations that describe the application's computational actions. The reading from and writing to Kahn channels represent the communication behavior of a process within the application model. By executing the KPN model, each Kahn process records its actions in order to generate its own trace of application events. These application events are an abstract representation of the workload that is imposed on the architecture, and drive the underlying architecture model. Application events typically are coarse grained, such as Execute(DCT), which represents the execution of a Discrete Cosine Transform (DCT), or Read(channel_id,pixel-block), which represents the reading of a pixel block from communication channel channel_id.

An architecture model simulates the performance consequences of the computation and communication events generated by an application model. To this end, each architecture model component is parameterized with an event table containing the latencies of the application events it can execute (illustrated for Processor 1 in Fig. 1, where the labels X, Y and Z refer to application event names). The event table entries could, for example, specify the latency of anExecute(DCT) event, or the latency of a memory access in the case of a memory component. The latency values are usually initialized using performance numbers from literature, and can be calibrated using measurements on available hardware or via lower-level simulations of architecture components [17]. This trace-driven simulation approach allows to quickly assess different HW/SW partitionings by experimenting with the latency parameters of processing components in the architecture model: a low computational latency refers to a HW implementation while a high latency mimics a SW implementation.

To bind application tasks to resources in the architecture model, Sesame provides an intermediate mapping layer. This layer controls the mapping of Kahn processes (i.e. their event traces) onto architecture model components by dispatching application events to the correct architecture model component. The mapping also includes the mapping of Kahn channels onto communication resources in the architecture model.

Extending the Sesame framework to also support power modeling could be done fairly easily by adding power consumption numbers to the event tables. So, this means that a component in the architecture model not only accounts for the timing consequences of an incoming application event, but also accounts for the power that is consumed by the execution of this application event (which is specified in the event tables now). The power numbers that need to be stored in the event tables can, of course, be retrieved from lower-level power simulators or from (prototype) implementations of components. However, simply adding fixed power numbers to the event tables would be a rigid solution in terms of DSE: these numbers would only be valid for the specific implementation used for measuring the power numbers. Therefore, we propose a high-level power estimation method based on so-called event signatures that allows for more flexible power estimation in the scope of system-level DSE.[1] As will be explained in the next sections, signature-based power estimation provides an abstraction of processor activity in comparison to traditional ISS-based power models, while still incorporating an explicit micro-architecture model and thus being able to perform micro-architectural DSE.

EVENT SIGNATURES

An event signature is an abstract execution profile of an application event that describes the computational complexity of the application event (in the case of computational events) or provides information about the data that is communicated (in the case of communication events). Hence, it can be considered as meta-data about an application event. In this paper, we purely focus on signatures for computational application events (i.e., Execute() events). The signatures for these events describe computational complexity in a (micro-)architecture independent fashion using an Abstract Instruction Set (AIS). Currently, our AIS is based on a load-store architecture and consists of only a small set of abstract instructions, or ratherinstruction classes, such as ‘Simple Integer Arithmetic’, ‘Simple Integer Arithmetic Immediate’, ‘Integer Multiply’, ‘Branch’, ‘Load’, and ‘Store’. The high level of abstraction of the AIS should allow for capturing the computational behavior for a wide range of processors with different instruction-set architectures. To construct the signatures, the real machine instructions that embody an application event (derived from an ISS as will be explained below) are first mapped onto the various AIS instruction classes, after which a compact execution profile is made. This means that the resulting signature is a vector containing the instruction counts of the different AIS instruction classes. Here, each index in this vector specifies the number of executed instructions of a certain AIS class in the application event. We note that the generation of signatures for each application event is a one-time effort, unless e.g. an algorithmic change is made to an application event’s implementation.

In Fig. 2, signature-based power modeling is illustrated. The Kahn application process for which a power estimation needs to be performed, is simulated using Sim-Cache, which is part of the SimpleScalar simulator suite [2]. Using this relatively fast simulator, the event signatures are constructed—by mapping the executed machine instructions onto the AIS as explained above—for every computational application event that can be generated by the Kahn process in question. The event signatures act as input to our parameterized microprocessor power model, which will be described in detail in the next section. For each signature, Sim-Cache (or any other ISS that could be used to construct signatures) may also provide the power model with some additional micro-architectural information, such as cache missrates, branch misprediction rates, etc. In our case, only instruction and data cache missrates are used. The microprocessor power model also uses a micro-architecture description file in which the mapping of AIS instructions to usage counts of microprocessor components is described. An example fragment of this mapping description is shown in Fig. 3. It specifies that for every AIS instruction (indicated by the ‘ALL’ tag), the instruction cache (il1) is read, the register update unit (RUU) is read and written, and branch prediction is performed. Furthermore, it specifies that for the AIS instruction ‘LOAD’, the ALU is used (to calculate the address), the level-1 data cache (dl1) is accessed, and that the integer register file (irf) is read and written. With respect to the latter, it takes register and immediate addressing modes into account by assuming 1.5 read operations to the irf on average. In addition, the micro-architecture description file also contains the parameters for our power model, such as e.g. the dimensions and organization of memory structures (caches, register file, etc.) in the microprocessor, clock frequency, and so on. Clearly, this micro-architecture description allows for easily extending the AIS and facilitates the modeling of different micro-architecture implementations.

The above ingredients (the event signatures, additional micro-architectural information per signature such as cache statistics, and the micro-architecture description of the processor) allow the power model to produce power consumption estimates for each computational application event. These estimates can be performed on-the-fly during a Sesame system simulation: application events generated by an application model, together with their signatures and additional micro-architectural information, are used at the architecture model level by the power model to dynamically estimate the power consumption. Alternatively, the power consumption of the different types of application events can also be estimated off-line (i.e., statically) after which these estimates are stored in Sesame’s event tables at the architecture model level (as discussed in Section 2).

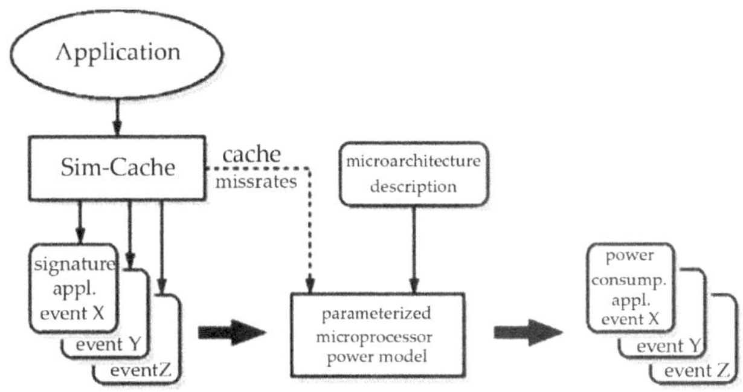

Figure 2. Signature-based power modeling.

```
<node name="ALL" class="AIS">
  <port name="bpred" dir="in">
    <property name="execute" value="1" />
  </port>
  <port name="il1" dir="in">
    <property name="read" value="1" />
  </port>
  <port name="RUU" dir="in">
    <property name="read" value="1" />
    <property name="write" value="1" />
  </port>
</node>
<node name="LOAD" class="AIS">
  <port name="dl1" dir="in">
    <property name="read" value="1" />
  </port>
  <port name="irf" dir="in">
    <property name="read" value="1.5" />
    <property name="write" value="1" />
  </port>
  <port name="alu" dir="in">
    <property name="execute" value="1" />
  </port>
</node>
```

Figure 3 Mapping AIS instructions to usage counts of the microprocessor components.

We note that the generation of event signatures can also be performed either statically or dynamically. In static signature generation, Sim-Cache measures the average instruction execution behavior of code fragments that represent application events and constructs the signatures based on these averages. So, in this case, the signature generation takes place entirely off-line. In dynamic signature generation, the signatures are constructed on-the-fly for every application event. This means that the signatures of the same type of application events may change over time due to e.g. data dependent execution behavior inside the code of these events. Another consequence of dynamic signature generation is that Sim-Cache must be co-simulated together with Sesame. The above reasoning also holds for the additional micro-architecture information, like cache missrates, that can be provided by Sim-Cache to the power model. This can also be done statically (i.e., average based) or dynamically (i.e., exact based).

MICROPROCESSOR POWER MODEL

The microprocessor model that underlies our power model is depicted in Fig. 4. It is a dynamic pipelined machine, consisting of one arithmetic logical unit, one floating point unit, a multiplier and two levels of caches. For the sake of simplicity, it does not use a hardware branch prediction module. This means that we currently assume taken, not-taken or perfect branch prediction. The level 1 caches are virtually addressed and each have a TLB for address translation. The level 2 cache is unified. All requests for this cache are issued to a load/store queue. The same is true for level-2 cache requests to the main memory.

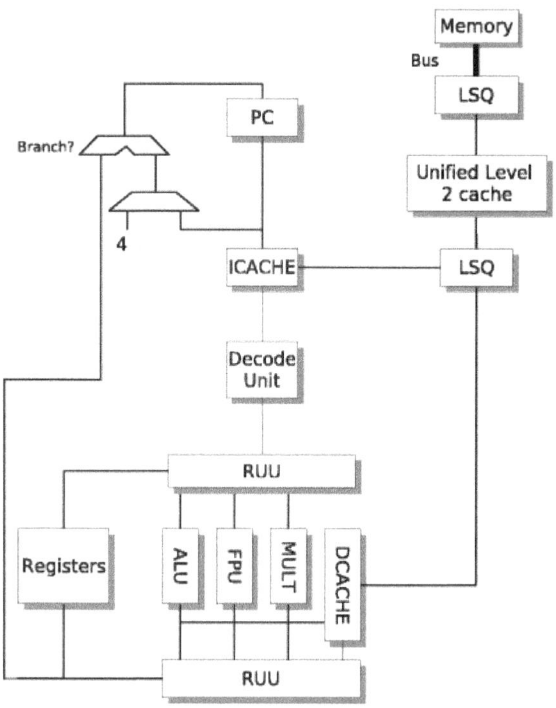

Figure 4 The underlying microprocessor model for our power models.

The dynamic pipeline is implemented using a RUU. This unit controls the dispatching of instructions to the functional units and, after execution, the RUU takes care of the (in-order) instruction commitment. This means that in Fig. 4 the two displayed RUUs are one and the same. We believe that the above microprocessor model that underlies our power model can reflect the behavior of popular embedded processors (like ARM10/11 or MIPS4000 based processors) in a fairly realistic fashion, albeit at a high level of abstraction.

The various components that constitute our microprocessor power model—which are of course based on our underlying microprocessor model as depicted in Fig. 4—are shown in Fig. 5. The power consumption of an application event is calculated by accumulating the power consumption in each of these components. More specifically, the first step to calculate an application event's power consumption is to map its signature (using the micro-architecture description file, as explained in the previous section) to usage counts of the various processor components. So, here it is determined how often e.g. the ALU (in 'other units' in Fig. 5), the register file and the level-1 instruction and data caches are accessed during the execution of an application event. For the memory components (level 1 and 2 caches, register file, etc.), we use Cacti 4.2 [22] to determine the power consumption of read and write accesses to these structures. These power estimates include leakage power. Moreover, we use the cache missrate information provided by Sim-Cache to derive the access counts for the level-2 cache, load/store queue and bus components. Here, we note that although we recently also incorporated Cacti 5.0 into our tool flow, which allows for modeling of DRAMs, this paper does not consider the modeling of the main memory.

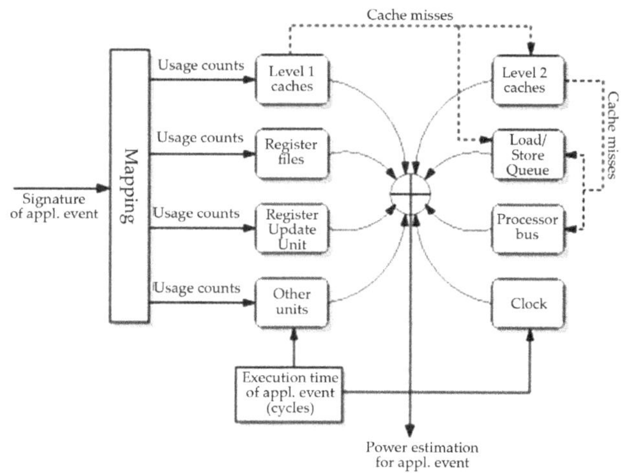

Figure 5 The different components in our microprocessor power model.

The non-memory components in our power model ('other units', 'bus' and 'clock' components in Fig. 5) are activity based. That is, they estimate power using the common power equation for CMOS technology:

$$P=\alpha CV2f \qquad (1)$$

where C specifies the capacitance, V the voltage, and f the frequency. The activity, which is defined as the percentage of transistors which make a switch on each clock cycle, is represented by the parameter α. For the bus component, which represents the processor's front-side bus, we use a simple model which abstracts the bus to a set of wires, without any logic, with input and output pins. Currently, we use an I/O pin capacitance of 5 pF per pin, a wire capacitance of 2.15 pF per inch, and a wire length of 3 inches. These numbers were taken from [11]. Further, we assume an activity α of 0.5 (half of the wires perform a state switch). For the models of the ALU and multiplier units (in the box 'other units' in Fig. 5), we use the capacitance numbers from Wattch [6]. The activity α for these units is calculated by dividing (an estimation of) the number of cycles in which a unit is active, which is derived from the usage counts of the unit (see Fig. 5), by the total number of cycles.

For the power model of the clock component, we base ourselves on the models used in [19, 21]. The model recognizes three sub-components: the clock distribution wiring, the clock buffering and the clocked node capacitance. We assume a H-tree based clock network using a distributed driver scheme (i.e. applying clock buffers).

To determine the wiring capacitance of the clock network, the wire length of the network has to be calculated. This is done in a number of steps. First, the length of a single straight wire at a specific level in the clock tree has to be calculated. The length of the first straight wire at the root level is equal to half of the width of the chip (where $\sqrt{A_{die}}$ is the width of the chip). Then, because of the H-tree form of the clock network, going to a next level in the clock tree halves the length of a straight wire. This means that the wire length at tree level i equals to:

$$Wirelength_i = \frac{1}{2^{\lceil i/2 \rceil - 1}} \times \sqrt{A_{die}} \tag{2}$$

With $Wires_i = 2^{i-1}$, which is the number of straight wires at tree level i, the total wire length can be calculated. However, this total length cannot directly be multiplied by the capacitance of the wire, as the capacitance of the wire is not uniform in the network. Starting form the end points in the network, the capacitance of the wire doubles at every T-junction.

$$FactorCap_i = 2^{N_{tree} - i} \tag{3}$$

This gives for the total capacitance of the distribution wiring the following equation:

$$C_{wiring} = C_{wire} \times \sqrt{A_{die}} \times 2^{N_{tree} - 1} \times \sum_{i=1}^{N_{tree}} \frac{1}{2^{\lceil i/2 \rceil}} \tag{4}$$

where we retrieve the wire capacitance (C_{wire}) and chip area A_{die} (determined by accumulating all cache and register-file areas) from Cacti 4.2, and calculate the depth of the clock tree (N_{tree}) using:

$$N_{tree} = \sqrt{A_{die} \frac{R_{wire} \times C_{wire}}{Skew_{clock}}} + 1 \tag{5}$$

where resistance R_{wire} is also retrieved from Cacti, and $Skew_{clock}$ is the maximum clock skew that is allowed.

The capacitance consumed by the buffers is modeled to be a fraction of the capacitance consumed by the wiring network. This fraction is dependent on the number of buffers, which is calculated by first taking the ratio of the capacitance of the wiring network and the capacitance of a single buffer. Over this the fourth root is taken, where the value four is actually a parameter, the optimal stage ratio, but this value is fixed within our model.

$$buffers = \sqrt[4]{\frac{C_{wiring}}{C_{single\_buffer}}} \tag{6}$$

$$C_{buffers} = C_{wiring} \times \frac{1}{1 - \left(\frac{1}{buffers}\right)} \tag{7}$$

For the clocked node capacitance, only memory components are considered. Here, we use the number of read and write ports and the blocksize to calculate the capacitance:

$$C_{clocked} = ports \times blocksize \times C_{trans} \tag{8}$$

The capacity for switching a port is acquired from Cacti, and is equal to the capacitance of a transistor. The clocked node capacitance of each memory structure is summed to the total clocked node capacitance.

For several components in our power model, the execution time of an application event is needed in order to calculate the activity parameter α (i.e., the time that a unit is active during the execution of an application event divided by the total execution time of the application event). These event execution times may be derived from the event tables in Sesame's architecture model (e.g. in the case signatures are generated statically), or they may be generated dynamically using e.g. our trace calibration co-simulation technique [23].

EXPERIMENTS

To evaluate our signature-based power estimation, we use three benchmark applications from the MiBench benchmark suite [8]: cjpeg (jpeg compression), susan (edge detection), and string search. We compare our results to those from Wattch [6] and Sim-Panalyzer [1], which are widely-used ISS-based power analysis tools. To compare against Wattch, we use a (in-order issue) PowerPC microprocessor model, with a voltage of 2.0 V and a frequency of 600 MHz. For comparison to Sim-Panalyzer, we use an (in-order issue) ARM microprocessor model, with a voltage of 1.8 V and a frequency of 233 MHz. For both models, we assume a 180 nm technology.

In the first experiment, we have varied the sizes of the level-1 instruction (il1) and data (dl1) caches as well as of the unified level-2 cache (ul2). This is done by increasing the number of sets in the caches. The level-1 caches are increased in size from 4 Kb to 512 Kb, and the level-2 cache from 32 Kb to 4 Mb. In Fig. 6a, the power estimation results from Wattch are shown, while Fig. 6b shows the results from our own power model. Clearly, the absolute power predictions differ significantly between Wattch and Sesame. As will be demonstrated later on, this is mostly due to the differences in the underlying power models. In Wattch, the power consumption is dominated by the level-1 instruction and data caches and, to a lesser extent, the clock network. In Sesame, the power is mostly dominated by the clock network, followed by the level-1 caches. The large discrepancy in absolute power estimations is mainly caused by two differences in the power models of Sesame and Wattch: First, Wattch has a more extensive model of the clocked node capacitance. For example, not only the clocked node capacitance of memory components are modeled, but also the datapath and other components are included. Second, whereas Sesame (in line with Cacti) only models the power consumption related to the critical path inside memory structures, Wattch directly relates the power consumption of a memory structure to the size of the memory. The latter can be clearly seen in Fig. 6a where the power consumption of the level-1 caches rapidly increases when increasing the number of cache sets.

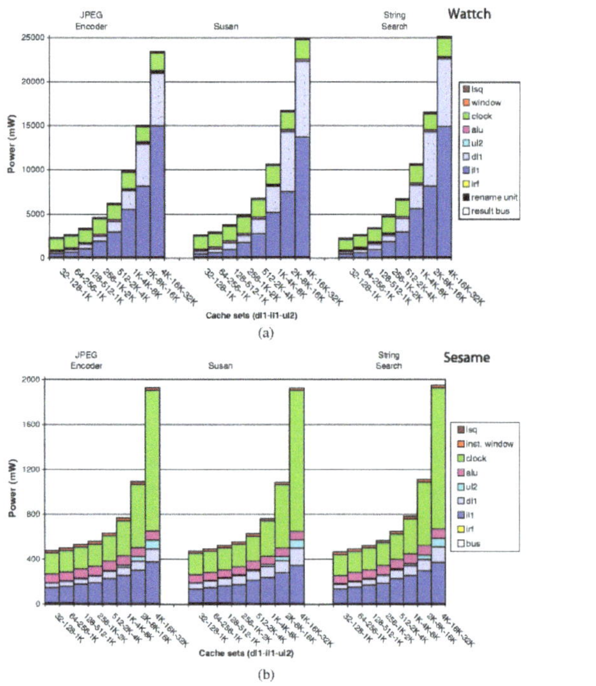

Figure 6 Power consumption estimation for Wattch (**a**) and Sesame (**b**) when varying the cache sizes.

Another observation that can be made is that the power consumption of the clock network in Wattch does not appear to increase for larger cache structures, while there is a significant increase of power consumed by the clock in Sesame. This is due to the fact that Sesame dynamically calculates the die area (using Cacti) for the power modeling of the clock network, while Wattch uses a fixed die area which evidently does not scale with the cache size.

In Fig. 7a, the power estimation results from Sim-Panalyzer for the same experiment are shown, while Fig. 7b again shows the results from our own power model but now configured to model the ARM processor. Like with Wattch, the absolute power predictions differ significantly between Sim-Panalyzer and Sesame. But maybe more surprising is the large increase of power consumption of the clock when scaling the cache sizes for Sim-Panalyzer. With the largest cache sizes in our experiment, the fraction of clock power consumption to the overall power consumption is equal to 99%. This significant increase of clock power in Sim-Panalyzer is mostly due to the fact that its model for clocked node capacitance is directly dependent on the number of sets in cache structures. In Sesame on the other hand, correspondingly to Eq. 8, the clocked node capacitance is only dependent on the block size of memory structures and the number of ports.

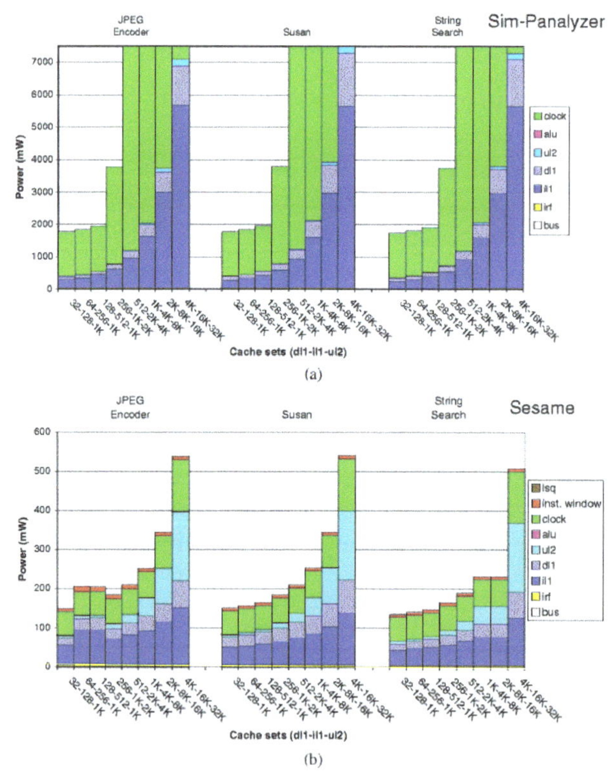

Figure 7 Power consumption estimation for Sim-Panalyzer (**a**) and Sesame (**b**) when varying the cache sizes.

Also, the increase of power consumption by the caches is much larger in Sim-Panalyzer than in Sesame. As is the case with Wattch, this is again due to the fact that Sim-Panalyzer directly relates the power consumption of a memory structure to its size, rather than relating the power consumption to the critical path inside memory structures like Sesame does. We further note that the discrepancies in Sesame's results for e.g. cjpeg for cache configurations between 64-256-1K and 512-2K-4K are caused by the reduced activity at the lower levels of the memory hierarchy due to the increasing cache sizes. In these cases, the increased hit-rate seems to amortize the higher dynamic and static power consumption of the larger caches.

From the above experiments, we learn that although Sesame can predict the total power consumption trend reasonably well (e.g. when compared to Wattch), its power model still needs to be improved considerably to better reflect the correct absolute power consumptions of the various architectural components. Moreover, the current differences in the underlying power models of Sesame, Wattch and Sim-Panalyzer make it impossible to actually evaluate the signature-based power modeling technique and the consequences (no notion of separate instructions nor of the data that they use) that come with it. For this reason, we also present an experiment in which we use the access power consumptions from Wattch for the various components in our power model. This way, we try to align both power models such that we can actually measure the effects of the high abstraction level at which Sesame operates. Figure 8shows the power consumption estimates for both Sesame and Wattch for all three benchmark applications. Here, we have executed cjpeg with three different input pictures of varying complexity: Cats (1.15 MB), Rain (150 KB) and Water (40 KB).

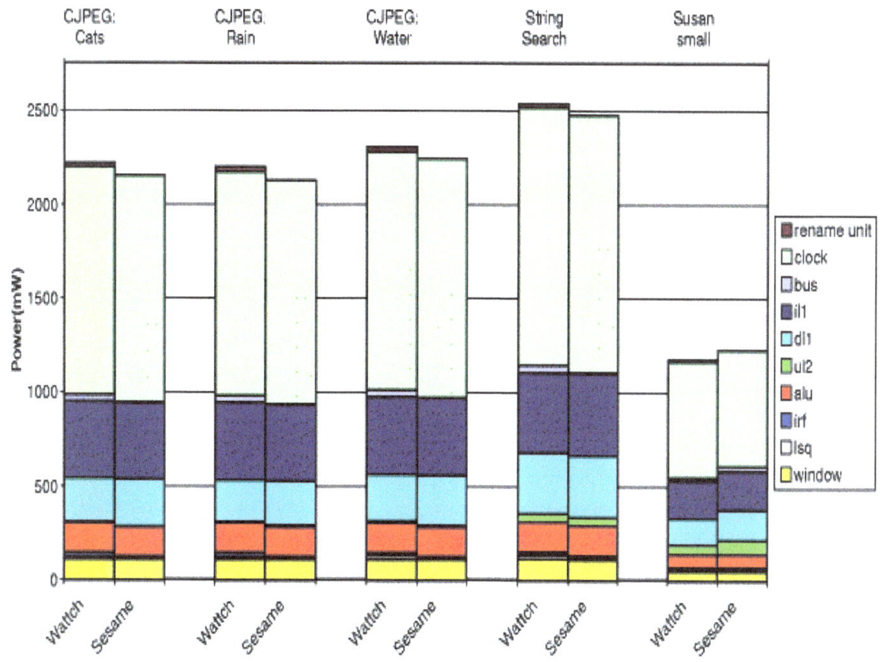

Figure 8.Comparing Sesame and Wattch when the same underlying power models are applied.

Using these aligned power models, the mean difference between Sesame and Wattch is only 3.1%, with a worst case of 4.2%. Individual components which are data dependent may have a larger mean error (and standard deviation σ), as indicated by Table 1. For example, the difference in power estimation for the instruction register file is 11% and for the level-2 cache even 22.6%. However, these components only account for a small fraction of the total power consumption. More importantly, for the level-1 caches—currently modeled with double-ended bitlines in both Sesame and Wattch, making their access energies largely data independent—the mean error is relative small: 2.1% for the level-1 instruction cache and 6.1% for the level-1 data cache. From these numbers, we can conclude that our high-level method may yield power estimations that are fairly close to those from tools such as Wattch.

Table 1 The mean difference in power estimation between Wattch and Sesame at component level.

Component	Mean difference	σ
Window	1.7%	1.6
irf	11%	7.5
il1	2.1%	2.6
dl1	6.1%	2.9
ul2	22.6%	22.9
alu	1.9%	1.7

We should note that Wattch and Sim-Panalyzer are also relatively high-level power simulators (in comparison to circuit-level power simulators such as Spice) and thus suffer from inaccuracies as well. Typically, they can be accurate within 10% [1, 5, 6]. But, as explained and demonstrated in [5], DSE studies can tolerate some error because relative accuracy is more important than absolute accuracy. Nevertheless, it would have been interesting to also compare our signature-based power estimates to power consumption numbers from real ARM or PowerPC processors. This is however considered as future work.

Since our initial aim was to speed up the power estimations in comparison to traditional ISS-based power simulators, we also measured the execution times of both the Sesame and Wattch[2]frameworks. Here, we should note that for Sesame we used a set-up in which Sim-Cache was co-simulated together with Sesame to provide our power model with exact cache missrates for application events (see Section 3). Decoupling Sesame and Sim-Cache (i.e., using average cache missrates for application events in Sesame rather than exact ones) would result in even much better performance of our power estimations. For the studied benchmark applications, Sesame is on average 10.5 times faster than Wattch. The largest speed-up we measured was 25. Given the fact that—with an appropriate power model—good power estimates can be achieved (see Fig. 8), this is a very promising result.

RELATED WORK

High-level microprocessor power modeling techniques range from analytical methods [4, 15], based on e.g. statistical or information-theoretic techniques, to micro-architecture level instruction set simulators (ISSs) such as Wattch [6], Sim-Panalyzer [1] and SimplePower [25]. Clearly, within this range, there is a trade-off between accuracy and estimation performance. A fair number of efforts also address microprocessor power estimation at a level that is in between analytical and ISS-level models. Many of these efforts estimate power based on a-priori knowledge about instructions or segments of instructions. For example, the power consumption of separate instructions (or instruction pairs [14]) can be measured (using a real processor or a low-level simulator) after which the power consumption of an entire application involves the accumulation of these per-instruction power consumptions [20]. Such measurement-based power estimation can also be performed at a coarser granularity such as at the level of entire functions [18].

Another interesting high-level microprocessor power estimation approach is functional-level power analysis (FLPA) [12, 13]. In FLPA, a processor is modeled using a relatively small number of functional blocks, for which parameterized analytical power models are determined. These power models, which only have a few high-level parameters (like internal/external data access rate and average number of processing units used per cycle), are determined using linear regression with a set of power measurements in which the aforementioned high-level parameters are varied. For these measurements, real processors or lower-level simulators are used together with a set of benchmark programs (scenarios) that stress the different functional blocks in the processor. In [3], a hybrid FLPA and instruction-level microprocessor power estimation approach is proposed. They have shown that by incorporating instruction dependent behavior in the FLPA models, the accuracy of the models can be further improved.

In terms of abstraction level, our signature- based power estimation technique is also in between analytical and ISS-based models. Compared to the methods described above, we abstract from single instructions while still being able to apply an explicit micro-architecture model to perform power estimation. This allows us, in contrast to e.g. instruction-level power modeling and FLPA, to perform micro-architectural DSE. Also, unlike instruction-level power modeling and FLPA, we do not need to perform power measurements to determine the power consumption of (blocks of) instructions or, in the case of FLPA, functional blocks in the processor.

CONCLUSIONS

In this paper, we have presented a technique for high-level power estimation of microprocessors. This technique, which is based on abstract execution profiles called 'event signatures', operates at a higher level of abstraction than commonly-used instruction-level power simulators and should thus be capable of achieving good evaluation performance. The signature-based power modeling technique has been integrated in our Sesame system-level simulation and DSE framework and will eventually be extended to allow for system-level power modeling of an entire MP-SoC (i.e., also support the power modeling of the interconnect, shared memory, dedicated IP blocks, etc.).

We compared the results from our signature-based power modeling to those from the instruction-level simulators Wattch and Sim-Panalyzer. Here, it was shown that although the underlying power model we applied is able to show approximately the correct overall trends, it needs to be improved considerably to show the correct absolute power consumptions of the various architectural components. However, we also demonstrated that with a good underlying power model, the signature-based power modeling technique can yield accurate estimations (a mean error of 3.1% compared to Wattch in our experiments). Important to note here is that the power estimations based on our event signature technique are at least an order of magnitude faster than with Wattch (being the fastest of the two used instruction-level simulators).

REFERERENCES

1. Austin, T., Mudge, T., & Grunwald, D. Sim-panalyzer. http://www.eecs.umich.edu/~panalyzer/.
2. Austin, T. M., Larson, E., & Ernst, D. (2002). Simplescalar: an infrastructure for computer system modeling. Computer, 35(2), 59–67, Feb.
3. Blume, H., Becker, D., Rotenberg, L., Botteck, M., Brakensiek, J., & Noll, T. G. (2007). Hybrid functional- and instruction-level power modeling for embedded and heterogeneous processor architectures. Journal of Systems Architecture, 53(10), 689–702, Oct.
4. Brandolese, C., Fornaciari, W., Salice, F., & Sciuto, D. (2000). An instruction-level functionally-based energy estimation model for 32-bits microprocessors. In: Proc. of the Design Automation Conference, pp. 346–351.
5. Brooks, D., Bose, P., & Martonosi, M. (2004). Power-performance simulation: design and validation strategies. SIGMETRICS Performance Evaluation Review, 31(4), 13–18.
6. Brooks, D., Tiwari, V., & Martonosi, M. (2000). Wattch: A framework for architectural-level power analysis and optimizations. In: Proc. of the Int. Symposium on Computer Architecture (ISCA), June.
7. Erbas, C., Pimentel, A. D., Thompson, M., & Polstra, S. (2007). A framework for system-level modeling and simulation of embedded systems architectures. EURASIP Journal on Embedded Systems, doi:10.1155/2007/82123.
8. Guthaus, M. R., Ringenberg, J. S., Ernst, D., Austin, T. M., Mudge, T., & Brown, R. B. (2001). Mibench: A free, commercially representative embedded benchmark suite. In: Proc. of the IEEE Workshop on Workload Characterization, pp. 3–14.
9. Jaddoe, S., & Pimentel, A. D. (2008). Signature-based calibration of analytical system-level performance models. In: Proc. of Int. Symposium on Systems, Architectures, MOdeling and Simulation (SAMOS), pp. 268–278, July.
10. Kahn, G. (1974). The semantics of a simple language for parallel programming. In: Proc. of the IFIP Congress 74.
11. Sung Kim, N., Kgil, T., Bertacco, V., Austin, T., & Mudge, T. (2004). Microarchitectural power modeling techniques for deep sub-micron microprocessors. In: Proc. of the International Symposium on Low Power Electronics and Design, pp. 212–217.
12. Laurent, J., Julien, N., Senn, E., & Martin, E. (2004). Functional level power analysis: An efficient approach

for modeling the power consumption of complex processors. In: Proc. of the conference on Design, Automation and Test in Europe (DATE), pp. 666–668, Feb.

13. Laurent, J., Senn, E., Julien, N., & Martin, E. (2001). High level energy estimation for dsp systems. In: Proc. of the Int. Workshop on Power And Timing Modeling, Optimization and Simulation (PATMOS), pp. 311–316.
14. Lee, M. T.-C., Fujita, M., Tiwari, V., & Malik, S. (1997). Power analysis and minimization techniques for embedded DSP software. IEEE Transactions on Very Large Scale Integration Systems, 5(1), 123–135.
15. Macii, E., Pedram, M., & Somenzi, F. (1998). High-level power modeling, estimation, and optimization. IEEE Transactions on CAD of Integrated Circuits and Systems, 17(11), 1061–1079.
16. Pimentel, A. D., Erbas, C., & Polstra, S. (2006). A systematic approach to exploring embedded system architectures at multiple abstraction levels. IEEE Transactions on Computers, 55(2), 99–112.
17. Pimentel, A. D., Thompson, M., Polstra, S., & Erbas, C. (2008). Calibration of abstract performance models for system-level design space exploration. Journal of Signal Processing Systems for Signal, Image, and Video Technology, 50(2), 99–114.
18. Qu, G., Kawabe, N., Usami, K., & Potkonjak, M. (2000). Function-level power estimation methodology for microprocessors. In: Proc. of the Design Automation Conference, pp. 810–813.
19. Vijaykrishnan, N., Chen, R., & Irwin, M. J. (1999). Clock power issues in system-on-chip designs. In: Proc. of IEEE CS Workshop on VLSI, pp. 48–53.
20. Russel, J. T., & Jacome, M. F. (1998). Software power estimation and optimization for high performance, 32-bit embedded processors. In: Proc. of the Int. Conference on Computer Design (ICCD), pp. 328–333, Oct.
21. Sung Kim, N., Austin, T., Mudge, T., & Grunwald, D. (2001). Challenges for architectural level power modeling. In: Power Aware Computing, pp. 48–53.
22. Tarjan, D., Thoziyoor, S., & Jouppi, N. P. (2006). Cacti 4.0. Technical Report HPL-2006-86, Hewlett-Packard.
23. Thompson, M., Pimentel, A. D., Polstra, S., & Erbas, C. (2006). A mixed-level co-simulation method for system-level design space exploration. In: Proc. of the IEEE Workshop on Embedded Systems for Real-Time Multimedia (ESTIMedia'06), pp. 27–32, Oct.
24. van Stralen, P., & Pimentel, A. D. (2007). Signature-based microprocessor power modeling for rapid system-level design space exploration. In: Proc. of the IEEE Workshop on Embedded Systems for Real-Time Multimedia (ESTIMedia '07), pp. 33–40, Oct.
25. Ye, W., Vijaykrishna, N., Kandemir, M., & Irwin, M. J. (2000). The design and use of simplepower: A cycle-accurate energy estimation tool. In: Proc. of the Design Automation Conference (DAC), June.

CHAPTER 2

TIME-PREDICTABLE COMPUTER ARCHITECTURE

Martin Schoeberl

INTRODUCTION

Standard computer architecture is driven by the following paradigm: *Make the common case fast and the uncommon case correct* [1]. However, this design approach leads to architectures where the worst-case execution time (WCET) is high and hard to predict by static analysis. For real-time systems, we have to design architectures with the following paradigm: *Make the worst case fast and the whole system easy to analyze*.

Classic enhancements in computer architectures are pipelining, instruction and data caching, dynamic branch prediction, out-of-order execution, speculative execution, and fine-grained chip multithreading. These features are increasingly harder to model for the low-level WCET analysis. Execution history is the key to performance enhancements, and also the main issue for WCET analysis. Thus, we need techniques to manage the execution history.

Pipelines should be simple, with minimum dependencies between instructions. It is agreed that caches are mandatory to bridge the gap between processor speed and memory access time. Caches in general, and particularly data caches, are usually hard to analyze statically. Therefore, we are introducing caches that are organized to speed up execution time and provide tight WCET bounds. We propose three different caches: (1) an instruction cache for full methods, (2) a stack cache, and (3) a small, fully associative buffer for heap access. Furthermore, the integration of a program—or compiler—managed scratchpad memory can help to tighten bounds for hard-to-analyze memory access patterns.

Out-of-order execution and speculation result in processor models that are too complex for WCET analysis. We discuss that the transistors are better used on chip multiprocessors (CMPs) with simple in-order pipelines. Real-time systems are naturally multithreaded and thus map well to the explicit parallelism of chip multiprocessors.

We propose a multiprocessor model with one processor per thread. Thread switching and schedulability analysis for each individual core disappears, but the access to the shared resource main memory still needs to be scheduled.

We have implemented most of the proposed concepts for evaluation in a Java processor. The Java processor JOP [2] is intended for real-time and safety critical applications written in a modern object-oriented language. It has to be noted that all concepts can also be applied to a standard RISC processor.

The following list points out the key arguments for a time-predictable computer architecture.

- There is a mismatch between performance-oriented computer architectures and worst-case analyzability.
- Complex features result in increasingly complex models.
- Caches, a very important feature for high performance, need new organization.
- Thread level parallelism is natural in embedded systems. Exploration of this parallelism with simple chip multiprocessors is a valuable option.
- One thread per processor obviates the classic schedulability analysis and introduces scheduling of memory access.

Citation: Martin Schoeberl, Time-Predictable Computer Architecture, DOI: 10.1155/2009/758480

Catching up with WCET analysis of features that enhance the average-case performance is not an option for future real-time systems. We need a sea change and should take the constructive approach by designing computer architectures where predictable timing is a first-order design factor.

The contributions of the paper are twofold: (1) an extensive overview is given of processor features that make WCET estimation difficult, (2) solutions for a time-predictable architecture that can be implemented in RISC, CISC, or VLIW style processors are provided. The implementations of some of the proposed concepts in the context of a Java processor, as described in Section 5, have been previously published in [3, 4].

The paper is organized as follows. Section 2 presents related work on real-time architectures. In Section 3, we describe the main issues that hamper tight WCET estimates of actual processors. We propose solutions for these issues in Section 4. In Section 5, we evaluate the proposed time-predictable computer architecture with an implementation of a Java processor in an FPGA. Section 6 concludes the paper.

RELATED WORK

Bate et al. [5] discuss the usage of modern processors in safety critical applications. They compare commercial off-the-shelf (COTS) processors with a customized processor developed specifically for the safety critical domain. While COTS processors benefit from a large user base and the resulting maturity of the design process, customized processors provide the following advantages:

- (design in conjunction with the safety argument,
- design for good worst-case performance,
- using only features that can be easily analyzed,
- the processor can be treated as a *white box* during verification and testing.

Despite these advantages, few research projects exist in the field of WCET-optimized hardware. Thiele and Wilhelm [6] discuss that a new research discipline is needed for time-predictable embedded systems to «match implementation concepts with techniques to improve analyzability.»

Similarly, Edwards and Lee discussed as «It is time for a new era of processors whose temporal behavior is as easily controlled as their logical function» [7]. A first simulation of their PRET architecture is presented in [8]. PRET implements the SPARC V8 instruction set architecture (ISA) in a six-stage pipeline and performs chip level multithreading for six threads to eliminate data forwarding and branch prediction. Scratchpad memories are used instead of instruction and data caches. The shared main memory is accessed via a TDMA scheme, called memory wheel, similar to the TDMA-based arbiter used in the JOP CMP system [9]. The SPARC ISA is extended with a*deadline* instruction that stalls the current thread until the deadline is reached. This instruction is used to perform time-based, instead of lock-based, synchronization for access to shared data.

Berg et al. identify the following design principles for a time-predictable processor: «recoverability from information loss in the analysis, minimal variation of the instruction timing, noninterference between processor components, deterministic processor behavior, and comprehensive documentation» [10]. The authors propose a processor architecture that meets these design principles. The processor is a classic five-stage RISC pipeline with minimal changes in the instruction set: it handles function calls with an explicit instruction for simpler reconstruction of the control flow graph and construction of 32-bit immediate values with two instructions to avoid immediate values in the code segment. The memory system has to be organized in Harvard-style with dedicated busses to the FLASH memory for the code and the SRAM memory for the data. The replacement strategy of caches has to be least-recently used (LRU).

Heckmann et al. provide examples of problematic processor features in [11]. The most problematic features found are the replacement strategies for set-associative caches. A pseudo-round-robin replacement strategy of the 4-way set-associative cache in the ColdFire MCF 5307 effectively renders the associativity useless for WCET analysis. The use of a single 2-bit counter for the whole cache destroys age information within the cache sets. The analysis of that cache results in effectively modeling only a quarter of the cache as a direct-mapped cache. Similarly, a pseudo-LRU replacement strategy for an 8-way set-associative cache of the PowerPC 750/755 uses an age counter for each set. Here, only half of the cache is modeled by the analysis. Slightly more complex pipelines, with branch prediction and out-of-order execution, need an integrated pipeline and cache analysis to provide useful WCET bounds. Such integrated analysis is complex and also demanding with respect to the computational effort. In conclusion, Heckmann et al. suggest the following restrictions for time-predictable processors: (1) separate data and instruction caches, (2) locally deterministic update strategies for caches, (3) static branch prediction, and (4) limited out-of-order execution. The authors discuss for restriction of processor features of actual processors (of the time) for embedded systems, but do not provide suggestions for *additional* or *alternative* features for a time-predictable processor.

The VISA approach [12] adapts a complex simultaneous multithreading processor that can be reconfigured to a simple single-issue pipeline. The complexity of the processor can be dynamically disabled at runtime. WCET analysis is performed for the simple pipeline. A task is divided into subtasks and each subtask is assigned a checkpoint. The task is executed on the complex pipeline and only if the checkpoint is missed, the processor is switched to the simple mode. The checkpoint is inserted early enough to complete the subtask on the simple pipeline before the deadline. The available slack time, when the task is executed on the fast, complex pipeline, is utilized for energy saving.

Puschner and Burns discuss for a single-path programming style [13] that results in a constant execution time. In that case, WCET can easily be measured. However, this programming paradigm is quite uncommon and restrictive. Single-path programming can be inefficient when the control flow is data-dependent. A processor, called SPEAR [14], was especially designed to evaluate the single-path programming paradigm. A single predicate bit can be set with a compare instruction whereby several instructions (e.g., move, arithmetic operations) can be predicated. The SPEAR implements a three-stage in-order pipeline and uses onchip memories for instruction and data instead of caches.

Complex hardware and software architectures hinder hierarchical timing analysis [15]. A radical simplification of the whole system to avoid unwanted timing interactions is proposed—single path programming, execution of a single task/thread per core, simple in-order pipelines, and statically scheduled access to shared memory in CMPs.

Whitham discusses that the execution time of a basic block has to be independent of the execution history [16]. As a consequence, his MCGREP architecture reduces pipelining to two stages (fetch and execute) and avoids caches all together. To reduce the WCET, Whitham proposes to implement the time-critical functions in microcode on a reconfigurable function unit (RFU). The main processor implements an RISC ISA as a microprogrammed, sequential processor. The interesting approach in MCGREP is that the RFUs implement the same architecture and microcode as the main CPU. Therefore, mapping a sequence of RISC instructions to microcode for one or several RFUs is straightforward. With several RFUs, it is possible to explicitly extract instruction level parallelism (ILP) from the original RISC code in a similar way to VLIW architectures.

Whitham and Audsley extend the MCGREP architecture with a trace scratchpad [17]. The trace scratchpad caches microcode and is placed after the decode stage. It is similar to a trace cache found in newer Pentium CPUs to cache the translated micro-operations. The differences from a cache are that the execution from the trace scratchpad has to be explicitly started and the scratchpad has to be loaded under program control. The authors extract ILP at the microcode level and schedule the instructions statically—similar to a VLIW architecture.

WCET ANALYSIS ISSUES

The WCET of tasks is the necessary input for schedulability analysis. Measuring the WCET is not a safe option. Only static WCET analysis can provide safe upper bounds of execution times.

WCET analysis can be separated into high-level and low-level analysis. The high-level analysis is a mature research topic [18, 19, 20]. An overview of WCET-related research can be found in [21, 22]. The main issues that need to be solved are in the low-level analysis. The processors that can be analyzed are usually several generations behind actual architectures [11, 23, 24] (e.g., Thesing models, in his Ph.D. thesis [25], the MPC755 variant of the PowerPC 750). The PowerPC 750 was introduced in 1997 and the MPC755 was *not recommended for new designs* in 2006.

The main issues in low-level analysis are features that increase average performance. All these features, such as multilevel caches, branch target buffer, out-of-order execution, and speculation, include a state that heavily depends on a large execution history. This caching of the execution history is actually fundamental for performance enhancements. However, it is the history which is hard to model for WCET analysis. A long history leads to a state explosion for the final WCET calculation. Low-level WCET analysis thus usually performs simplifications and uses conservative estimates. One example of this conservative estimate is to classify a cache access as a miss, if the outcome of the cache access is unknown.

Lundqvist and Stenström have shown that this intuitive assumption can be wrong on dynamically scheduled microprocessors [26]. They provide an example of such a timing anomaly in which a cache hit can cause a longer execution time than a cache miss. The principles behind these timing anomalies are further elaborated in [27].

Pipeline Dependencies

Simple pipelines, similar to the original Berkeley/Stanford RISC design [28], are easy to model for WCET analysis. In a nonstalled pipeline, the execution time latency corresponds to the length of the pipeline. The effective execution time itself is only a single cycle. What makes pipeline analysis necessary are stalls introduced by dependencies within the pipeline. Those stalls are introduced by

- data dependencies between instructions,
- control dependencies between instructions.

In one of the first RISC designs, the MIPS [29], these dependency hazards are explicitly exposed to the compiler. They have to be resolved by the compiler with instruction scheduling for delayed branches and for the single cycle delay between a memory load and the data use. Therefore, these effects are also recognized by the WCET tool. More advanced pipelines avoid exposing stalls from the ISA in order to avoid too many (compiler) target variations and retain binary compatibility between processor versions. Nevertheless, this information is needed for WCET analysis.

Dependencies within a basic block can be easily modeled. The challenge is to merge the effects from different basic blocks and across function boundaries. In [30], the *timing schema* [31] is extended to include the pipeline information. Timing schema is a tree-based WCET analysis. After the determination of basic block execution times, the control flow graph is processed in a bottom-up manner until a final WCET bound is available. Branches are merged with the higher WCET bound as result. For the extension, the pipeline is represented by reservation stations, and the state at the head and tail of a basic block is considered when basic blocks are merged.

Pipelines with timing dependencies can result in an unbounded effect, called *long timing effect* (LTE) [32]. This means that an instruction far back in the history (longer than the pipeline length) influences the execution time of the current instruction. These LTEs can be negative or positive. A positive LTE means longer execution time. An instruction with a possible positive LTE needs a safe approximation of that effect for the pipeline analysis.

More complex pipelines can be analyzed with abstract interpretation, but the analysis time can become impractical. Berg et al. [10] report that up to 1000 states per instruction are needed for the model of the PowerPC 755. This processor was introduced in 1998 and we expect a considerable growth of the states that need to be tracked by abstract interpretation for newer processors.

Instruction Fetch

The instruction fetching is often decoupled from the main memory or the instruction cache by a prefetch unit. This unit fills the prefetch queue with instructions independently of the main pipeline. This form of prefetching is especially important for a variable length instruction set as the x86 ISA or the bytecode instructions of the Java virtual machine (JVM). The fill status of the prefetch queue depends on the history of the instruction stream. The possible length of this history is unbounded. To model this queue for a WCET tool, we need to cut the maximum history and assume an empty queue at such a cut point.

In [33], the authors model the 4-byte-long prefetch queue of an Intel 80188. Even for this simple prefetch queue, the authors have to perform some simplifications in their approach to handle the resulting complexity due to the interaction between the instruction execution and the instruction prefetch (the consuming and the producing ends of the queue).

Caches

Between the middle of the 1980s and 2002, CPU performance increased by around 52% per year, but memory latency decreased only by 9% [1]. To bridge this growing gap between CPU and main memory performance, a memory hierarchy is used. Several layers with different tradeoffs between size, speed, and cost form that memory hierarchy.

A typical hierarchy consists of

(1) register file,
(2) per-processor level 1 instruction and data cache,
(3) onchip, shared unified level 2 cache,
(4) offchip level 3 cache,
(5) main memory,
(6) hard disc for virtual memory.

The only layer under the control of the compiler is the register file. The rest of the memory hierarchy is usually not visible—it is not part of the ISA abstraction. Placement of data in the different layers is performed automatically by the hardware for caches and by the OS for virtual memory management. The access time for a word located in a memory block paged out by the OS is several orders of magnitude higher than a level 1 cache hit. Even the access times to the level 1 cache and to the main memory differ by two orders of magnitudes.

Cache memories for the instructions and data are classic examples of the *make the common case fast* paradigm. Avoiding or ignoring this feature in real-time systems, due to its unpredictable behavior, results in a very pessimistic WCET bound. Much effort has been expended on research to integrate the instruction cache into the timing analysis of tasks [34, 35], on the cache's influence on task preemption [36, 37], and on integration of the cache analysis with the pipeline analysis [38]. The influence of different cache architectures on WCET analysis is described in [11].

A unified cache for data and instructions can easily destroy all the information on abstract cache states. Access to n unknown addresses in an n-way set-associative cache results in the state *unknown* for all cache lines. Modern processors usually have separate instruction and data caches for the level 1 cache. However, the level 2 cache is usually shared. Most CMP systems also share the level 2 cache between the different cores. The possible interactions between concurrent threads running on different cores are practically impossible to model.

Data caches are considerably harder to analyze than instruction caches. For some data accesses, especially for data allocated on the heap, the addresses cannot be predicted. However, access to the stack can be predicted statically. A data cache that caches heap and stack content suffers from the same problem as a unified instruction and data cache: an unknown address for a heap access will evict one block from all sets in the abstract cache state and will increase the age of all cache blocks.

In a recent paper, Reineke et al. analyzed the predictability of different cache replacement policies [39]. It is shown that LRU performs best with respect to predictability. Pseudo-LRU and FIFO perform similarly, as both perform considerably worse than LRU. In an 8-way set-associative setting, pseudo-LRU and FIFO take more than twice as long as LRU to recover from lost information.

Branch Prediction

Accurate branch prediction is of utmost importance to keep long pipelines filled. The penalty of a wrongly predicted conditional branch is typically almost as long as the pipeline length. Modern branch predictors guess the outcome primarily from results of earlier branches. They heavily rely on the execution history, an effect we want to avoid for a tight worst-case prediction. Global branch predictors and caches have a similar issue: as soon as a single index into the branch history is unknown, the whole information of branch prediction is lost for the analysis at that point.

Two-level branch predictors are not suitable for time-predictable architectures [40]; for example, on the Pentium III, Pentium 4, and UltraSparc III, a decrease in the number of loop iterations can actually result in an increase of the execution time. This is another form of timing anomaly [26].

Branch prediction also interferes with cache contents. When the analysis cannot anticipate the outcome of the prediction, both branch directions need to be considered for cache analysis.

Instruction Level Parallelism

Some microprocessors try to extract ILP from the instruction stream, that is, execute more than one instruction per clock cycle. ILP extractions can be done either statically by the compiler or dynamically by the hardware.

Processors with static-scheduled ILP are known as very long instruction word (VLIW) processors. The main issue of VLIW processors is that the pipeline details are exposed at the ISA. The compiler has to group parallel instructions and needs to consider pipeline constraints. Some processors rely on the compiler to resolve data dependencies and do not stall the pipeline. Therefore, each new generation of VLIW processors needs a new compiler back end. However, this issue is actually an advantage for low-level WCET analysis, as these details are needed for the pipeline analysis.

Dynamically scheduled, superscalar microprocessors combine several parallel execution units with an out-of-order execution to extract the ILP from the instruction stream. In current processors, about hundred instructions (e.g., 128 in the Pentium 4 [1]) can be in flight at each cycle. Analysis of a realistically sized application with an accurate processor model is thus (almost) impossible. Even modeling the pipeline states for basic blocks leads to a state space explosion; and modeling only basic blocks would result in very long penalties for the branches—on a later version of the Pentium 4, a simple instruction takes at least 31 clock cycles from fetch to retire [1].

Despite this complexity, in [41], a hypothetical out-of-order executing microprocessor is modeled for WCET analysis. Verification of the proposed approach on a real processor is missing. We think modeling out-of-order processors is practically not feasible.

Chip Multithreading

Dynamic extraction of ILP is limited to about two instructions per cycle on current processors, such as Pentium 4 and AMD Opteron [1]. Another path to speed up multithreaded workloads is the extraction of thread-level parallelism (TLP). The concept of TLP in a single processor is quite old—it was used in the CDC 6600, a supercomputer from the 1960s—but is now being reconsidered in all desktop and server processors. Fine-grained multithreading can hide the latency of load/use hazards or a cache miss for one thread by the execution of other threads.

The main issue with multithreading in real-time systems arises when the execution time of one thread depends on the state of a different thread. The main source of timing interactions in a CMP comes from shared caches and shared main memory. In the worst case, all latency hiding has to be ignored by the analysis and the sum of the execution times of several threads is the same as the serial execution on a single-threaded CPU. In addition, multithreaded processors usually share the level 1 caches. Therefore, each thread invalidates the abstract cache state of the other threads.

Dynamic ILP and TLP can be combined for simultaneous multithreading (SMT). With this technique, independent threads can be active in the same pipeline stage. This results in a higher utilization of processor resources that are already available for the ILP extraction. Modeling the fine-grained interaction of different SMT threads for WCET analysis seems, at least to the author, an intractable problem.

Chip Multiprocessors

Due to the power wall [1], CMP systems are now state-of-the-art in desktop and server processors. There are three different CMP systems: (1) multicore versions of superscalar architectures (Intel/AMD), (2) multicore chips with simple RISC processors (Sun Niagara), and (3) the cell architecture.

Mainstream desktop processors from Intel and AMD include two or four out-of-order executing processors. These processors are replications of the original, complex cores that share a level 2 cache and the memory bus. Cache coherence protocols on the chip keep the level 1 caches coherent and consistent. Furthermore, these cores also support SMT, sometimes also called hyperthreading.

Sun took a completely different approach with their Niagara T1 [42] by abandoning their superscalar architecture. The T1 contains 8 simple RISC cores, each supporting 4 threads, scheduled round-robin. When a thread stalls due to a cache miss or a load-use dependency, it is skipped in the schedule. The first version of the chip contains a single floating point unit which is shared by all 8 processors. Each core implements a six-stage, single-issue pipeline similar to the original five-stage RISC pipeline. Such a simple pipeline brings WCET analysis back into consideration.

The cell multiprocessor [43, 44, 45] takes an approach similar to a distributed memory multiprocessor. The cell contains, beside a PowerPC microprocessor, 8 synergistic processors (SPs). The SPs contain 256 KB onchip memory that is incoherent with the main memory. The PowerPC, the 8 SPs, and the memory interface are connected via a network consisting of four independent rings. Communication between the cores in the network has to be set up explicitly. All memory management, for example, transfer between SPs or between onchip memory and main memory, is under program control, resulting in a new programming model. The time-predictable memory access to the onchip memory and the in-order pipeline of the SPs should be a reasonable target for WCET analysis. The challenge is to include the explicit memory transfers between the cores and the main memory into the analysis.

Intel recently announced a CMP system named Larrabee [46]. Larrabee is intended as a replacement for graphic processing units from other vendors. It is notable that Intel uses several dual issue, in-order x86 cores. They discuss that for some workloads, in-order pipelines are more power efficient than out-of-order cores. The design is based on the first Pentium processor, enhanced with multithreading support and vector instructions.

The main source of timing interactions in a CMP comes from the shared level 2 (and probably level 3) cache and the shared main memory. The shared memory provides an easy-to-use programming model at the cost of unpredictable access time to the data. A shared level 2 cache is practically not analyzable due to the interthread interference. This is the same issue as with multithreading with a shared level 1 cache.

Cache coherence protocols (bus snooping or directory based) enforce a coherent and consistent view of the main memory. These protocols exchange the cache information between all cores on each memory access and introduce a high variability of the cache access time even when the access is a cache hit.

Yan and Zhang analyze a shared instruction cache on a dual core system that executes two threads [47]. To restrict the set of conflicting cache blocks, they introduce the category *always-except one hit* for level 2 cache blocks. Assuming threads A and B, a cache block c is classified as *always-except one hit* for thread A when c is part of a loop in thread A, c conflicts with a block used by thread B, and the conflicting block in thread B is not part of a loop in thread B. However, the approach has two drawbacks: (1) for n threads/cores, several categories (up to $n - 1$) need to be introduced; (2) *not in a loop* is not a proper model for real-time threads as these are usually periodic.

The memory arbitration algorithm determines the worst-case access time to the main memory. Any fairness-based arbitration is, at least, difficult to integrate into WCET analysis. Priority-based arbitration can only guarantee access time for the core with the highest priority because lower priority cores can be blocked indefinitely.

Documentation

To model the processor for the low-level analysis, an accurate documentation of the processor internals is needed. However, this information is often not available or sometimes simply wrong [32]. For actual processors, the documentation of the internals is usually not disclosed. Over time, due to reverse engineering and less competition with other processors, more information becomes available. This is probably another reason why WCET analysis is about 10 years behind the processor technology.

Summary

While conventional techniques in designing processor architectures increase average throughput, they are not feasible for real-time systems. The influence of these architectural enhancements is at best hardly WCET analyzable. From a survey of the literature, we found that modeling a new version of a microprocessor and finding all undocumented details is usually worth a full Ph.D. thesis.

We discuss that trying to catch up on the analysis side with the growing complexity of modern computer architectures is not feasible. A paradigm shift is necessary. Computer architecture has to be redefined or adapted for real-time systems. Predictable and *analyzable* execution time is of primary importance.

TIME-PREDICTABLE ARCHITECTURE

We propose a computer architecture designed especially for real-time applications. We do not want to restrict features only, but we also want to actively add features that enhance performance and are time-predictable.

Figure 1 illustrates the aim of a time-predictable architecture, showing the distribution of the different execution times for a task: they are best-case execution time (BCET), average-case execution time (ACET), worst-case execution time (WCET), and the bound of the WCET that an analysis tool can provide. The difference between the actual WCET and the bound is caused by the pessimism of the analysis resulting from two factors: (a) certain information, for example, infeasible execution paths, not being known statically and (b) the simplifications to make the analysis computationally practical. For example, infeasible execution paths may significantly impact the WCET bound because the static analysis cannot prove that these paths may never be executed. Similarly, dynamic features such as speculative execution and pipelining often need to be modeled conservatively to prevent an explosion of the analysis complexity.

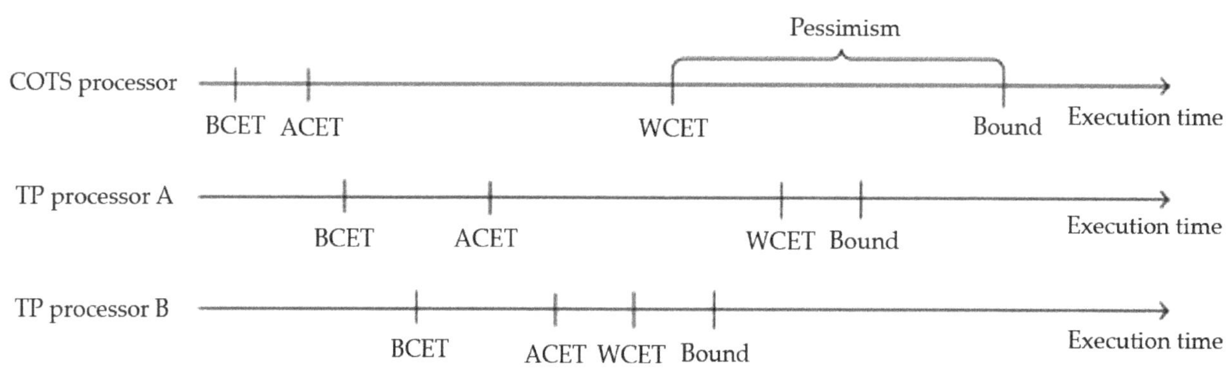

Figure 1 Distribution of the best-case, average-case, and worst-case execution times and the WCET bound of a task on different architectures. A time-predictable processor has less pessimism and the average-case execution time is not important.

The first time line shows the distribution of the execution times for a commercial off-the-shelf (COTS) processor. The other two time lines show the distribution for two different time-predictable processors.

Variant A depicts a time-predictable processor with a higher BCET, ACET, and WCET than a standard processor. Although the WCET is higher than the WCET of the standard processor, the pessimism of the analysis is lower and the resulting WCET bound is lower as well. Even this type of processor is a better fit for hard real-time systems than today›s standard processors.

Processor B shows an architecture where the BCET and ACET are further increased, but the WCET and the WCET bound are decreased. Our goal is to design an architecture with a low WCET bound. For hard real-time systems, the likely increase in the ACET and BCET is acceptable because the complete system needs to be designed to reduce the WCET. It should be noted that a processor designed for low WCET will never be as fast in the average case as a processor optimized for ACET. Those are two different design optimizations. We define a time-predictable processor as «under the assumption that only feasible execution paths are analyzed, a time-predictable processor›s WCET bound is close to the real WCET.»

In the following, we propose time-predictable solutions or replacements, if possible, for the issues we identified in the Section 6. Table 1 summarizes the issues of standard processors for WCET analysis and the proposed architectural solutions.

Table 1 Architectural issues for WCET analysis of standard processors and proposed architectural solutions.

	Standard processor WCET issues	Time-predictable processor
Pipeline	Dependencies and shared state	Simple pipeline
Instruction fetch	Unbounded timing effects	Avoid prefetch queue, use double buffer
Caches	Replacement policy, abstract cache state destruction by unknown addresses	Method cache, stack cache, and a highly associative, small heap cache
Branch prediction	Long history in dynamic predictors	Static branch prediction
Superscalar architectures	Timing anomalies	Avoid, instead use CMP and/or VLIW
Chip-multithreading	Interthread interference (cache, pipeline)	Avoid, instead use CMP
Chip-multiprocessors	Intercore interference via shared memory, cache	TDMA scheduled memory access

Pipeline Dependencies

Pipelining is a major architectural feature to speed up program execution. Different stages of an instruction are overlapped and, therefore, executed in parallel. The theoretical throughput of a scalar pipeline is one instruction per clock cycle.

In contrast to Whitham [16], we think that a time-predictable architecture should be pipelined. The pipeline should be simple and dependencies between instructions avoided, or at least minimized, to avoid unbounded timing effects.

Instruction Fetch

To avoid a prefetch queue, with probably unbounded execution-time dependencies over a stream of instructions, a fixed-length instruction set is recommended. Variable length instructions can complicate instruction cache analysis because an instruction can cross a block boundary. The method cache, as proposed in the following section, avoids this issue. Either all instructions of a function, independent of their length, are in the cache, or none of them.

Fetching variable-sized instructions from the method cache can be performed in a single cycle. The method cache is split into two interleaved memories banks. Each of the two cache memories needs a read port wide enough for a maximum-sized instruction. Accessing both memories concurrently with a clever address calculation overcomes the boundary issue for variable-sized instruction access.

Caches

To reduce the growing gap between the clock frequency of the processor and memory access times, multilevel cache architectures are commonly used. Since even a single-level cache is problematic for WCET analysis, more levels in the memory architecture are practically not analyzable. The additional levels also increase the latency of the memory access on a cache miss.

For the cache analysis, the addresses of the memory accesses need to be predicted. The addresses for the instruction fetch are easy to determine, and access to stack allocated data, for example, function arguments and local variables, is also quite regular. The addresses can be predicted when the call tree is known.

The addresses for heap-allocated data are very hard to predict statically—the addresses are only known during runtime (we found no publication that describes analysis of the data cache for heap-allocated data). Without knowing the address, a single access influences all sets in the cache.

To avoid corruption of the abstract cache state in the analysis by data accesses, separate instruction and data caches are mandatory [11]. Furthermore, we propose to split the data cache into a cache for stack-allocated data and a cache for global- or heap-allocated data. As stack allocated data is guaranteed thread local, the stack cache can be further simplified for CMP systems.

For the instruction cache, we propose a new form of organization where whole functions are loaded on a miss on call or return. Figure 2 shows the proposed organization of the three caches.

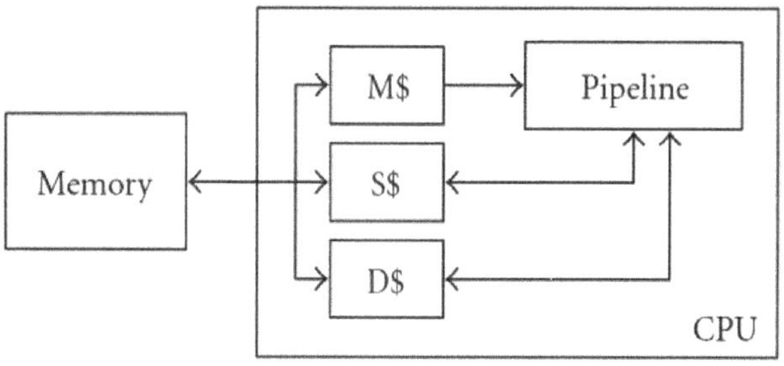

Figure 2 Cache configuration for the time-predictable architecture. The method cache M$ caches instructions of a full method/function. The data cache D$ is augmented by a stack cache S$ to avoid cache trashing of stack allocated data with heap allocated data.

The Instruction Cache

We propose a new form of organization for the instruction cache: the *method cache* [3], which has a novel replacement policy. A whole function or method is loaded into the cache on a call or return. This cache fill strategy pools all the cache misses of a function. All instructions except call and return are guaranteed cache hits. Only the call tree needs to be analyzed during the cache analysis. With the proposed cache organization, the cache analysis can be performed independently of the pipeline analysis.

Filling the cache on call and return only removes another source of interference: there is no competition for the main memory access between instruction cache and data cache. In traditional architectures, there is a subtle dependency between the instruction cache and memory access for a load or store instruction. For example, a load or store at the end of the processor pipeline competes with an instruction fetch that results in a cache miss. One of the two instructions is stalled for additional cycles by the other instruction. With a data cache, this situation can be even worse. The worst-case scenario for the memory stall time for an instruction fetch or a data load is two miss-penalties when both cache reads are a miss.

The main restriction of the method cache is that a whole method needs to fit into the cache. For larger methods, software- and hardware-based options are possible to resolve this issue. The compiler can split large methods into several shorter methods. At the hardware level, there are two options for methods that are too large: the cache can be disabled or the method cache can be switched into a direct-mapped mode.

If we avoid absolute jumps within a method, we can use a relative program counter within the method and place a method at each position within the cache. This property is fulfilled with Java bytecode, but can also be enforced by the compiler for C code.

For a full method load into the cache, we need to know the length of the method. This information is available in the Java class file. For compiled C code, this information can be provided in the executable. A simple convention, implemented in the linker, is to store the method length one word before the actual method starts. In order to use the method cache in an RISC processor, the ISA is extended with a prefetch instruction to force the cache load. The prefetch instruction can be placed immediately before the call or return instruction. It can also be scheduled earlier to hide the cache load latency.

The Stack Cache

Access patterns to stack allocated data are different from heap- or static-allocated data. Addresses into the stack are easy to predict statically because the allocation addresses of stack frames can be predicted by the analysis of the call tree. Furthermore, a new stack frame for a function call does not need to be cache-consistent with the main memory. The involved cache blocks need no cache fill from the main memory?

To benefit from these properties for WCET analysis, we propose to split the data cache into a stack cache and a cache for static- and heap-allocated data (it is possible to further split the data cache into a cache for static data and heap data). The organization of the cache for static- and heap-allocated data, further referred to as data cache, will be proposed in the following section.

The regular access pattern to the stack cache will not benefit from set associativity. Therefore, the stack cache is a simple direct-mapped cache. The stack contains local variables and the write frequency is higher than for other memory areas. The high frequency mandates a write back organization.

A stack cache is similar to a windowed register file as implemented in the Berkeley RISC processor [28]. A stack cache can be organized to exchange data with the main memory on a stack frame basis. When the cache overflows, which happens only during a call, the oldest frame or frames have to be moved to the memory. A frame needs to be loaded from the memory only when a function returns. Exchange with the main memory can be implemented in hardware, microcode, or with compiler visible machine instructions.

If the maximum call depth results in a stack that is smaller than the stack cache, all accesses will be a cache hit. A write back occurs first when the program reaches a call depth resulting in a wraparound within the cache. A cache miss can occur only when the program goes up in the call tree and needs access to a cache block that was evicted by a call down in the call tree.

Figure 3 shows the call and return behavior of a program over time and the changing stack cache window. The stack grows downwards in the figure. The dashed box shows a possibility to enforce a write back at some program point. The following stack changes fit into the enforced stack window and no memory transactions are necessary.

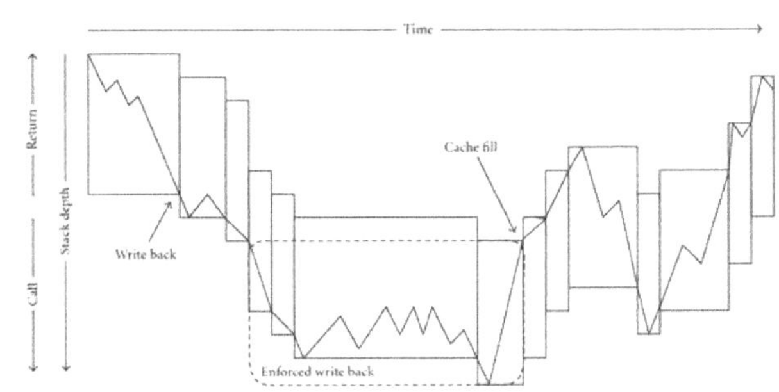

Figure 3 Stack usage for call and return and the resulting stack cache window. When the window overflows on a call, a write back of old frames is necessary. The stack cache fill is caused by an underflow after a return. Enforcing a write back of the whole stack cache can guarantee hits for subsequent, more deeply nested functions.

On a return, the previously used cache blocks can be marked empty because function local data is not accessible after the return (it could be accessed in C by returning a pointer to the stack data; however, this is undefined and considered poor programming practic). As a result, cache lines will never need to be written back on a cache wrap around after return. The stack cache activity can be summarized in the following way.

- A cache miss can only occur after a return. The first miss is at least *one cache size* away from a leaf in the call tree.
- Cache write back can only occur after a function call. The first write back is *one cache size* away from the root of the call tree.

We can make the misses and write backs more predictable by forcing them to occur at explicit points in the call tree. At these points, the cached stack frames are written back to the main memory and the whole stack cache is marked empty. If we place the flush points at function calls in the call tree that are within *one cache size* from the leaf functions, all cache accesses into that area are guaranteed hits. This algorithm can actually improve WCET because most of the execution time of a program is spent in inner loops further down the call tree.

Stack data is usually not shared between threads and no cache coherence and consistence protocol—the major bottleneck for CMP scaling—needs to be implemented for a CMP system.

The Data Cache

For conservatively written programs with statically allocated data, the address of the data is known after program linking. Value analysis results in a good prediction of read and write addresses. The addresses are the input for the cache analysis. In [48], control tasks, from a real-time benchmark provided by Airbus, were analyzed. For this benchmark, 90% of the memory accesses were predicted precisely.

In a modern object-oriented language, data is usually allocated on the heap. The address for these objects is only known at runtime. Even when using such a language in a conservative style, where all data is allocated during an initialization phase, it is not easy to predict the resulting addresses. The order of the allocations determines the addresses of the objects. When the order becomes unknown at one point in the initialization phase, the addresses for all following allocations cannot be determined precisely.

It is possible to analyze local cache effects with unknown addresses for an LRU set-associative cache. For an n -way associative cache, the history for n different addresses can be tracked. Because the addresses are unknown, a single access influences *all* sets in the cache. The analysis reduces the effective cache size to a single set.

The local analysis for the LRU-based cache is illustrated by a small example with a four-word cache. The example cache allocates a cache block on a write. Table 2 shows a code fragment with access to heap-allocated data (objects `a` , `b` , `c` , and `d`). The cache state after the load or store instruction is shown in the right section of the table. The leftmost column of the cache state represents the youngest element, the rightmost column the oldest (the LRU element). We assume a 4-way set-associative cache for the example. Therefore, we can locally track four different and *unknown* addresses. After the first two constant assignments, we know that `a.v` and `b.v` are in the cache. The following load of `b.v` is trivially a hit and the store into `c.v` changes the cache content and the age of `a.v` and `b.v` . All following loads are hits and only change the age ordering of the cache elements. In this small example we dealt with four different and unknown addresses, but could classify all read accesses as hits for a four-word cache.

Table 2 An example of analyzable accesses to three heap-allocated objects with a four-word LRU cache. The cache content after the execution of a statement is depicted in the right section of the table.

		Cache state			
Instruction	**Memory**	**Youngest**		**Oldest**	
a.v = 123;	store a.v	a.v	—	—	—
b.v = 456;	store b.v	b.v	a.v	—	—
c.v = b.v;	load b.v	b.v	a.v	—	—
	store c.v	c.v	b.v	a.v	—
d.v = b.v;	load b.v	b.v	c.v	a.v	—
	store d.v	d.v	b.v	c.v	a.v
b.v = a.v;	load a.v	a.v	d.v	b.v	c.v
	store b.v	b.v	a.v	d.v	c.v

We propose to implement the cache architecture exactly as it results from this analysis—a small, fully associative cache with an LRU replacement policy. This cache organization is similar to the victim cache [49], which adds associativity to a direct-mapped cache. A small, fully associative buffer holds discarded cache blocks. The replacement policy is LRU.

LRU is difficult to calculate in hardware and only possible for very small sets. Replacement of the oldest block gives an approximation of LRU. The resulting FIFO strategy can be used for larger caches. To offset the less predictable behavior of the FIFO replacement [39], the cache has to be much larger than an LRU-based cache.

The Scratchpad Memory

A common method for avoiding data caches is an onchip memory called scratchpad memory, which is under program control. This program managed memory entails a more complicated programming model although it can be automatically partitioned [50, 51]. A similar approach for time-predictable caching is to lock cache blocks. The control of the cache locking [52] and the allocation of data in the scratchpad memory [53, 54] can be optimized for the WCET. A comparison between locked cache blocks and a scratchpad memory with respect to the WCET can be found in [55].

Exposing the scratchpad memory at the language level can further help to optimize the time-critical path of the application.

Branch Prediction

As the pipelines of current general-purpose processors become longer to support higher clock rates, the penalty of branches also increases. This is compensated by branch prediction logic with branch target buffers. However, the upper bound of the branch execution time is the same as without this feature.

Simple static branch prediction (e.g., backward branches are assumed taken, forward branches not taken) or compiler-generated branch predictions are WCET-analyzable options. One-level dynamic branch predictors can be analyzed [56]. The branch history table has to be separate from the instruction cache to allow independent modeling for the analysis.

Instruction Level Parallelism

Statically scheduled VLIW processors are an option for a time-predictable architecture. The balance between the VLIW width and the number of cores in a CMP system depends on the application domain. For control-oriented applications, we assume that a dual-issue VLIW is a practical architecture. DSP-related applications can probably fill more instruction slots with useful instructions.

Dynamically scheduled superscalar architectures are not considered as an option for a time-predictable architecture. The amount of hardware that is needed to extract ILP from a single thread is better spent on a (VLIW-based) CMP system.

Chip Multithreading

Fine-grained multithreading within the pipeline is in principle not an issue for WCET analysis. The scheduling algorithm of the threads needs to be known and must not depend on the state of the threads. Round-robin scheduling is a time-predictable option. The execution time for simple instructions simply increases by a factor equal to the number of threads. The benefit of hiding pipeline stalls due to data dependencies or branches results in a lower factor for these instructions. Execution of n tasks on an n -way multithreading pipeline takes less (predictable) time than executing these tasks serially on a single threaded processor. However, cache misses, even if a single cache miss could be hidden, result in interference between the different threads because the memory interface is a shared resource.

Fine-grained multithreading resolves the data dependencies for a thread within the pipeline: the thread is only active in a single pipeline stage. Therefore, the forwarding network can be completely removed from the processor. This is an important simplification of the pipeline because the forwarding multiplexer is often part of the critical path that restricts the maximum clock frequency.

To avoid cache thrashing, each thread needs—in addition to its own register file—its own instruction and data cache, which reduces the effectively shared transistors to the pipeline itself. We think that the cost is too high for the small performance enhancement. Therefore, also duplicating the pipeline—resulting in a CMP solution—will result in a better performance/cost factor.

SMT is not an option as the interaction between the threads is too complex to model.

Chip Multiprocessors

Embedded applications need to control and interact with the *real*-world, a task that is inherently parallel. Therefore, these systems are good candidates for CMPs. We discuss that the transistors required to implement superscalar architectures are better used on complete replication of simple cores.

CMP systems share the access bandwidth to the main memory. To build a time-predictable CMP system, we need to schedule the access to the main memory in a predictable way. A predictable scheduling can only be time-based, where each core receives a fixed time slice. This scheduling scheme is called time division multiple access (TDMA). The time slices do not need to be of equal size. The execution time of un-cached loads and stores and the cache miss-penalty depend on this schedule and, therefore, for accurate WCET analysis, the complete schedule needs to be known.

Assuming that enough cores are available, we propose a CMP model with a single thread per processor. In that case, thread switching and schedulability analysis for each individual core disappears. Since each processor executes only a single thread, the WCET of that thread can be as long as its deadline. When the period of a thread is equal to its deadline, 100% utilization of that core is feasible. For threads that have enough slack time left, we can increase the WCET by decreasing their share of the bandwidth on the memory bus. Other threads with tighter deadlines can, in turn, use the freed bandwidth and run faster. The usage of the shared resource main memory is adjusted by the TDMA schedule. The TDMA schedule itself is the input for WCET analysis for all threads. Finding a TDMA schedule, where all tasks meet their deadlines, is thus an iterative optimization problem.

Figure 4 shows the analysis tool flow for the proposed time-predictable CMP with three tasks. First, an initial arbiter schedule is generated, for example, one with equal time slices. That schedule and the tasks are the input of WCET analysis performed for each task individually. If all tasks meet their deadline with the resulting WCETs, the system is schedulable. If some tasks do not meet their deadline and other tasks have some slack time available, the arbiter scheduler is adapted accordingly. WCET analysis is repeated, with the new arbiter schedule until all tasks meet their deadlines or no slack time for an adaption of the arbiter schedule is available. In the latter case, no schedule for the system is found.

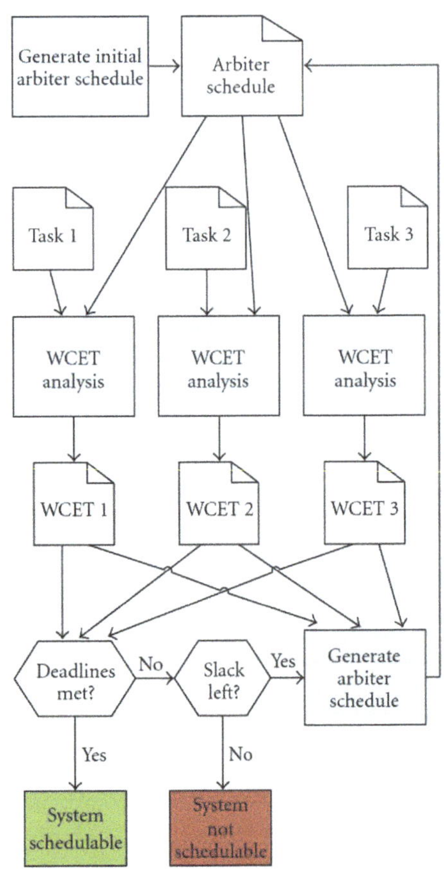

Figure 4 Tool flow for a CMP-based real-time system with one task per core and a static arbiter schedule. If the deadlines are not met, the

arbiter schedule is adapted according to the WCETs and deadlines of the tasks. After the update of the arbiter schedule, the WCET of all tasks needs to be recalculated.

Documentation

The hardware description language VHDL was originally developed to document the behavior of digital circuits. Today, digital hardware can be synthesized from a VHDL description. Therefore, the VHDL code for the processor is the ideal form of documentation. VHDL code can also be simulated and all interactions between different components are observable.

An open-source design enables the WCET tool provider to check the real processor when the documentation is missing; documentation errors are also easier to find. Sun provides the Verilog source of their Niagra T1 [42] as open-source under the GNU GPL (http://www.opensparc.net/opensparc-t1/downloads.html).

EVALUATION

In this section, we evaluate some of the proposed time-predictable architectural features with JOP [2], an implementation of a Java processor. We have chosen to natively support Java as it is the language which will be used for future safety critical systems [57, 58]. Java's intermediate representation, the Java class file, is analysis friendly and the type information can be reconstructed from the class file. Executing bytecodes—the instruction set of the Java virtual machine (JVM)—directly in the hardware allows WCET analysis at the bytecode level. The translation step from bytecode to machine code, which introduces timing inaccuracies, can be avoided.

The Java Processor JOP

The major design goal of JOP is the time-predictable execution of Java bytecodes [59]. All functional units, and especially the interactions between them, are carefully designed to avoid any timing dependency between bytecodes.

JOP dynamically translates the Java bytecodes to a stack-based microcode that can be executed in a three-stage pipeline. The translation takes exactly one cycle per bytecode. Compared to other forms of dynamic code translation, the scheme used in JOP does not add any variable latency to the execution time and is, therefore, time-predictable.

JOP contains a simple execution stage with the two topmost stack elements as discrete registers. No write back stage or forwarding logic is needed. The short pipeline (four stages) results in short conditional branch delays; a difficult to analyze branch prediction logic or a branch target buffer can be avoided.

All microcode instructions have a constant execution time of one cycle. No stalls are possible in the microcode pipeline. Loads and stores of object fields are handled explicitly. The absence of timing dependencies between bytecodes results in a simple processor model for the low-level WCET analysis.

The proposed architecture is open-source and all design files are available (http://www.jopdesign.com/). The instruction timing of the bytecodes is documented.

Method Cache

JOP contains the proposed method cache. The default configuration is 4 KB, divided into 16 blocks of 256 Bytes. The replacement strategy is FIFO.

WCET analysis of the method cache and of standard instruction caches is currently under development. Therefore, we perform only average-case measurements for a comparison between a time-predictable cache organization and a standard cache organization. With a simulation of JOP, we measure the cache misses and miss-penalties for different configurations of the method cache and a direct-mapped cache. The miss-penalty and the resulting effect on the execution time depend on the main memory system. Therefore, we simulate three different memory technologies: static RAM (SRAM), synchronous DRAM (SDRAM), and double data rate (DDR) SDRAM. For the SRAM, a latency of 1 clock cycle and an access time of 2 clock cycles per 32 bit word are assumed. For the SDRAM, a latency of 5 cycles (3 cycles for the row address and 2 cycles for the CAS latency) is assumed. The SDRAM delivers one word (4 bytes) per cycle. The DDR SDRAM has a shorter latency of 4.5 cycles and transfers data on both the rising and falling edges of the clock signal.

The resulting miss-cycles are scaled to the bandwidth consumed by the instruction fetch unit. The result is the number of cache fill cycles per fetched instruction byte: in other words, the average main memory access time in cycles per instruction byte. A value of 0.1 means that for every 10 fetched instruction bytes, one clock cycle is spent to fill the cache.

Table 3 shows the result for different configurations of a direct-mapped cache. For the evaluation, we used an adapted version of the real-time application Kfl (the Kfl benchmark is also used in Section 5.4), which is a node in a distributed control application. As the embedded application is quite small (1366 LOC), we simulated small instruction caches. The best performing configuration depends on the relationship between memory bandwidth and memory latency. The data in bold emphasize the best block size for the different memory technologies. As expected, memories with a higher latency and bandwidth perform better with larger block sizes. For small block sizes, the latency clearly dominates the access time. Although the SRAM has half the bandwidth of the SDRAM and a quarter of the DDR SDRAM, it is faster than the SDRAM memories with a block size of 8 byte. In most cases, a block size of 16 bytes is fastest.

Table 3: Direct-mapped cache, average memory access time.

Cache size	Block size	SRAM	SDRAM	DDR
1 KB	8 B	**0.18**	0.25	0.19
1 KB	16 B	0.22	**0.22**	0.16
1 KB	32 B	0.31	0.24	**0.15**

2 KB	8 B	**0.11**	0.15	0.12
2 KB	16 B	0.14	**0.14**	**0.10**
2 KB	32 B	0.22	0.17	0.11

Table 4 shows the average memory access time per instruction byte for the method cache. Because we load full methods, we have chosen larger block sizes than for a standard cache. All configurations benefit from a memory system with a higher bandwidth. The method cache is less latency sensitive than the direct-mapped instruction cache. For the small 1 KB cache, the access time is almost independent of the block size. The capacity misses dominate. From the 2 KB configuration, we see that smaller block sizes result in less cache misses. However, smaller block sizes result in more hardware for the hit detection since the method cache is in effect fully associative. Therefore, we need a balance between the number of blocks and the performance.

Table 4 Method cache, average memory access time.

Cache size	Block size	SRAM	SDRAM	DDR
1 KB	16 B	0.36	0.21	0.12
1 KB	32 B	0.36	0.21	0.12
1 KB	64 B	0.36	0.22	0.12
1 KB	128 B	0.41	0.24	0.14
2 KB	32 B	0.06	0.04	0.02
2 KB	64 B	0.12	0.08	0.04
2 KB	128 B	0.19	0.11	0.06
2 KB	256 B	0.37	0.22	0.13

The cache conflict is high for the small configuration with 1 KB cache. The direct-mapped cache, backed up with a low-latency main memory, performs better than the method cache. When high-latency memories are used, the method cache performs better than the direct-mapped cache. This is expected as the long latency for a transfer is amortized when more data (the whole method) is filled in one request.

A small block size of 32 bytes is needed in the 2 KB method cache to outperform the direct-mapped cache with the low-latency main memory as represented by the SRAM. For higher latency memories (SDRAM and DDR), a method cache with a block size of 128 bytes outperforms the direct-mapped instruction cache.

The comparison does not show if the method cache is more easily predictable than other cache solutions. It shows that caching full methods performs similarly to standard caching techniques.

Stack Cache

In JOP, a simplified version of the proposed stack cache is implemented. The JVM uses the stack not only for the activation frame and for local variables but also for operands. Therefore, the two top elements of the stack are implemented as registers [4]. With this configuration, we can avoid the write-back pipeline stage.

The fill and spill between the stack cache and the main memory is simplified. The cache content is exchanged only on a thread switch. Therefore, the maximum call depth is restricted by the onchip cache size. In a future version of JOP, we intend to relax this limitation. The cache fill will be performed on a return and the write back on invoke when necessary. A stack analysis tool will add a marker to the methods where a full cache write back will be performed and the stack access in methods deeper in the call tree will be guaranteed hits. Heap-allocated data and static fields are not cached in the current implementation of JOP.

Branch Prediction

In JOP, branch prediction is avoided. This results in pressure on the pipeline length. The microprogrammed core processor has a pipeline length of as little as three stages resulting in a branch execution time of three cycles in microcode. The two slots in the branch delay can be filled with instructions or *nop*. With the additional bytecode fetch and translation stage, the overall pipeline is four stages and results in a four-cycle execution time for a bytecode branch.

WCET analysis

Bytecode instructions that do not access memory have a constant execution time. Most simple bytecodes are executed in a single cycle. Table 5 shows example instructions and their timing. The access time to object, array, and class fields depends on the timing of the main memory. With a memory with r_{ws} wait states for a read access, the execution time for, for example, getfield, is

$$t_{getfield} = 11 + 2r_{ws}. \quad (1)$$

Table 5 Execution time of simple bytecodes in cycles.

Instruction	Cycles	Funtion
iconst_0	1	load constant 0 on TOS
bipush	2	load a byte constant on TOS
iload_0	1	load local variable 0 on TOS
iload	2	load a local variable on TOS
dup	1	duplicate TOS
iadd	1	integer addition
isub	1	integer subtraction
ifeq	4	conditional branch

To demonstrate that JOP is amenable to WCET analysis, we have built an IPET-based WCET analyzer [60]. While loop bounds are annotated at the source level, the analysis is performed at the bytecode level. Without dependencies between bytecodes, the pipeline analysis can be omitted. The execution time of basic blocks is calculated simply by adding the execution time of individual bytecodes. For the method cache, we have implemented a simplified analysis where only leaf nodes in the call tree are considered. A return from such a leaf node is a guaranteed hit. (The maximum method size is restricted to half of the cache size.) Invocation of a leaf node in a tight loop (without invocations of other methods) is classified as a miss for the first iteration and a hit for the following iterations. For small benchmarks, the overestimation of the WCET is around 5%. For two real applications (Lift and Kfl) the analysis resulted in an overestimation of 56% and 116%. It should be noted that the overestimation is calculated by comparison with measurement-based WCET estimation, which is not a safe approach.

Another indication that JOP is a WCET *friendly* design is that other real-time analysis projects use JOP as the primary target platform. Harmon has developed a tree-based WCET analyzer for interactive back-annotation of WCET estimates into the program source [61]. Bøgholm et al. have developed an integrated WCET and scheduling analysis tool based on model checking [62].

Comparison with Picojava

We compare the time-predictable JOP design with picoJava [63, 64], a Java processor designed for average-case performance. Simple bytecodes are directly supported by the processor. Most of them execute in a single cycle. More complex bytecodes trap to a software routine. However, the invocation time of the trap depends on the cache state and is between 6 cycles in the best case and 426 cycles in the worst case—a factor in the order of two magnitudes. Some of the trapped instructions (e.g., invokevirtual) can be replaced at runtime by a *quick* version (e.g.,invokevirtual_quick). This replacement results in different execution times for the first execution of some code and following executions.

To speed up sequences of stack operations, picoJava can fold several instructions into an RISC style three register operation, for example, the sequence: load , load , add , store . This feature compensates for the inefficiency of a stack machine. However, the folding unit depends on a 16-byte instruction buffer with all the resulting unbounded timing effects of a prefetch queue.

However, picoJava implements a 64-word stack buffer as discrete registers. Spill and fill of that stack buffer is performed in background by the hardware. Therefore, the stack buffer closely interacts with the data cache. The interference between the folding unit, the instruction buffer, the instruction cache, the stack buffer, the data cache, and the memory interface causes complications in modeling picoJava for WCET analysis.

Also, picoJava is about 8 times larger than JOP and can be clocked at less than half of the frequency of JOP in the same technology [65]. Therefore, the small size of a time-predictable architecture naturally leads to a CMP system.

Performance

One important question remains: is a time-predictable processor slow? We evaluate the average-case performance of JOP by comparing it with other embedded Java systems: Java processors from industry and academia as well as two just-in-time (JIT) compiler-based systems. For the comparison, we use JavaBenchEmbedded , (available athttp://www.jopwiki.com/JavaBenchEmbedded) a set of open-source Java benchmarks for embedded systems. However, Kfl and Lift are two real-world applications adapted with a simulation of the environment to run as stand-alone benchmarks. Udplp is a simple client/server test program that uses a TCP/IP stack written in Java.

Table 6 shows the raw data of the performance measurements of different embedded Java systems for the three benchmarks. The numbers are iterations per second whereby a higher value represents better performance. Figure5 shows the results scaled to the performance of JOP.

Table 6 Application benchmark performance on different Java systems. The table shows the benchmark results in iterations per second—a higher value means higher performance.

	Kfl	**UdpIp**	**Lift**
Cjip	176	91	
Jamuth	3400	1500	
EJC	9893	2882	
SHAP	11570	5764	12226
aJ100	14148	6415	
JOP	18275	8467	18649
picoJava	23813	11950	25444
CACAO/YARI	39742	17702	38437

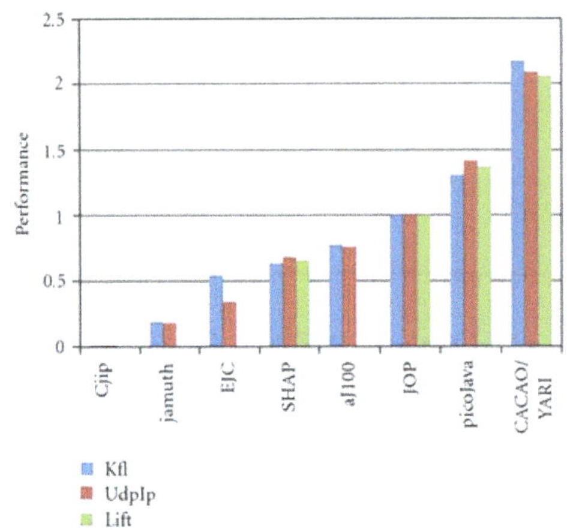

Figure 5 Performance comparison of different Java systems with embedded application benchmarks. The results are scaled to the performance of JOP.

The numbers for JOP are taken from an implementation in the Altera Cyclone FPGA [66], running at 100 MHz. JOP is configured with a 4 KB method cache and a 1 KB stack cache.

Cjip [67] and aJ100 [68] are commercial Java processors, which are implemented in an ASIC and clocked at 80 and 100 MHz, respectively. Both cores do not cache instructions. However, aj100 contains a 32 KB onchip stack memory. jamuth [69] and SHAP [70] are Java processors that are implemented in an FPGA. Also, jamuth is the commercial version of the Java processor Komodo [71], a research project for real-time chip multithreading; and it is configured with a 4 KB direct-mapped instruction cache for the measurements. The architecture of SHAP is based on JOP and enhanced with a hardware object manager. Also, SHAP implements the method cache [72]. The benchmark results for SHAP are taken from the SHAP website (http://shap.inf.tu-dresden.de/, accessed July, 2008); and it is configured with a 2 KB method cache and 2 KB stack cache.

Also, picoJava [73] is a Java processor developed by Sun. picoJava is no longer produced and the second version (picoJava-II) was available as open-source Verilog code. Puffitsch implemented picoJava-II in an FPGA (Altera Cyclone-II) and the performance numbers are obtained from that implementation [65]. picoJava is configured with a direct-mapped instruction cache and a 2-way set-associative data cache. Both caches are 16 KB.

EJC [74] is an example of a JIT system on an RISC processor (32-bit ARM720T at 74 MHz). The ARM720T contains an 8 KB unified cache. To compare JOP with a JIT-based system in exactly the same hardware, we use the research JVM CACAO [75] on top of the MIPS compatible soft-core YARI [76]. YARI is configured with a 4-way set-associative instruction cache and a 4-way set-associative write-through data cache. Both caches are 8 KB.

The measurements do not provide a clear answer to the question of whether a time-predictable architecture is slow. JOP is about 33% faster than the commercial Java processor aJ100. However, picoJava is 36% faster than JOP and the JIT/RISC combination is about 111% faster than JOP. (The numbers of CACAO/YARI are from [76]. In the mean time, YARI has been enhanced and outperforms JOP by a factor of 2.8.) We conclude that a time-predictable solution will never be as fast in the average case as a solution optimized for the average case.

Hardware Area and Clock Frequency

Table 7 compares the resource consumption and maximum clock frequency of a time-predictable processor (JOP), a standard MIPS architecture (YARI), and a complex Java processor (picoJava), when implemented in the same FPGA. The streamlined architecture of JOP results in a small design: JOP is half the size of the MIPS core YARI, and compared to picoJava consumes about 12% of the resources. JOP's size allows implementing a CMP version of JOP even in a low-cost FPGA. The simple pipeline of JOP achieves the highest clock frequency of the three designs. From the frequency comparison, we can estimate that the maximum clock frequency of JOP in an ASIC will also be higher than a standard RISC pipeline in an ASIC.

Table 7 Resource consumption and maximum operating frequency of JOP, YARI, and picoJava.

Soft-core	Logic cells	Memory	Frequency
JOP	3,300	7.6 KB	100 MHz
YARI	6,668	18.9 KB	75 MHz
picoJava	27,560	47.6 KB	40 MHz

Jop CMP System

We have implemented a CMP version of JOP with a fairness-based arbiter [77]. All cores are allotted an equal share of the memory bandwidth. Each core has its own method cache and stack cache. Heap-allocated data is not cached in this design.

When comparing a JOP CMP system against the complex Java processor picoJava, a dual core version of JOP is about 5% slower than a single picoJava core, but consumes only 22% of the chip resources. With four cores, JOP outperforms picoJava by 30% with size of 43% of picoJava.

A configurable TDMA arbiter for a time-predictable CMP system and the integration of the arbitration schedule into the WCET tool [60] is presented in [9].

Summary

A model of a processor with accurate timing information is essential for tight WCET analysis. The architecture of JOP and the microcode are designed with this in mind. Execution time of bytecodes is known cycle accurately [59]. It is possible to analyze the WCET at the bytecode level [78] without the uncertainties of an interpreting JVM [79] or generated native code from ahead-of-time compilers for Java.

CONCLUSION

In this paper, we discuss a time-predictable computer architecture for embedded real-time systems that supports WCET analysis. We have identified the problematic microarchitecture features of standard processors and provided alternative solutions when possible.

Dynamic features, which model a large execution history, are problematic for WCET analysis. Especially interferences between different features result in a state space explosion for the analysis. The proposed architecture is an in-order pipeline with minimized instruction dependencies. The cache memory consists of a method cache containing whole methods and a data cache that is split for stack-allocated data and heap-allocated data. The pipeline can be extended to a dual-issue pipeline when the instructions are compiler scheduled. For further performance enhancements, we propose a CMP system with time-sliced arbitration of the main memory access. Running each task on its own core in a CMP system eliminates scheduling, and the related cache thrashing, from the analysis. The schedule of the memory access becomes an input for WCET analysis. With nonuniform time slices, the arbiter schedule can be adapted to balance the utilization of the individual cores.

The concept of the proposed architecture is evaluated by a real-time Java processor, called JOP. We have presented a brief overview of the architecture. A simple four-stage pipeline and microcoded implementation of JVM bytecodes result in a time-predictable architecture. The proposed method and stack caches are implemented in JOP. The resulting design makes JOP an easy target for the low-level WCET analysis of Java applications.

We compared JOP against several embedded Java systems. The result shows that a time-predictable computer architecture does not need to be slow. A streamlined, time-predictable processor design is quite small. Therefore, we can regain performance by the exploration of thread level parallelism in embedded applications with a replication of the processor in a CMP architecture.

The proposed processor has been used with success to implement several commercial real-time applications [80]. JOP is open-source under the GNU GPL, and all design files and the documentation are available athttp://www.jopdesign.com/.

We plan to implement some of the suggested architectural enhancements in an RISC-based system in the future. We will implement the proposed stack cache and the method cache in YARI [76], an open-source, MIPS ISA-compatible RISC implementation in an FPGA.

A scratchpad memory for JOP is implemented and the integration into the programming model is under investigation. We will add a small fully associative data cache to JOP. This cache will also serve as a buffer for a real-time transactional memory for the JOP CMP system. We will investigate whether a standard cache for static data is a practical solution for Java.

REFERENCES

1. Hennessy J, Patterson D: Computer Architecture: A Quantitative Approach. 4th edition. Morgan Kaufmann, San Francisco, Calif, USA; 2006.
2. Schoeberl M: A Java processor architecture for embedded real-time systems. Journal of Systems Architecture 2008,54(1-2):265-286. 10.1016/j.sysarc.2007.06.001
3. Schoeberl M: A time predictable instruction cache for a Java processor. In Proceedings of the On the Move to Meaningful Internet Systems: Workshop on Java Technologies for Real-Time and Embedded Systems (JTRES '04), October 2004, Agia Napa, Cyprus, Lecture Notes in Computer Science. Volume 3292. Springer; 371-382.
4. Schoeberl M: Design and implementation of an efficient stack machine. Proceedings of the 12th IEEE Reconfigurable Architecture Workshop (RAW '05), April 2005, Denver, Colo, USA
5. Bate I, Conmy P, Kelly T, McDermid J: Use of modern processors in safety-critical applications. The Computer Journal 2001,44(6):531-543. 10.1093/comjnl/44.6.531

6. Thiele L, Wilhelm R: Design for timing predictability. Real-Time Systems 2004,28(2-3):157-177.
7. Edwards SA, Lee EA: The case for the precision timed (PRET) machine. In Proceedings of the 44th ACM/IEEE Annual Conference on Design Automation (DAC '07), June 2007, San Diego, Calif, USA. ACM Press; 264-265.
8. Lickly B, Liu I, Kim S, Patel HD, Edwards SA, Lee EA: Predictable programming on a precision timed architecture. In Proceedings of the International Conference on Compilers, Architectures and Synthesis for Embedded Systems (CASES '08), October 2008, Atlanta, Ga, USA. Edited by: Altman ER. ACM Press; 137-146.
9. Pitter C: Time-predictable memory arbitration for a Java chip-multiprocessor. Proceedings of the 6th International Workshop on Java Technologies for Real-Time and Embedded Systems (JTRES '08), September 2008, Santa Clara, Calif, USA 115-122.
10. Berg C, Engblom J, Wilhelm R: Requirements for and design of a processor with predictable timing. In Perspectives Workshop: Design of Systems with Predictable Behaviour, 2004, Schloss Dagstuhl, Germany, Dagstuhl Seminar Proceedings. Volume 03471. Edited by: Thiele L, Wilhelm R. Internationales Begegnungsund Forschungszentrum für Informatik (IBFI);
11. Heckmann R, Langenbach M, Thesing S, Wilhelm R: The influence of processor architecture on the design and the results of WCET tools. Proceedings of the IEEE 2003,91(7):1038-1054. 10.1109/JPROC.2003.814618
12. Anantaraman A, Seth K, Patil K, Rotenberg E, Mueller F: Virtual simple architecture (VISA): exceeding the complexity limit in safe real-time systems. In Proceedings of the 30th Annual International Symposium on Computer Architecture (ISCA '03), June 2003, San Diego, Calif, USA. ACM Press; 350-361.
13. Puschner P, Burns A: Writing temporally predictable code. In Proceedings of the 7th IEEE International Workshop on Object-Oriented Real-Time Dependable Systems (WORDS '02), January 2002, San Diego, Calif, USA. IEEE Computer Society; 85-94.
14. Delvai M, Huber W, Puschner P, Steininger A: Processor support for temporal predictability—the SPEAR design example. Proceedings of the 15th Euromicro Conference on Real-Time Systems (ECRTS '03), July 2003, Porto, Portugal 169-176.
15. Puschner P, Schoeberl M: On composable system timing, task timing, and WCET analysis. Proceedings of the 8th International Workshop on Worst-Case Execution Time (WCET) Analysis, July 2008, Prague, Czech Republic
16. Whitham J: Real-time processor architectures for worst case execution time reduction, Ph.D. thesis. University of York, York, UK; 2008.
17. Whitham J, Audsley N: Using trace scratchpads to reduce execution times in predictable real-time architectures. Proceedings of the 14th IEEE Real-Time and Embedded Technology and Applications Symposium (RTAS '08), April 2008, St. Louis, Mo, USA 305-316.
18. Li Y-TS, Malik S: Performance analysis of embedded software using implicit path enumeration. In Proceedings of the ACM SIGPLAN Workshop on Languages, Compilers, & Tools for Real-Time Systems (LCTES '95), November 1995, La Jolla, Calif, USA. ACM Press; 88-98.
19. Puschner PP, Schedl AV: Computing maximum task execution times—a graph-based approach. Real-Time Systems 1997,13(1):67-91. 10.1023/A:1007905003094
20. Gustafsson J: Analyzing execution-time of object-oriented programs using abstract interpretation, Ph.D. thesis. Uppsala University, Uppsala, Sweden; 2000.
21. Puschner P, Burns A: Guest editorial: a review of worst-case execution-time analysis. Real-Time Systems 2000,18(2-3):115-128.
22. Wilhelm R, Engblom J, Ermedahl A, et al.: The worst-case execution-time problem-overview of methods and survey of tools. Transactions on Embedded Computing Systems 2008,7(3):1-53.
23. Engblom J, Ermedahl A, Södin M, Gustafsson J, Hansson H: Worst-case execution-time analysis for embedded real-time systems. International Journal on Software Tools for Technology Transfer 2003,4(4):437-455. 10.1007/s100090100054
24. Nilsen KD, Rygg B: Worst-case execution time analysis on modern processors. ACM SIGPLAN Notices 1995,30(11):20-30. 10.1145/216633.216650
25. Thesing S: Safe and precise worst-case executiontime prediction by abstract interpretation of pipeline models, Ph.D. thesis. University of Saarland, Saarland, Germany; 2004.
26. Lundqvist T, Stenström P: Timing anomalies in dynamically scheduled microprocessors. In Proceedings of the 20th IEEE Real-Time Systems Symposium (RTSS '99), December 1999, Phoenix, Ariz, USA. IEEE Computer Society; 12-21.
27. Wenzel I, Kirner R, Puschner P, Rieder B: Principles of timing anomalies in superscalar processors. In Proceedings of the 5th International Conference on Quality Software (QSIC '05), September 2005, Melbourne, Australia. IEEE Computer Society Press; 295-303.
28. Patterson DA: Reduced instruction set computers. Communications of the ACM 1985,28(1):8-21. 10.1145/2465.214917
29. Hennessy JL: VLSI processor architecture. IEEE Transactions on Computers 1984,33(12):1221-1246.
30. Lim S-S, Bae YH, Jang GT, et al.: An accurate worst case timing analysis for RISC processors. IEEE Transactions on Software Engineering 1995,21(7):593-604. 10.1109/32.392980
31. Shaw AC: Reasoning about time in higher-level language software. IEEE Transactions on Software Engineering 1989,15(7):875-889. 10.1109/32.29487
32. Engblom J: Processor pipelines and static worst-case execution time analysis, Ph.D. thesis. Uppsala University, Uppsala, Sweden; 2002.

33. Zhang N, Burns A, Nicholson M: Pipelined processors and worst case execution times. Real-Time Systems 1993,5(4):319-343. 10.1007/BF01088834

34. Arnold R, Mueller F, Whalley D, Harmon M: Bounding worst-case instruction cache performance. Proceedings of the Real-Time Systems Symposium (RTSS '94), December 1994, San Juan, Puerto Rico, USA 172-181.

35. Healy CA, Whalley DB, Harmon MG: Integrating the timing analysis of pipelining and instruction caching. Proceedings of the 16th Real-Time Systems Symposium, December 1995, Pisa, Italy 288-297.

36. Lee C-G, Hahn J, Seo Y-M, et al.: Analysis of cache-related preemption delay in fixed-priority preemptive scheduling. IEEE Transactions on Computers 1998,47(6):700-713. 10.1109/12.689649

37. Busquets-Mataix JV, Serrano JJ, Ors R, Gil P, Wellings A: Adding instruction cache effect to schedulability analysis of preemptive real-time systems. In Proceedings of the 2nd IEEE Real-Time Technology and Applications Symposium (RTAS '96), June 1996, Brookline, Mass, USA. IEEE Computer Society Press; 204-212.

38. Healy CA, Arnold RD, Mueller F, Whalley DB, Harmon MG: Bounding pipeline and instruction cache performance. IEEE Transactions on Computers 1999,48(1):53-70. 10.1109/12.743411

39. Reineke J, Grund D, Berg C, Wilhelm R: Timing predictability of cache replacement policies. Real-Time Systems 2007,37(2):99-122. 10.1007/s11241-007-9032-3

40. Engblom J: Analysis of the execution time unpredictability caused by dynamic branch prediction. In Proceedings of the 9th IEEE Real-Time and Embedded Technology and Applications Symposium (RTAS '03), May 2003, Toronto, Canada. IEEE Computer Society; 152-159.

41. Li X, Roychoudhury A, Mitra T: Modeling out-of-order processors for WCET analysis. Real-Time Systems 2006,34(3):195-227. 10.1007/s11241-006-9205-5

42. Kongetira P, Aingaran K, Olukotun K: Niagara: a 32-way multithreaded sparc processor. IEEE Micro 2005,25(2):21-29. 10.1109/MM.2005.35

43. Hofstee HP: Power efficient processor architecture and the cell processor. Proceedings of the 11th International Symposium on High-Performance Computer Architecture (HPCA '05), February 2005, San Francisco, Calif, USA 258-262.

44. Kahle JA, Day MN, Hofstee HP, Johns CR, Maeurer TR, Shippy D: Introduction to the cell multiprocessor. IBM Journal of Research and Development 2005,49(4-5):589-604.

45. Kistler M, Perrone M, Petrini F: Cell multiprocessor communication network: built for speed. IEEE Micro 2006,26(3):10-23.

46. Seiler L, Carmean D, Sprangle E, et al.:Larrabee: a many-core https://static-content.springer.com/image/art%3A10.1155%2F2009%2F758480/MediaObjects/13639_2008_Article_205_IEq44_HTML.gif 86 architecture for visual computing. ACM Transactions on Graphics 2008,27(3):1-15.

47. Yan J, Zhang W: WCET analysis for multi-core processors with shared L2 instruction caches. Proceedings of the 14th IEEE Real-Time and Embedded Technology and Applications Symposium (RTAS '08), April 2008, St. Louis, Mo, USA 80-89.

48. Ferdinand C, Heckmann R, Langenbach M, et al.: Reliable and precise WCET determination for a real-life processor. In Proceedings of the 1st International Workshop on Embedded Software (EMSOFT '01), October 2001, Tahoe City, Calif, USA, Lecture Notes in Computer Science. Volume 2211. Edited by: Henzinger TA, Kirsch CM. Springer; 469-485.

49. Jouppi NP: Improving direct-mapped cache performance by the addition of a small fully-associative cache and prefetch buffers. Proceedings of the 17th Annual International Symposium on Computer Architecture (ISCA '90), May 1990, Seattle, Wash, USA 364-373.

50. Angiolini F, Benini L, Caprara A: Polynomial-time algorithm for on-chip scratchpad memory partitioning. In Proceedings of the International Conference on Compilers, Architecture, and Synthesis for Embedded Systems (CASES '03), October-November 2003, San Jose, Calif, USA. ACM Press; 318-326.

51. Verma M, Marwedel P: Overlay techniques for scratchpad memories in low power embedded processors. IEEE Transactions on Very Large Scale Integration (VLSI) Systems 2006,14(8):802-815.

52. Puaut I: WCET-centric software-controlled instruction caches for hard real-time systems. Proceedings of the 18th Euromicro Conference on Real-Time Systems (ECRTS '06), July 2006, Dresden, Germany 2006: 217-226.

53. Wehmeyer L, Marwedel P: Influence of memory hierarchies on predictability for time constrained embedded software. Proceedings of the Conference on Design, Automation and Test in Europe (DATE '05), March 2005, Munich, Germany 1: 600-605.

54. Suhendra V, Mitra T, Roychoudhury A, Chen T: WCET centric data allocation to scratchpad memory. In Proceedings of the 26th IEEE International Real-Time Systems Symposium (RTSS '05), December 2005, Miami, Fla, USA. IEEE Computer Society; 223-232.

55. Puaut I, Pais C: Scratchpad memories vs locked caches in hard real-time systems: a quantitative comparison. In Proceedings of the Conference on Design, Automation and Test in Europe (DATE '07), April 2007, Nice, France. EDA Consortium; 1484-1489.

56. Colin A, Puaut I: Worst case execution time analysis for a processor with branch prediction. Real Time Systems 2000,18(2-3):249-274.

57. Wellings A: Is Java augmented with the RTSJ a better realtime systems implementation technology than Ada 95. Ada Letters 2003,23(4):16-21.

58. Java Expert Group : Java specification request JSR 302: Safety critical Java technology. http://jcp.org/en/jsr/detail?id=302

59. Schoeberl M: A time predictable Java processor. Proceedings of the Conference on Design, Automation

and Test in Europe (DATE '06), March 2006, Munich, Germany 1: 800-805.

60. Schoeberl M, Pedersen R: WCET analysis for a Java processor. In Proceedings of the 4th International Workshop on Java Technologies for Real-Time and Embedded Systems (JTRES '06), October 2006, Paris, France. Volume 177. ACM Press; 202-211.

61. Harmon T, Klefstad R: Interactive back-annotation of worst-case execution time analysis for Java microprocessors. Proceedings of the 13th IEEE International Conference on Embedded and Real-Time Computing Systems and Applications (RTCSA '07), August 2007, Daegu, Korea 209-216.

62. Bøgholm T, Kragh-Hansen H, Olsen P, Thomsen B, Larsen KG: Model-based schedulability analysis of safety critical hard real-time Java programs. In Proceedings of the 6th International Workshop on Java Technologies for Real-Time and Embedded Systems (JTRES '08), September 2008, Santa Clara, Calif, USA. ACM Press; 106-114.

63. Sun : picoJava-II Microarchitecture Guide. Sun Microsystems, March 1999

64. Sun : picoJava-II Programmer's Reference Manual. Sun Microsystems, March 1999

65. Puffitsch W: picoJava-II in an FPGA, M.S. thesis. University of Technology, Vienna, Austria; 2007.

66. Altera : Cyclone FPGA Family Data Sheet. ver. 1.2, April 2003

67. Imsys Im1101c (the Cjip) technical reference manual, v0.25, 2004

68. aJile : aj-100 real-time low power Java processor. preliminary data sheet, 2000

69. Uhrig S, Wiese J: Jamuth: an IP processor core for embedded Java real-time systems. In Proceedings of the 5th International Workshop on Java Technologies for Real-Time and Embedded Systems (JTRES '07), September 2007, Vienna, Austria. ACM Press; 230-237.

70. Zabel M, Preußer TB, Reichel P, Spallek RG: Secure, real-time and multi-threaded general-purpose embedded Java microarchitecture. Proceedings of the 10th Euromicro Conference on Digital System Design Architectures, Methods and Tools (DSD '07), August 2007, Lübeck, Germany 59-62.

71. Kreuzinger J, Brinkschulte U, Pfeffer M, Uhrig S, Ungerer Th: Real-time event-handling and scheduling on a multithreaded Java microcontroller. Microprocessors and Microsystems 2003,27(1):19-31. 10.1016/S0141-9331(02)00082-0

72. Preußer TB, Zabel M, Spallek RG: Bump-pointer method caching for embedded Java processors. In Proceedings of the 5th International Workshop on Java Technologies for Real-Time and Embedded Systems (JTRES '07), September 2007, Vienna, Austria. ACM Press; 206-210.

73. O'Connor JM, Tremblay M: picoJava-I: the Java virtual machine in hardware. IEEE Micro 1997,17(2):45-53. 10.1109/40.592314

74. EJC : The ejc (embedded Java controller) platform. http://www.embedded-web.com/index.html

75. Krall A, Grafl R: CACAO—a 64-bit JavaVM just-in-time compiler. In Proceedings of the Workshop on Java for Science and Engineering Computation (PPoPP '97), June 1997, Las Vegas, Nev, USA. Edited by: Fox GC, Li W. ACM Press;

76. Brandner F, Thorn T, Schoeberl M: Embedded JIT compilation with CACAO on YARI. Institute of Computer Engineering, Vienna University of Technology, Vienna, Austria; June 2008.

77. Pitter C, Schoeberl M: Performance evaluation of a Java chip-multiprocessor. Proceedings of the 3rd International Symposium on Industrial Embedded Systems (SIES '08), June 2008, La Grande Motte, France 34-42.

78. Bernat G, Burns A, Wellings A: Portable worst-case execution time analysis using Java byte code. Proceedings of the 12th Euromicro Conference on Real-Time Systems (ECRTS '00), June 2000, Stockholm, Sweden 81-88.

79. Bate I, Bernat G, Murphy G, Puschner P: Low-level analysis of a portable Java byte code WCET analysis framework. Proceedings of the 7th International Conference on Real-Time Computing and Applications (RTCSA '00), December 2000, Cheju Island, Korea 39-48.

80. Schoeberl M: Application experiences with a real-time Java processor. Proceedings of the 17th IFAC World Congress, July 2008, Seoul, Korea

CHAPTER 3

"POWER AND ENERGY CONSUMPTION OF NONVOLATILE MICROCONTROLLERS"

Dahoo Kim, Itaru Hida, Eric Shun Fukuda, Tetsuya Asai and Masato Motomura

Keywords: Embedded System, Microcontroller, Instruction Cache, Nonvolatile, Low-Power Design

INTRODUCTION

In recent years, sensor networks have been widely studied as a fundamental technology to help realize the con- cept of the "SmartSociety" [1] . In order to implement sensor networks in various fields of application, sensor nodes that can operate for a long time with a small energy source are required. Therefore, it is necessary to re- duce power consumption of microcontrollers that operate sensor nodes.

Meanwhile, NV microcontrollers (microcontrollers integrated with on-chipnonvolatile memories) are widely used due to their convenience in helping develop embedded system software. However, the power consumption of the nonvolatile memory dominates the total power consumption of the microcontroller [2]. Furthermore, it is difficult to reduce the power consumption of nonvolatile memories in microcontrollers. The purpose of our work is to reduce power and energy consumption by introducing an instruction cache to the microcontroller, which will reduce the frequency of access of the nonvolatile memory.

As shown in Table 1, traditional cache architecture research has attempted to improve microprocessor per- formance by introducing a high-speed cache memory, known as static random-access memory (SRAM), be- tween the main memory and the datapath by reducing memory access time [3] . On the other hand, in the case of NV microcontrollers used for real-time applications in embedded systems, the program execution time can be calculated in advance. Furthermore, the change of execution due to cache misses should be avoided, since such a change can cause problems in the system such as real-time applications. Thus, it is necessary to introduce an in- struction cache that does not cause cache miss penalties and leaves the speed of memory access during cache hits unchanged.

Therefore, in this paper, we attempt to reduce power and energy consumption instead of improving performance. To accomplish this, we introduce a "transparent" instruction cache that does not change the cycle-level timing of existing NV microcontrollers. In particular, we evaluate power and energy reduction by introducing the transparent instruction cache to a base microcontroller that is integrated with a flash memory, which is the most popular nonvolatile memory.

The rest of this paper is organized as follows: Section 2 describes the features of our research. Section 3 describes the architecture of the proposed instruction cache employed with the NV microcontroller. Section 4 dis- cusses the evaluation results of the instruction cache's effects of reducing power and energy consumption. Section 5 concludes this paper.

FEATURES OF OUR RESEARCH

The features of our research are as follows:

1) Examination of instruction cache suitable for microcontroller deployments: As described in Section 1, in the case of a microcontroller, it is more important to prolong the battery run time than to reduce the processing time. Thus, the instruction cache that we propose in this paper is intended to reduce power consumption rather than to improve processing speed.
2) Evaluation based on a realistic microcontroller architecture: We evaluated the power and energy con- sumption of our system, which was built on a base

Citation: Kim, D. , Hida, I. , Fukuda, E. , Asai, T. and Motomura, M. (2014) Power and Energy Consumption of Nonvolatil, doi:10.4236/cs.2014.511027.

microcontroller, Renesas Electronics Corporation's 78K0R. Since not all of the 78K0R's specifications are open, we implemented the base microcontroller using only publicly available information [4] -[6] .

3) Power evaluation with high precision: We evaluated the power and energy consumption using the chip design data generated from the RTL description. This allowed us to obtain the results with high precision.

Table 1. Concept of this work.

<table>
<tr><th>Traditional Cache Architecture Research</th><th>This work</th></tr>
<tr><td>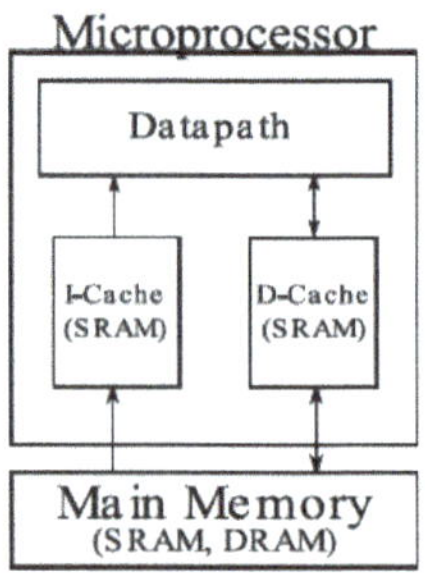
</td><td>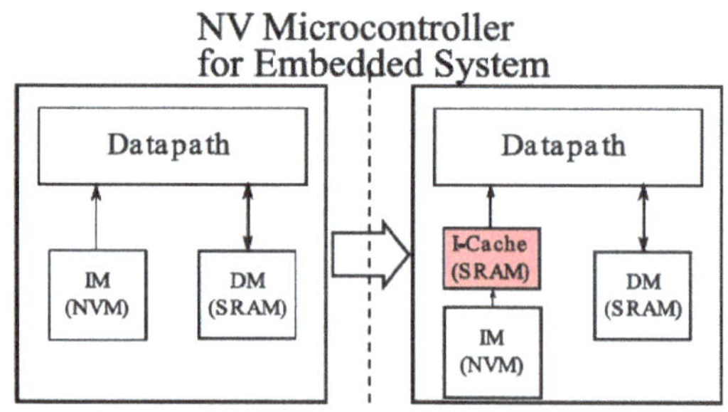
</td></tr>
<tr><td>High Performance</td><td>Low power and low energy transparent (i.e., doesn't change cycle level timing)</td></tr>
</table>

ARCHITECTURE

Base Microcontroller

Our microcontroller is based on the 78K0R, which is used in various industrial fields [7] . The block diagram is shown as a part of Figure 1. The 78K0R has an on-chip flash memory as its NV instruction memory, and a SRAM as its data memory [4] .

This architecture has a three-stage pipeline structure (IF stage, ID stage, MEM stage) [5] . In the IF stage, the microcontroller provides an address from the program counter (PC) to the instruction memory (flash memory), and fetches an instruction sequence from the instruction memory. This instruction sequence is stored in an in- struction queue (I-Queue). In the ID stage, the microcontroller decodes the instruction that has been fetched in the IF stage, and extracts data memory (SRAM) and register file (RF) addresses to be accessed. In the MEM stage, the microcontroller retrieves the data from the data memory and executes the instruction.

In addition, the base microcontroller uses complex instruction set computing (CISC) architecture [5] . The number of instructions of the base microcontroller is 915, and the instruction length is 1 byte to 5 bytes. The base microcontroller has a 4-byte (1 word) instruction queue that contains an instruction sequence fetched from the instruction memory. Therefore, if a valid instruction exists in the I-Queue, there is no need to access the flash memory.

1-Word-Per-Line Instruction Cache

Most (on-chip) NV microcontrollers, including the 78K0R, access the NV memory in one cycle [5] . Therefore, ourbase microcontroller fetches one word from the instruction sequence in the flash memory in one cycle [5] . Thus, as the first step to introduce the instruction cache to the base microcontroller, we designed a 1-word-per- line instruction cache architecture, as shown in Figure 1. We assumed the operation timing of the instruction cache to be the same as the timing of the base microcontroller's access to the flash memory: The instruction sequence is read from flash memory in the subsequent cycle of the access to the flash memory.

Operation of the instruction cache is as follows:

1) In the case of a cache miss, the microcontroller accesses the flash memory and fetches 1 word from the in- struction sequence.
2) In the case of a cache hit, the microcontroller accesses the instruction cache (I-Cache) and fetches 1 word from the flash memory. The word is written to the instruction cache. Since the bit widths of the instruction cache and the flash memory are the same, there is no penalty for writing to the instruction cache.

Since the proposed instruction cache does not allow for a cache miss penalty, there is a problem in that a read access and a write access to the instruction cache can occur coincidentally. For example, as shown in Figure 2,

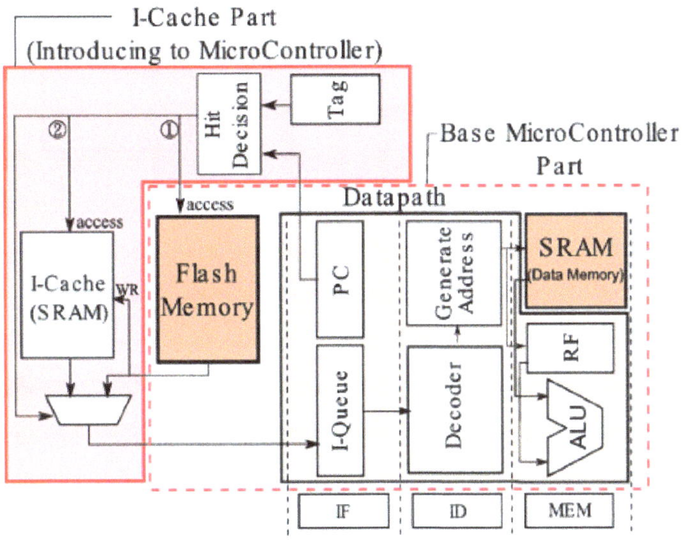

Figure 1. Concept of proposed 1-word-per-line instruction cache architecture.

when a cache miss occurs at PC1, the microcontroller reads the flash memory. Then, in the next cycle, the in- struction sequence is read from the flash memory. Thus, the instruction sequence for the cache miss (F-DATA1) must be written to the instruction cache in this cycle (the next cycle of the cache miss). In the cycle immediately after the cache miss, when a cache hit occurs at PC2, an instruction is fetched from the instruction cache. In this case, a read access and a write access collide in the instruction cache.

There are two methods for solving this problem:

1) Clock cycle dividing method: On collision of read access and write access in the instruction cache, this method bisects the clock cycle and allows write access in the first half and read access in the second half.

2) Intermediate buffer insertion method: This method delays the write access to the cache by inserting an intermediate buffer to which the instruction sequence in the cache miss is written between the instruction cache and the flash memory. Figure 3 shows its architecture. In this method, the timing of writing the instruction se- quence to the instruction cache is a cycle that does not access the instruction cache. If a valid instruction is still in the I-Queue of the datapath as described in the base microcontroller architecture, and if a cache miss occurred, the microcontroller does not have access to the instruction cache.

In particular, when a cache miss occurs, the instruction sequence that has been present in the intermediate buffer is written to the instruction cache, and the instruction sequence that has been read from the flash memory is newly written to the intermediate buffer.

In the case of the clock cycle dividing method, controlling the divided clock cycle is complicated. In addition, because it is necessary to generate a write/read access signal in one cycle, the operating frequency of the micro- controller must be lowered unless another clock is provided. Therefore, in this paper, we adopt the intermediate buffer insertion method, which simplifies the control scheme.

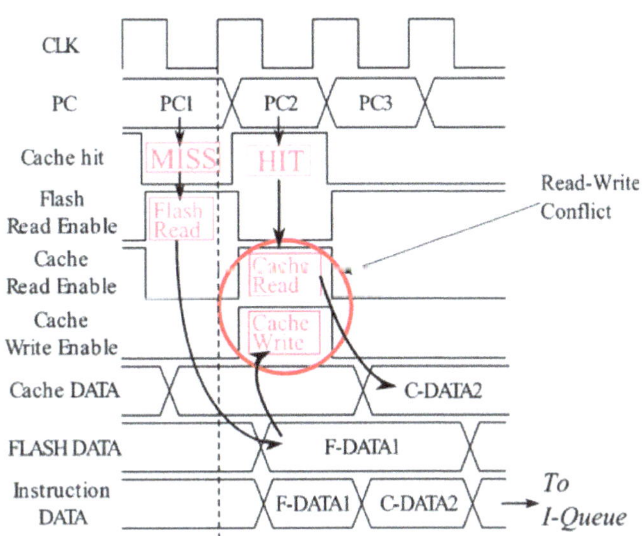

Figure 2. Instruction memory access timing in case of cache hit immediately after cache miss.

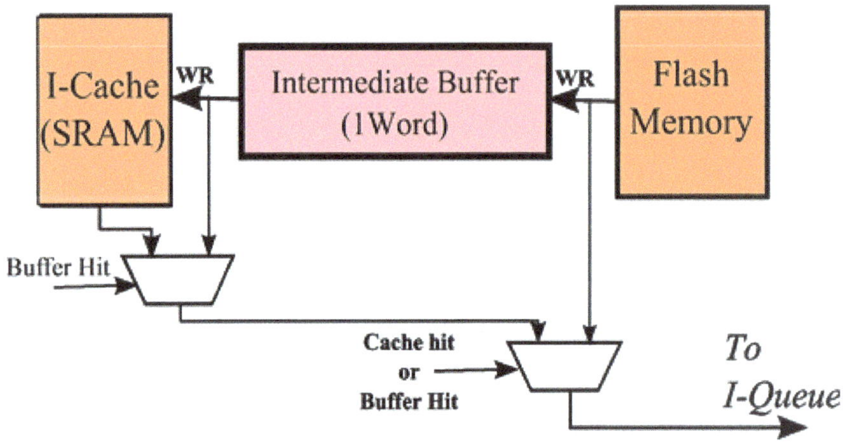

Figure 3. Intermediate buffer insertion method for 1-word-per- line instruction cache architecture.

4-Word-Per-Line Instruction Cache

For the second step to integrate the instruction cache to the base microcontroller, we designed a 4-word-per-line instruction cache architecture that can take advantage of spatial locality.

The flash memory (NV memory) of the base microcontroller (the existing NV microcontroller) reads one word from the instruction sequence in one cycle. Therefore, in order to implement the 4-word-per-line instruction architecture, a buffer for storing four words is required. We implemented a 4-word-per-line instruction cache by extending the number of words in the intermediate buffer, as shown in Figure 4. The instruction se- quence that is read from the flash memory in one cycle is stored in the intermediate buffer at one word per cycle. When all four entries of the intermediate buffer are filled, the contents are sent to the 4-word-per-line instruction cache memory.

For example, when a miss occurs in the first cycle, the instruction sequence that is read from the flash memory is stored to one of the entries of the intermediate buffer by referring to the lower 2 bits of the instruction address (held in the PC). In the next cycle, if the flash memory is not accessed (in other words, if there are cache hits, buffer hits, or I-Queue hits), the next instruction sequence is read from the flash memory by incrementing the PC, and is stored to the next entry of the intermediate buffer. Four words of the instruction sequence will be written to the instruction cache memory.

However, when a miss occurs before the four words are collected in the intermediate buffer, it is necessary to store the instruction sequence that has been read from the flash memory to the intermediate buffer. In this case, the existing instruction sequences in the intermediate buffer are written to the instruction cache, and the instruction sequence that has been read from the flash memory on a miss is stored to the intermediate buffer. For ex- ample, as shown in Figure 5, if a miss occurs when only three words of an instruction sequence are stored in the intermediate buffer, the three words are written to the instruction cache, invalidating the 0th word. In this paper, we extend the 4-word-per-line tag memory's valid bits to 4 bits, and make them indicate the valid word of each line.

However, when writing the line that has invalid words to the instruction cache, such as the case shown in Figure 5, it is expected that there will be a case of a low hit rate, due to the deletion of the valid word that ex- isted in one line of the instruction cache. In Section 4, we will evaluate our system in this regard.

Associativity

As the third step to integrate the instruction cache with the base microcontroller, we increased the associativity from 1 to 2 and 4. It is expected that the power consumption of the control unit of the tag memory will increase

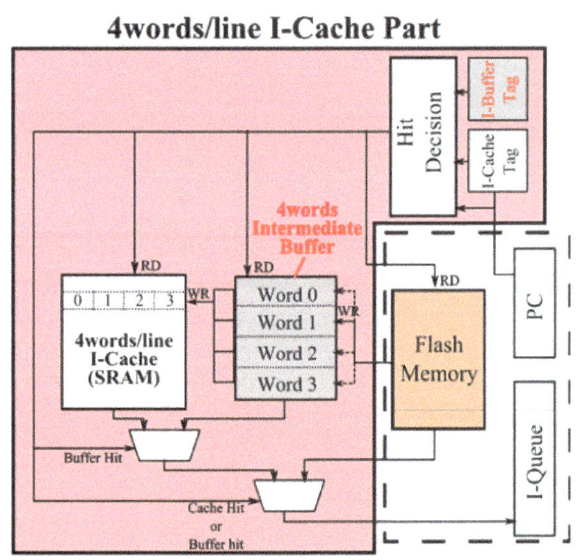

Figure 4. Proposed 4-word-per-line instruction cache architecture.

because the control unit of the tag memory becomes complicated by increasing the associativity. Yet there is a possibility that the hit rate will rise. We will evaluate our system in this regard as well. In addition, we adopt a pseudo least-recently-used (LRU) replacement algorithm.

EVALUATION

Method of Evaluation

Figure 6 shows our evaluation method. In the first step, we made a RTL description of the base microcontroller based on the public specifications of the 78K0R, and attached the instruction cache architecture to the RTL description. In the second step, we generated the chip design data of the base microcontroller with and without the instruction cache. The chip design data was generated through a logic synthesis and a placement/routing, using the RTL description. The tools used for the logic synthesis and the placement/routing are shown in Table 2. In the third step, we evaluated the power and energy consumption of each instruction cache system by examining the results of a back-annotated simulation, using the chip design data generated in the second step.

We used the evaluation programs shown in Table 3. The method of estimating power consumption is as follows:

1) CPU logic: We estimated the power consumption from the results of the back-annotated simulation using chip design data with the Synopsys Power Compiler.
2) Data memory: We evaluated the power consumption of the base microcontroller's data memory (SRAM) using the SRAM's power consumption parameters, which are publicly available [8] .
3) Instruction cache memory: Similarly, we evaluated the power consumption from publicly available information about the SRAM's power consumption parameters. Further, we changed the size of the cache memory from 128 bytes to 4 Kbytes and evaluated it.
4) Tag memory: Tag memory was also evaluated using the SRAM information.

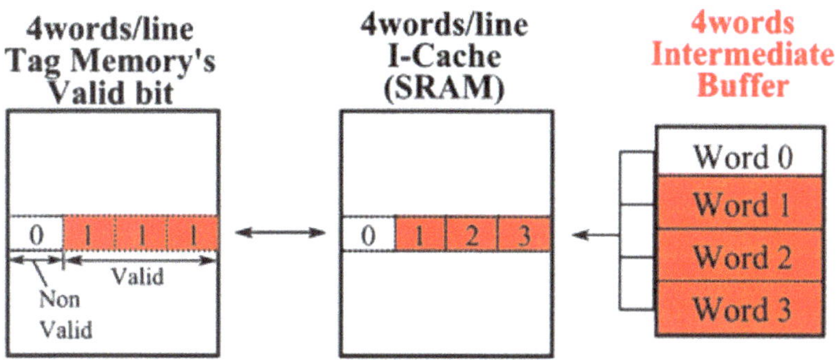

Figure 5. Writing operation of proposed 4-word-per-line instruction cache.

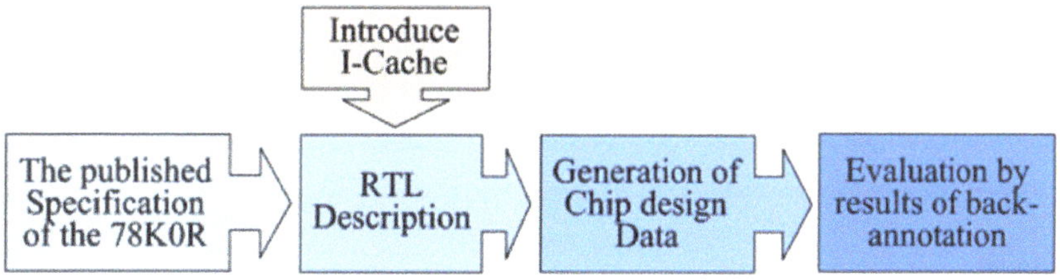

Figure 6. Method of evaluation for this work.

Table 2. Tools and process used in evaluation.

Phase	Tools and Process
Logic Synthesis	Synopsys Design Compiler (ver. 2013.03-SP3)
Placement and Routing	Cadence Encounter (ver. 10.13)
Estimated Power Consumption	Synopsys Power Compiler (ver. 2013.03-SP3)
Process	TSMC 0.18 μm

Table 3. Evaluation programs and hit rate.

Evaluation Program	Size (Byte)	1 Word/Line (%)	4 Word/Line (%)
Bubble Sort	614	98.69	98.76
Celsius to Fahrenheit	585	96.76	96.11
Checksum	535	96.92	98.46
Copy Verify	602	97.90	98.18
Factorial	606	98.63	95.67
EEMBC Coremark	11701	99.89	93.03

In addition, we used a 1.8 V power supply voltage and a 20 MHz operating frequency for evaluating power and energy consumption. Operation frequency was determined from the data arrival time of the critical path of its architecture, after logic synthesis. Furthermore, we evaluated energy consumption using the power consump- tion and delay time of the instruction cache, before and after its introduction.

Evaluation of Flash Memory (NV Memory)

In general, it is known that the power consumption of the flash memory access occupies a large part of the entire power consumption of a flash (NV) microcontroller. Moreover, the power consumption of flash memory differs according to the memory's capacity, the device technology, and circuit configurations. In addition, power con- sumption specifications of the NV memory are not publicly available.

In this paper, we evaluated the power consumption of the flash memory by parameterizing the ratio that the power consumption of flash memory occupies (i.e., 30%, 50%, and 70%). We selected these parameters to indi- cate the energy reduction in a wide range of practical NV microcontroller configurations. Thus, by realizing re- duced power and energy consumption of the proposed instruction cache, we believe we can provide useful in- formation to microcontroller designers.

Power Consumption

First, the hit rates for the evaluation programs are shown in Table 3. The cache size is 2 Kbytes. Because the programs are small in size, they (from the bubble sort program to the factorial program) showed high hit rates. Hereafter, we will evaluate power consumption in greater detail using the EEMBC Coremark programin order to discuss power consumption more realistically.

In the EEMBC Coremark program, the hit rate of each cache size is shown in Figure 7. For a large instruction cache (1 Kbytes or more), the 1-word-per-line instruction cache has a higher hit rate than the 4-word-per-line in- struction cache. This is because there is also a case where valid data that was in the instruction cache in the 4-word-per-line instruction cache has been discarded as described in the 4-word-per-line instruction cache ar- chitecture. In the case of an instruction cache of less than 1 Kbyte, the 4-word-per-line instruction cache has a higher hit rate. This is because the 4-word-per-line instruction cache can take advantage of spatial locality, even if the valid data has been discarded. Further, if the associativity was high, the hit rate was high for both methods. In particular, when the cache size was small, this effect became significant.

Next, we evaluated actual power consumption. Estimated power consumption results for an instruction cache of 512 bytes are shown in Figure 8. The results show that the proposed instruction cache reduces the power consumption of the flash memory by 66% with a 1-way, 1-word-per-line instruction cache, and by 90% with a 1-way, 4-word-per-line instruction cache. The larger effect on the 4-word-per-line instruction cache is a result of its higher hit rate.

In addition, for the 4-way instruction cache, the reduction of power consumption increased by 4% for the 1-word-per-line instruction cache, and by 2% for the 4-word-per-line instruction cache, compared to a 1-way instruction cache. However, the power consumption of the instruction cache memory and the tag memory shown in Table 4 is added after introducing the instruction cache.

For the instruction cache memory, since the bit width of the 4-word-per-line instruction cache is larger than that of the 1-word-per-line instruction cache, the power consumption for accessing the 4-word-per-line instruction

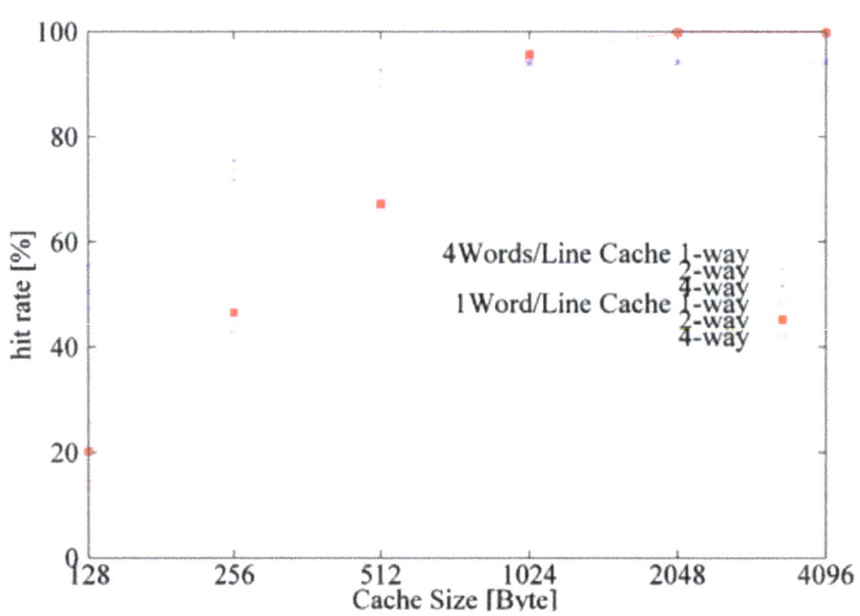

Figure 7. Cache hit rate by cache size. (Evaluation program: EEMBC Coremark)

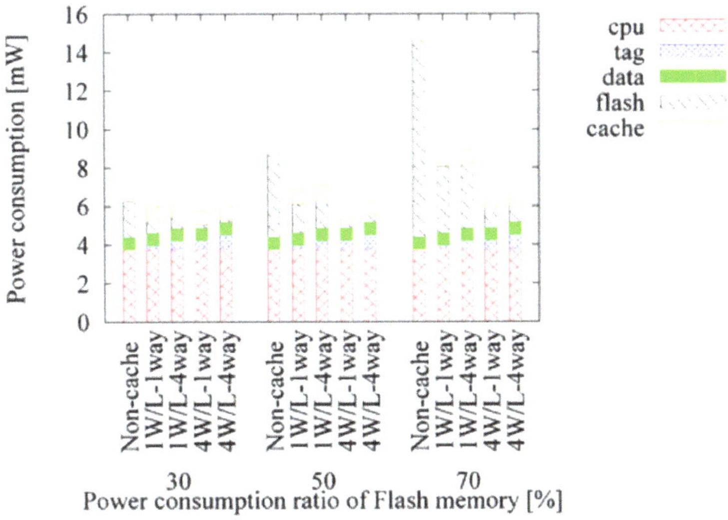

Figure 8. Power consumption by ratio of flash memory's power consumption to overall power consumption of microcontroller. (Cache size: 512 bytes)

Table 4. Power consumption of instruction cache and tag memory. (Cache size: 512 bytes)

Evaluation Program	Cache (mW)	1 Way Tag (mW)	4 Way Tag (mW)
1 Word/Line Instruction Cache	0.775	0.209	0.449
4 Words/Line Instruction Cache	0.775	0.461	0.749

cache memory does not increase significantly. This occurs because the architecture often has cache hits at the intermediate buffer, and the access rate to the cache memory is lower than that of the 1-word-per-line cache.

For the tag memory, power consumption increases as the width of the instruction cache line and the associativity increase, since the control circuit becomes complicated when the bit width of the tag memory increases.

From the above results, we note that the absolute amount of reduced power consumption by the proposed instruction cache becomes larger as the ratio of the flash memory's power consumption increases. In addition, when their capacities are 512 bytes, the 4-word-per-line instruction cache has a larger effect on reducing power consumption than the 1-word-per-line instruction cache. As associativity increases, power consumption de- creases for the 4-way instruction cache.

Next, we evaluated power consumption by changing the size of the instruction cache. The results are shown in Figure 9. The red line in this figure represents power consumption without the instruction cache. There is a case where total power consumption with the instruction cache increases when the size of the 1-word-per-line instruction cache is 128 bytes (1-way associative). This is because the reduced power consumption of the flash memory is small (because the hit rate for this cache size is low), and the power consumption of the instruction cache memory and tag memory has been added.

In addition, reflecting the behavior of the hit rate, the 4-word-per-line instruction cache is more effective at reducing power consumption for a small instruction cache (512 bytes or less). Similarly, in the case of a large (1 Kbyte or more) instruction cache, the 1-word-per-line instruction cache is more effective in reducing power consumption. In addition, as associativity increases, the reduction of power consumption becomes larger as the hit rate improves in the case of a small (128-byte) instruction cache. In other cases, the power consumption does not decrease as much because the power consumption of the tag memory increases.

Furthermore, we investigated the crossover point, which is the point where power consumption before and after the introduction of the instruction cache are the same. If the ratio of the power consumption of the flash memory exceeds the crossover point, power consumption can be expected to be reduced by introducing the instruction cache. The evaluation results of the crossover point are shown in Figure 10.

In the case of a 512 byte or larger instruction cache, the crossover point was approximately 20% to 30%. Accordingly, if the power consumption of the flash memory was approximately 20% or more, a reduction in power consumption can be expected by introducing a 512 byte or larger instruction cache. In the case of a 256 byte or smaller instruction cache, a reduction in power consumption cannot be expected in the 1-word-per-line instruction cache. However, in this case, the reduction in power consumption improved slightly by introducing a 4-way associative instruction cache. However, for other sizes of instruction cache, the crossover point increased because power consumption did not decrease as much.

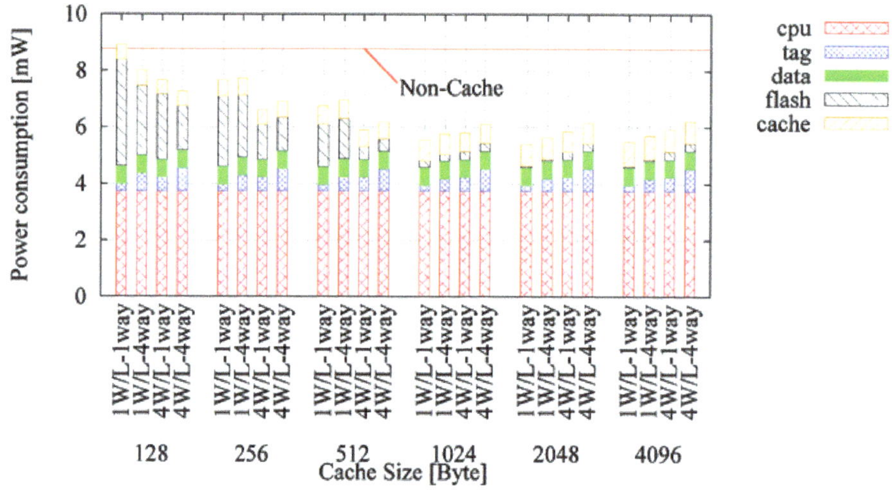

Figure 9. Power consumption by cache size. (Ratio of flash memory's power consumption: 50%)

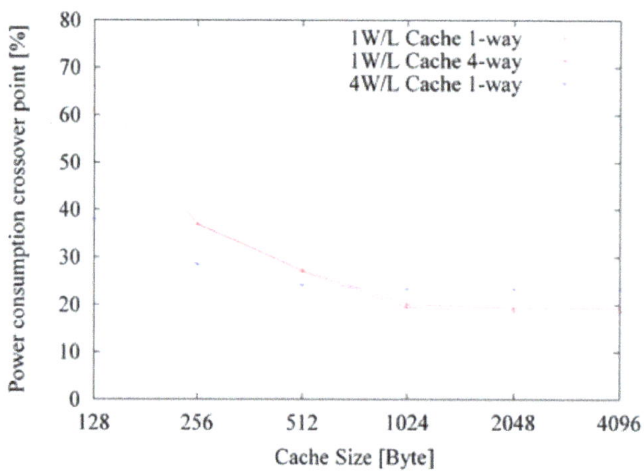

Figure 10. Change of crossover point by cache size for nor- mal voltage operation.

Reducing the Power Supply Voltage

In recent years, research in reducing the power consumption of logic circuits by lowering the power supply voltage has been reported [9] . However, lowering the power supply voltage of flash memories or other NV memories is difficult, owing to the circuit characteristics of NV memories [10] . Because the proposed instruction cache can work at a lower power supply voltage in accordance with the logic circuits, and because it can de- crease the frequency of yet high-power flash memory access, the power reduction should be more enhanced.

Thus, we examined the effect of reducing power consumption by introducing a proposed instruction cache with a decreased power supply voltage (to 0.9 V) for all components except for the flash memory. In addition, we used a 4 MHz operating frequency, which was estimated from a circuit simulation of a single gate.

Calculating with the following Formula (1), the ratio of the flash memory's power consumption at normal voltage is shown in Table 5. By decreasing the power supply voltage, the power consumption of the flash memory becomes dominant, compared to the power consumption during normal voltage operation.

$$P \propto C_L V_{dd}^{2} f_{clk} \tag{1}$$

The crossover points during low-voltage operation are shown in Figure 11. With a lower voltage, the cross- over point is reduced by approximately 20% from the crossover point during normal voltage operation. This is because the absolute amount of power consumption by the proposed instruction cache decreases more as the power consumption of the flash memory becomes dominant. Therefore, during low-voltage operation, a reduction in power consumption can be expected for an even smaller ratio of the flash memory's power consumption. When the crossover point in the instruction cache of 512 bytes or more becomes approximately 10%, a reduction in power consumption can be expected by introducing an instruction cache of 512 bytes or more, even if the power consumption of the flash memory occupies approximately 10% of the entire power consumption at nor- mal voltage.

Area and Critical Path

Using the proposed instruction cache, various increased areas of the microcontroller chip are shown in Table 6. These values indicate an area of the proposed instruction cache control logic and cache memory (SRAM). We estimated the area using the following method:

- Proposed instruction cache control logic: We calculated this area based on the chip design data of the post placement and routing.
- Cache memory (SRAM): We calculated this area using memory design data that was obtained by using MEMAKER by Faraday Technology.

The critical path of the NV microcontroller is generally a read-path of NV memory. This is because NV memory is slower than that of the other logic types [11] . In addition, operation frequency is determined by the critical path

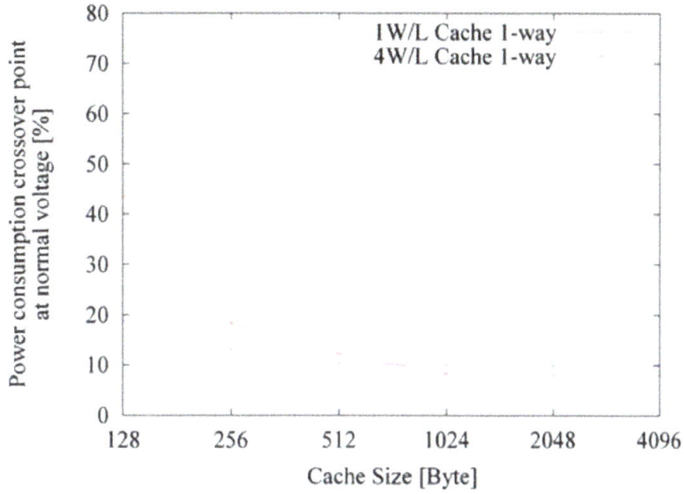

Figure 11. Change of crossover point by cache size for low-voltage operation.

(read-path of NV memory). By introducing the proposed instruction cache, the delay time of the logic path in- creases. This increase occurs because of the cache hit determination of the proposed instruction cache. In most cases, the increased delay time of the cache hit determination will not exceed the existing critical path of the NV microcontroller (read-path of NV memory). In most cases, the delay time of the NV memory read-path is higher than that of the logic path [11] . However, if the increased delay time of the logic path by cache hit determination exceeds the existing critical path, the delay time of the critical path will increase up to 2.91 ns for a 1-word-per- line instruction cache, and 3.62 ns for a 4-word-per-line instruction cache. These delay times are calculated using the chip design data of the post placement and routing.

Low-Energy Effect

In this section, we will show the low-power-consumption effects and the low-energy-consumption effects by examining the reduction rate of power consumption for the ratio of the power consumption of the flash memory, which is 30% to 70%.

First, we determine the low-power-consumption effects for 512 byte and 2 Kbyte instruction caches, as shown in Table 7. In the case of the 512 byte instruction cache, the effect of the 4-word-per-line instruction cache is larger than that of the 1-word-per-line instruction cache, and the obtained low-power-consumption effect is approximately 16.7% to 56.5%. For the 2 Kbyte instruction cache, the effect of the 1-word-per-line instruction cache is larger, and the obtained low-power-consumption effect is approximately 37.5% to 75.6%. In addition, a low-power effect was confirmed for low-voltage operation.

In the worst case, the delay time due to hit determination was approximately 2.91 ns for the 1-word-per-line instruction cache, and approximately 3.62 ns for the 4-word-per-line instruction cache. We determined the low- energy-consumption effects on the basis of these results, as shown in Table 8. Reflecting the delay time of the hit determination, the low-energy-consumption effect decreased by approximately 5% for the 1-word-per-line instruction cache, and approximately 6% for the 4-word-per-line instruction cache.

Table 5. Change of ratio of power consumption of flash memory.

Normal Voltage Operation	Low-Voltage Operation
30%	63%
50%	80%
70%	90%

Table 6. Area of proposed instruction cache.

Cache Size (Byte)	1-Word-per-Line Instruction Cache	4-Word-per-Line Instruction Cache
256	0.112, mm^2	0.241, mm^2
512	0.133, mm^2	0.265, mm^2
1024	0.167, mm^2	0.301, mm^2
2048	0.220, mm^2	0.369, mm^2

Table 7. Low-power-consumption effect.

	Cache Size: 512 Bytes (4 Words/Line)	Cache Size: 2 Kbytes (1 Word/Line)
Normal Voltage Operation	Approximately 16.7% - 56.5%	Approximately 37.5% - 75.6%
Low Voltage Operation	Approximately 23.1% - 64.3%	Approximately 47.4% - 84.1%

Table 8. Low-energy-consumption effect.

	Cache size: 512 Bytes (4 Words/Line)	Cache size: 2 Kbytes (1 Word/Line)
Normal Voltage Operation	Approximately 10.1% - 52.4%	Approximately 33.3% - 74.4%
Low Voltage Operation	Approximately 16.7% - 60.0%	Approximately 44.4% - 83.1%

CONCLUSIONS

In this paper, we proposed a transparent instruction cache architecture for reducing the power and energy consumption of a practical NV microcontroller architecture. Unlike traditional cache architecture, we intended to reduce power and energy consumption rather than improve processing speed. This is because, in the case of a NV microcontroller used for real-time applications in embedded systems, it is important to prolong the battery run time and not change the existing NV microcontroller's cycle-level execution time.

In addition, we presented the effects of reducing power and energy consumption by introducing proposed instruction caches. The proposed instruction cache reduces energy consumption by 10% to 52% for a ratio of the power consumption of the flash memory of 30% to 70%, if the size of the cache is one-twentieth of the program size. Meanwhile, for NV memories that include flash memories, it is generally difficult to lower the power supply voltage due to the circuit characteristics. Therefore, the proposed instruction cache is more effective in reducing energy consumption by lowering the power supply voltage, since the proposed instruction cache can work at a lower power supply voltage and can decrease the frequency of NV memory accesses. In this case, the proposed instruction cache reduces energy consumption by 16%, to 60%.

From the above evaluation results, we can see that the proposed instruction cache can successfully reduce the energy involved. In particular, if the power consumption of the flash memory occupied approximately 20% or more of normal voltage operation, a reduction in power consumption can be expected by introducing a 1 Kbyte or smaller instruction cache. In addition, microcontroller designers can lower a logic circuit's power supply voltage in order to reduce energy consumption. In this case, if the power consumption of flash memory is approximately 10% or more, a reduction in power consumption can be expected by introducing a 512-byteor smaller instruction cache.

Dahoo Kim, Itaru Hida, Eric Shun Fukuda, Tetsuya Asai, Masato Motomura Our future work involves devising a architecture for increasing the hit rate of the instruction cache, and to evaluate the plurality of a realistic large-size application. In addition, we intend to evaluate more precise power consumption using a prototype chip.

REFERENCES

1. Gungor, V.C., Lu, B. and Hancke, G.P. (2010) Opportunities and Challenges of Wireless Sensor Networks in Smart Grid. IEEE Transactions on Industrial Electronics, 57, 3557-3564. http://dx.doi.org/10.1109/TIE.2009.2039455
2. Lee, H.G. and Chang, N. (2003) Energy-Aware Memory Allocation in Heterogeneous Nonvolatile Memory Systems, Proceedings of the 2003 International Symposium on Low Power Electronics and Design, 25-27 August 2003, 420-423.http://dx.doi.org/10.1109/LPE.2003.1231941
3. Goodman, J.R. (1983) Using Cache Memory to Reduce Processor-Memory Traffic. Proceedings of the 10th Annual International Symposium on Computer Architecture, June 1983, 255-262. http://dx.doi.org/10.1145/285930.285984
4. Renesas Electronics Corporation, 78L0R/KE3 User's Manual.http://documentation.renesas.com/doc/DocumentServer/U17854EJ9V0UD00.pdf
5. Renesas Electronics Sorporation 78K0R Microcontroller User's Manual: Instructions.http://documentation.renesas.com/doc/DocumentServer/r01us0029ej0600_78k0r.pdf
6. Renesas Electronics Corporation, 78L0R/Kx3-A User's Manual: Hardware.http://documentation.renesas.com/doc/DocumentServer/r01uh0003ej0200_78k0rkx3a.pdf
7. Oba, K., Kawai, K., Matsushita, R., Ishihara, K. and Eto, K. (2009) Development of 16-Bit All Flash

Microcomputers "78K0R/Kx3-L" Featuring Ultralow Power Consumption. NEC Technical Journal, 4, 35-39. http://www.nec.com/en/global/techrep/journal/g09/n01/pdf/090109.pdf

8. Katevenis, M. (2003) On-Chip SRAM.http://www.csd.uoc.gr/~hy534/03a/s31_ram_bl.htm

9. Chandrakasan, A.P. and Brodersen, R.W. (1995) Minimizing Power Consumption in Digital CMOS Circuits. Proceed- ings of the IEEE, 83, 498-523.http://dx.doi.org/10.1109/5.371964

10. Bez, R., Camerlenghi, E., Modelli, A. and Visconti, A. (2003) Introduction to Flash Memory. Proceeding of the IEEE, 91, 489-502.http://dx.doi.org/10.1109/JPROC.2003.811702

11. Park, C., Seo, J., Seo, D., Kim, S. and Kim, B. (2003) Cost-Efficient Memory Architecture Design of NAND Flash Me- mory Embedded Systems. Proceedings of the 21st International Conference on Computer Design, 13-15 October 2003, 474-480.http://dx.doi.org/10.1109/ICCD.2003.1240943

CHAPTER 4

JITTER REDUCED SELF BIASED PLLS

J. Dhurga Devi

Keywords: Jitter, Dual Loop PLL, Self Biased PLL, Adaptive Bandwidth PLL

INTRODUCTION

The well proven self biased adaptive bandwidth PLL [1] considers supply and substrate noise as the dominant sources that lead to timing uncertainties of the output clock. Its architecture and loop parameters were optimally designed to make the output clock immune to these noise sources. The Voltage Controlled Oscillator (VCO) employed in this PLL exhibits high gain and facilitates wide operating frequency range for the PLL. Its loop parameters are process insensitive, and therefore the system also becomes process technology independent. These unique features of the self biased PLL ideally are suited for microprocessor clock generation and it has been used for this purpose for nearly two decades [2] - [4] .

However in today's scenario of tiled and multi core processers, this well proven PLL architecture operates in an environment of rapidly and widely varying combinations of clock frequencies and (low) supply voltages. This is needed in order to meet the overall power envelope constraints [2] - [4] . The processor cores when operated at very high frequencies and low supply voltages are subjected to increasingly more acute jitter constraints. In spite of these severe constraints, the choice of the self-biased PLL for microprocessor clock generation is considered "… unchallenged in the foreseeable future due to the wide frequency range required for product binning and power management" [5] .

In this work, a systematic design of these self biased PLLs for jitter reduction is described. The crucial functional blocks that decide the jitter performance, namely the charge pump and the VCO are designed in such a way that the overall jitter contribution is reduced. This design methodology is further extended to a self biased PLL incorporating dual loop scheme to attain significant jitter performance improvement.

Though there have been many papers appearing periodically in the literature addressing the issue of jitter reduction in PLLs, very few of these are in the context of self biased PLLs. In [5] , a delay interpolator based VCO design is proposed with a low gain VCO to provide low jitter while retaining the benefits resulting from bandwidth adaptivity and self biased nature. However this design employs an inverter based delay cell without the use of symmetric loads, and hence lacks the benefit of dynamic supply noise rejection. A calibration technique is proposed in [6] to operate the self biased adaptive bandwidth PLL at various frequency sub bands. In each operating sub band, the VCO gain is designed to be low; thereby this design ensures a low jitter performance. However a systematic approach to jitter reduction has not been addressed in [6] .

Dual loop techniques using conventional oscillators have recently been proposed in [7] - [9] reporting significant reductions in jitter (phase noise) and use a separate loop for coarse frequency acquisition. The present work carries out the systematic design of a similar dual loop scheme for jitter minimization, and closely parallels the work reported in [7] - [9] . But as opposed to using a ring oscillator based PLL as in [7] [8] or an LC oscillator as in [9] , the present work uses the self biased symmetric load VCO based adaptive PLL. Further for the chosen VCO, the appropriate selection of VCO gain for reduced jitter is also discussed. The dual loop technique proposed in [8] , attains jitter improvement by eliminating resistance in the loop filter, instead uses a Multiplying Delay

Citation: J. Dhurga Devi (2016) Jitter Reduced Self Biased PLLs. Circuits and Systems, 07,533-542. doi: 10.4236/cs.2016.75045.

Lock Loop (MDLL) to provide the feed forward path. This MDLL was made to act as fine loop that minimizes ripple and thermal noise generated by the loop filter resistance. This technique [8] discusses about optimization of PLL to PVT variations but a systematic procedure to choose its loop parameters for optimization of jitter performance is not suggested.

The present paper is organized as follows. Section 2 explains the design procedure of the self biased PLL with emphasis on jitter reduction. Detailed simulation results and overall system performance are presented in Section 3 and the conclusions of the present work are given in Section 4.

DESIGN OF REDUCED JITTER SELF BIASED PLL

The block diagram of the traditional self biased PLL [1] is shown in Figure 1 that includes a Phase Frequency Detector (PFD), Charge Pump (CP_1), Loop Filter (LPF), VCO and a Prescalar. It also includes a Bias Generator (BG) to provide bias currents to the charge pump, the PFD as well as the VCO, and this bias current generated adaptively, tracks the operating frequency. The VCO employed is a ring oscillator, and its delay element uses a linearized resistive load enabling the PLL to operate over a wide range of frequencies.

Impact of System Parameters on Jitter

The jitter performance of the traditional self biased PLL can be normally studied by considering the Noise Transfer Functions (NTFs) and the noise Power Spectral Densities (PSDs) of the various functional blocks of the

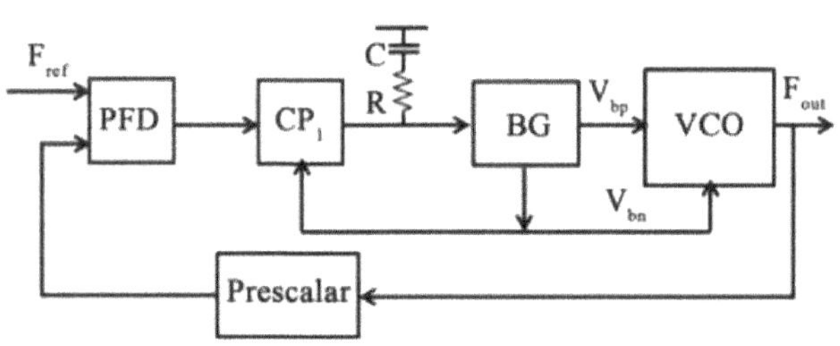

Figure 1. Block diagram of self biased PLL.

PLL. The overall PLL output noise PSD $\overline{\Phi^2_{out,n}}$ is given by (as given in [10] for e.g.),

$$\overline{\Phi^2_{out,n}} = \overline{\Phi^2_{cp,n}} + \overline{\Phi^2_{lpf,n}} + \overline{\Phi^2_{vco}} + \overline{\Phi^2_{div}}$$
$$= |H_{CP}|^2\, \overline{i^2_{cp,n}} + |H_{LPF}|\, \overline{v^2_{lpf,n}} + |H_{VCO}|\, \overline{\Phi^2_{vco,n}} + |H_{DIV}|\, \overline{\Phi^2_{div,n}} \tag{1}$$

In the above expression, H_{CP}, H_{LPF}, H_{VCO} and H_{DIV} represent the NTFs of the blocks CP_1, LPF, VCO and Prescalar blocks respectively (Figure 1). The quantities $\overline{i^2_{cp,n}}$, $\overline{v^2_{lpf,n}}$, $\overline{\Phi^2_{vco,n}}$ and $\overline{\Phi^2_{div,n}}$ represent the PSD's of the corresponding noise sources. The expressions for the NTFs of the various functional blocks as described in [10] are given in (2)-(5). The purpose of reproducing these equations here is to identify the key parameters that contribute to jitter and to suggest a methodology to reduce jitter.

$$H_{cp}(s) = \frac{N \cdot F_{lpf}(s) \cdot 2\pi K_{VCO}}{N \cdot s + I_{cp} \cdot F_{lpf}(s) \cdot K_{VCO}} \tag{2}$$

$$H_{LPF}(s) = \frac{N \cdot 2\pi K_{VCO}}{N \cdot s + I_{cp} \cdot F_{lpf}(s) \cdot K_{VCO}} \tag{3}$$

$$H_{VCO}(s) = \frac{N \cdot s}{N \cdot s + I_{cp} \cdot F_{lpf}(s) \cdot K_{VCO}} \tag{4}$$

$$H_{DIV}(s) = \frac{N \cdot I_{cp} \cdot F_{lpf}(s) \cdot K_{VCO}}{N \cdot s + I_{cp} \cdot F_{lpf}(s) \cdot K_{VCO}} \tag{5}$$

In the above expressions, I_{cp} represents the charge pump bias current and N represents the Prescalar divider ratio. $F_{lpf}(s)$ represents the transfer function of first order RC loop filter shown in Figure 1. If one considers the quantities K_{VCO}, I_{cp}, and the loop filter cutoff frequency, it is well known [10] [11] that they strongly influence the NTFs as well the noise PSDs (though the latter dependence is not explicitly shown). Further, the random noise contribution alone is captured in the above equations. Overall jitter also has a systematic component in it, and this is also a function of I_{cp} and K_{VCO} [12] . Further, these latter quantities themselves are intricately related through constraints on loop natural resonant frequency ω_n and damping constant ζ given by

$$\zeta = \frac{R}{2}\sqrt{\frac{I_{cp} \cdot K_{VCO} \cdot C}{N}} \tag{6}$$

$$\omega_n = \sqrt{\frac{I_{cp} \cdot K_{VCO}}{N \cdot C}} \tag{7}$$

Given all the above dependencies, the selection of optimum values of K_{VCO} and I_{cp} for minimizing overall jitter is a non trivial task. Further minimizing overall jitter implies reducing the noise PSDs as well as the NTFs of the various blocks.

Taking up the case of NTFs first, a careful consideration of all the dependencies leads to the following set of assumptions/observations which can aid in the systematic design of a low jitter self biased PLL.

1) Since the original PLL is wideband, it is assumed that the dominant contributor to the overall jitter is the noise originating from the charge pump [10] .
2) A reduction in K_{VCO} leads to a reduction in the charge pump contribution to overall jitter [7] .
3) While reducing K_{VCO}, due consideration must be given to the fact that other factors that are related to it through the PLL loop equations could mitigate the possible improvement in jitter.

4) One possibility for reducing jitter is to reduce K_{VCO} without altering the denominators in (1), and this can be accomplished with a corresponding increase in I_{cp}. This approach is followed in [10] . On the other hand, the work reported in [9] notes this approach as that "… there are practical difficulties which should be kept in mind when employing this strategy in multipath PLLs". After considering the systematic noise sources like capacitor leakage and charge pump mismatch currents, it is concluded in [9] that K_{VCO} cannot be reduced below a certain limit.

5) Alternatively, one can reduce K_{VCO} and also alter the denominator in (1), which in turn would imply that the system would operate with altered loop parameter values of ω_n and ζ. In return for a reduction in overall jitter, these loop parameters can be altered to the extent that the conditions of PLL loop stability and lock time are not degraded significantly. This is the approach followed in the present work and is iteratively designed through simulation so that the reduced jitter does not degrades settling time performance.

Next, taking up the case of the reduction of the noise PSDs of the individual blocks, the dominant noise sources within the key blocks like charge pump and the VCO have to be identified, and then suitably altered to the extant permissible without disturbing the PLL loop dynamics. Considering the VCO first, the circuit diagram of the delay element employed in the self biased PLL [1] is shown in Figure 2. Unlike the symmetric load transistors (M_4-M_7) and switch transistors (M_2-M_3), the tail transistor M_1 is generally considered to have negligible impact on the overall jitter [13] . Hence, given the overall PLLloop characteristics and NTFs, the sizes of the transistors (M_2-M_7) have to be chosen such that the VCO noise PSD is reduced while meeting the operating frequency range and VCO gain K_{VCO}. VCO gain of the symmetric load VCO as derived in [1] is given in the below expression (8). In the above equation k stands for the device trans conductance of M_4 or M_7. C_{eff} represents the effective capacitance that determines oscillator frequency.

$$K_{VCO} = \frac{d\omega_{out}}{dV_{ctrl}} = \frac{k}{2\pi \cdot C_{eff}} \qquad (8)$$

The circuit diagram of the charge pump employed in the self biased PLL is as shown in Figure 3. This architecture is a replica of the delay element, but with the differential switch transistors controlled by UP and DN signals generated by the PFD. Analogous to the delay element, the devices that adversely affect the overall jitter performance are the current source transistors (M_7-M_{14}) and switch transistors (M_3-M_6).

The dimensions of all the jitter sensitive transistors (M_2-M_7 of Figure 2 and M_3-M_{14}, of Figure 3) were chosen through simulations such that the respective block level noise PSDs were minimized while satisfying the overall PLL system level constraints (tuning range of VCO, K_{VCO} and bias currents in the charge pump). But a unique feature in self biased PLL is that the charge pump device dimensions are derived using the delay element device dimensions [1] . This constraint was imposed to mainly minimize charge pump mismatch currents, and implies that fixing the dimensions in the circuit of Figure 2 also fixes those in the circuit of Figure 3.

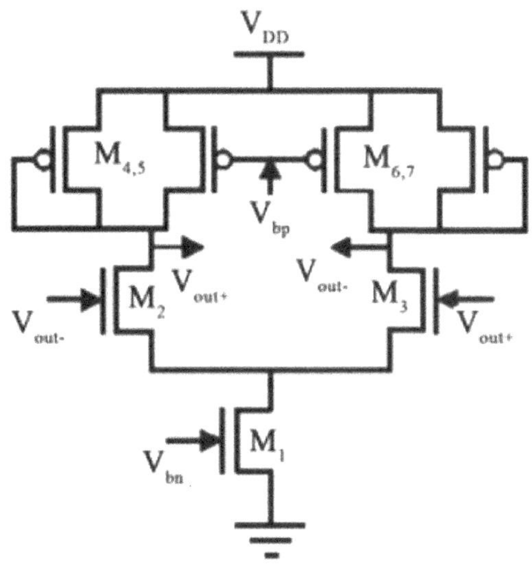

Figure 2. Circuit schematic of self biased VCO delay element.

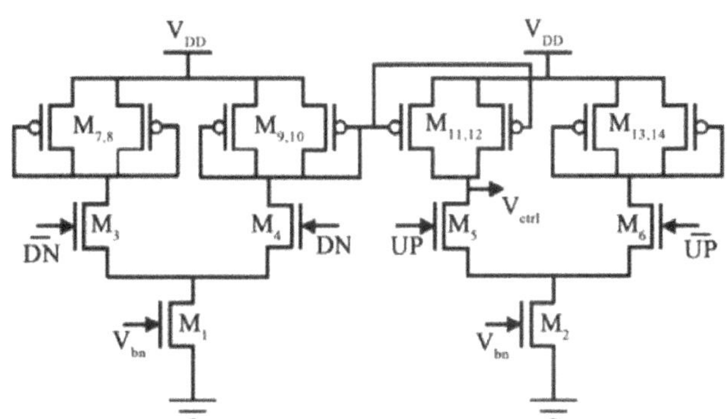

Figure 3. Circuit schematic of self biased VCO delay element.

For the traditional self biased PLL, the above mentioned considerations about NTFs and the noise PSDs of the various functional blocks will enable one to (a) size the devices in charge pump and VCO to minimize device noise contribution to jitter and (b) systematically reduce the K_{VCO} and study its impact on jitter. The quantitative results based on simulations will be presented in the next section.

While considering the reduction of K_{VCO}, one cannot reduce K_{VCO} without compromising the capture range and settling time of the traditional self biased PLL. Hence the dual loop scheme [14] is employed and allows one to reduce K_{VCO} without any trade off with capture range and settling time. The principle based on which the traditional single loop self biased PLL is modified into a dual loop scheme to reduce K_{VCO} is briefly discussed next.

Dual Loop Scheme for Jitter Minimization

The complete block diagram of the dual loop scheme is shown in Figure 4. The VCO of the self-biased PLL has two control inputs. One of the control inputs has a large conversion gain (coarse), provides wide tuning range, and forms part of the Frequency Lock Loop (FLL). The other control input has a low conversion gain (fine) and form part of the PLL. The loop constituting the FLL inFigure 4 comprises of a Quadricorrelator based Frequency Detector (FD), two charge pumps CP_1 and CP_2, and a bias generator BG_1. The charge pump block CP_2 in the FLL is used to emulate the resistance in LPF.

The PLL part in the dual loop scheme comprises of the PFD, charge pump CP_3, an LPF and a bias generator BG_2. The bias current of CP_3 is derived from the bias generator BG_2. The bias voltages from BG_1 and BG_2, V_{bn_FLL} and V_{bn_PLL} respectively are combined in a half buffer replica stage to generate the bias voltage V_{bp} for the ring oscillator.

The FLL acquires and brings the output frequency to within the PLL's capture range. While the FLL tracks the reference frequency, the PLL is disabled by a digital control circuit, and is later enabled when the frequency falls within its capture range. Therefore, under phase lock and very close to phase lock, the entire dual loop system operates only with the PLL being active, and the system reduces to a conventional self biased PLL with a reduced VCO conversion gain.

As was done above the single loop case, the dual loop scheme's device dimensions of VCO and CP of the PLL part of loop are chosen so as to minimize the respective noise PSDs while satisfying the respective block level specifications. The jitter performance of this systematically designed dual loop is compared with the systematically designed traditional self biased PLL and the results are discussed in the following section.

RESULTS AND DISCUSSIONS

All circuit simulations have been carried out using UMC 0.18 µm CMOS technology process libraries. The simulation results are presented in the following order. First, the traditional self biased PLL was designed with the optimum device dimensions of VCO and charge pump functional blocks determined using PNoise analysis in Cadence for reduced jitter performance. The choice of the optimum device dimensions from the PNoise plot is

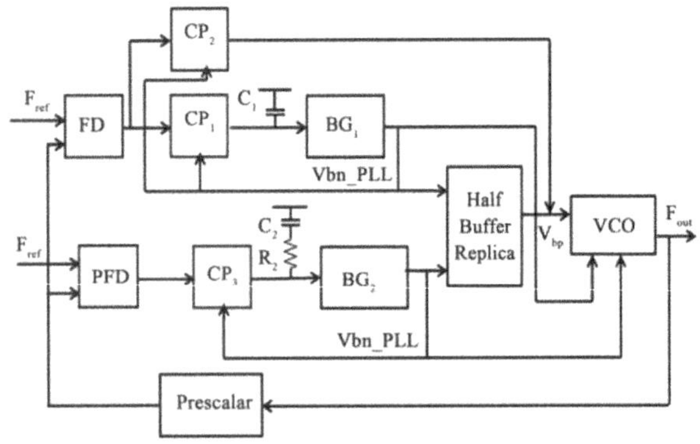

Figure 4. Block diagram of dual loop self biased PLL.

discussed. Next, the extant of possible improvement in the jitter performance obtained by systematically reducing K_{VCO} of the traditional self biased PLL, is presented. Based on this, the K_{VCO} of the VCO in the PLL is set and accordingly the dual loop self biased PLL is designed. The functionality of the dual loop scheme is verified and the capture transients are presented. The jitter performance of the dual loop is then compared with the jitter reduced traditional self biased PLL employing a high K_{VCO} for the required capture range.

The traditional self biased PLL was designed with the loop parameters ω_n/ω_{ref} as 1/15 and ζ chosen as 1 respectively. The bias generator was designed for a linear operating range of 0.3 V to 1.1 V that provides control voltage to VCO. For a VCO tuning range of 200 MHz to 3 GHz, K_{VCO} is computed to be 3.5 GHz/V. Based on the K_{VCO} expression given in (6) the ratio of device dimensions of M_{4-7} to C_{eff} is defined as given below.

$$(W/L)_{M4-7}/C_{eff} > 4\times10^{13} \tag{9}$$

With C_{eff} chosen as 210 fF, the constraint on dimensions $(W/L)_{M4-7}$ is set as given below.

$$(W/L)_{M4-7} > 1.5/0.18 \tag{10}$$

Based on chosen C_{eff}, the above constraint defines the device dimensions for $(W/L)_{M2-3}$ as 6/0.18 since capacitance of $(W/L)_{M4-7}$ and $(W/L)_{M2-3}$ together contributes C_{eff}.

The device dimensions of switch transistors $(W/L)_{M2-3}$ and symmetric load transistors $(W/L)_{M4-7}$ can be increased from the given constraint of (9)-(10) while satisfying the ratio required to meet the K_{VCO} value. Increasing device dimensions increases C_{eff}. Hence increasing the device dimensions beyond a range will affect its highest operating frequency. Thus the upper limit on the possible device dimensions that satisfies the operating frequency range is given by the constraint below.

$$(W/L)_{M4-7} < 6/0.18$$
$$(W/L)_{M2-3} < 26/0.18 \tag{11}$$

The permissible values of device dimensions of $(W/L)_{M2-3}$ and $(W/L)_{M4-7}$, that satisfies K_{VCO} and operating frequency range are given as below

$$1.5/0.18 < (W/L)_{M4-7} < 6/0.18$$
$$6/0.18 < (W/L)_{M2-3} < 26/0.18 \tag{12}$$

Choosing from these range of device dimensions, there VCO gain characteristics are plotted and are shown in Figure 5. Gain characteristic is plotted for two extreme values and one within the permissible range. The gain remains the same for the three characteristics but for the higher limit in the device dimensions, the highest operating frequency is observed to get limited. Therefore the constraints in (12) define the possible choice of device dimensions of the delay element.

Device dimension of M_1 of the delay element is chosen for the required bias current. At the highest frequency of operation the maximum bias current was computed to be 200 µA dependent on the gain and band width value decided in the bias generator. Dependent on this bias current and the respective bias voltage V_{bn}, the device dimension (W/L) of the tail transistor M_1 is computed to be 32/1.

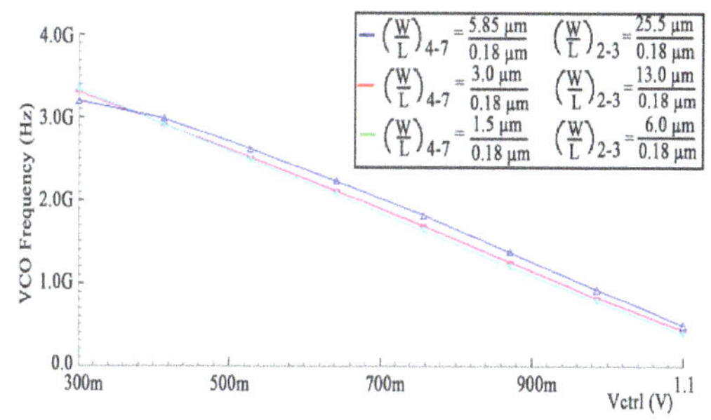

Figure 5. VCO gain characteristic for the three possible choices of device dimensions.

Since the charge pump circuit in the traditional self biased PLL is a replica of the delay element architecture. Therefore for every feasible set of device dimensions in VCO, there is a related device dimension set for the charge pump circuit, which is appropriately scaled by the factor x by which the charge pump current of the self-biased PLL is derived from the delay element bias current as described in [1] . The factor x was chosen as 0.2 that sets the ratio between ω_n/ω_{ref} as 1/15. The permissible range of device dimensions of charge pump circuit thus obtained is given below:

$$0.3/0.25 < (W/L)_{M7\text{-}14} < 0.8/0.25$$

$$6/0.18 < (W/L)_{M3\text{-}6} < 26/0.18 \tag{13}$$

In Figure 6, the VCO phase noise characteristic is plotted by progressively increasing the jitter sensitive device dimensions of M_2-M_7 (of Figure 2) within the permissible range as given in (12). The plots are shown for the extreme limits in the range and one within the permissible range. The noise PSD plots show improvement with increase in device dimensions. This obeys the fact that the resistance of the switch transistors and symmetric load transistors reduces with increase in device dimensions, and hence is the improvement in the noise PSD characteristics.

Similarly, in Figure 7, the noise PSDs for charge pump are plotted within the permissible range given in (13) and the plots are given for the extreme limits and one within the permissible range. From the noise PSD plots it can be observed that the lowest noise PSD characteristic corresponds to the lowest device dimensions from the feasible device dimension set. Since these minimum dimensions sets the current source transistors $M_{7\text{-}14}$ with minimum transconductance g_m, the output noise current PSD is smallest.

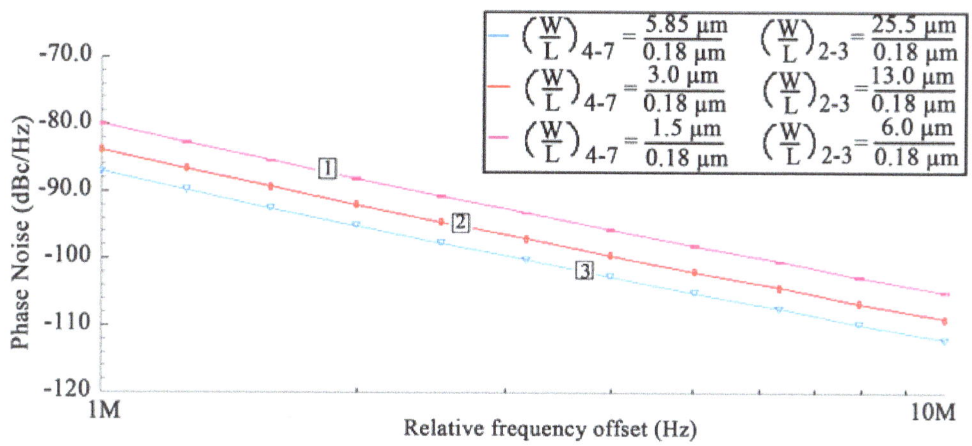

Figure 6. Phase noise plot of VCO at 2100 MHz operating frequency.

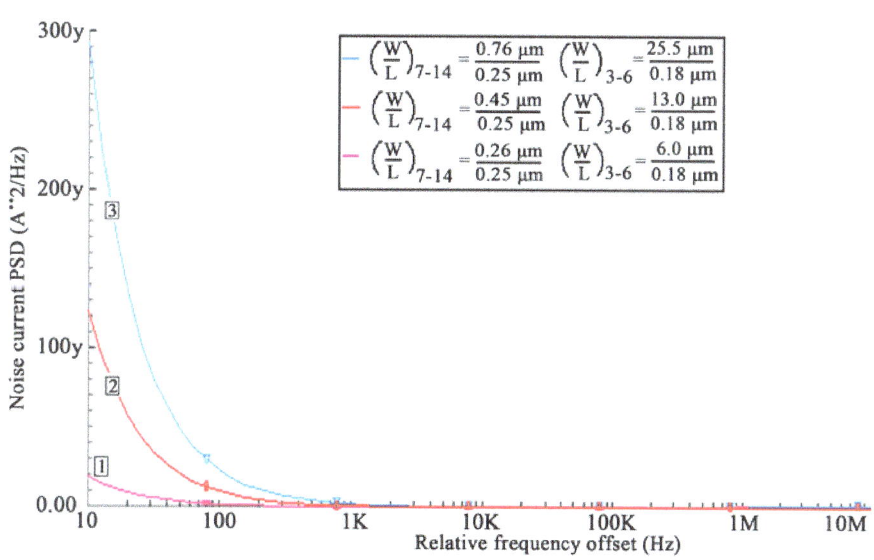

Figure 7. Charge pump noise current PSD.

It was observed through simulations that the device sizes corresponding to the plot labeled 1 in Figure 6 and Figure 7 led to the minimum jitter for the traditional single loop self biased PLL and were used for designing VCO and CP circuits while designing the traditional self biased PLL.

Considering the choice of reduced K_{VCO}, its impact on jitter performance was analyzed by simulating the traditional self biased PLL of dual loop system using VCOs with different K_{VCO} settings but with device dimensions (obtained from PNoise plot) that presents minimum noise PSDs. The jitter performance results are given below in Table 1. From the table, it can be noted that the dual loop PLL scales down its jitter performance by the factor by which the K_{VCO} is reduced. This follows with the fact that as K_{VCO} is reduced by certain factor, the impact of systematic noise on output phase noise is also reduced by the same factor. From the different values tabulated, K_{VCO} of 467 MHz/V was chosen for the design of dual loop PLL.

Figure 8 depicts the typical capture transients obtained for the resulting dual loop self biased PLL. The capture transients show the independent tracking nature of FLL and PLL incorporated in the dual loop system. The designed FLL is observed to settle in about 170 ns. The PLL is enabled as the frequency difference is reduced very close to the output frequency (within 4%) and is seen to settle at about 600 ns.

Next, in order to compare the jitter performances of the dual and traditional self biased PLLs, the eye diagram for both cases are plotted in Figure 9 for a nominal output frequency of 2.1 GHz. The RMS jitter computed from this plot was found to be 2.3 pS and 0.5 pS respectively for the traditional and dual loop self biased PLLs, and corresponds to improvement of 77% for the latter.

Two deviations need to be pointed out this stage. First, a K_{VCO} setting of 467 MHz/V is used as against the lowest possible K_{VCO} of 340 MHz/V as given in Table 1. The reason being, for the latter setting, the dual loop VCO settling time was degraded beyond 1 μs, and hence it was avoided. Second, the jitter reduction observed deviates from the prediction of Table 1. This is because Table 1 does not include the noise from the FLL, its bias circuits, and the buffer stages.

A summary of the performance of the present work is provided in Table 2 at the extreme operating frequency of 2700 MHz and a comparison is also provided with those of the dual loop schemes reported recently [6] - [8] .

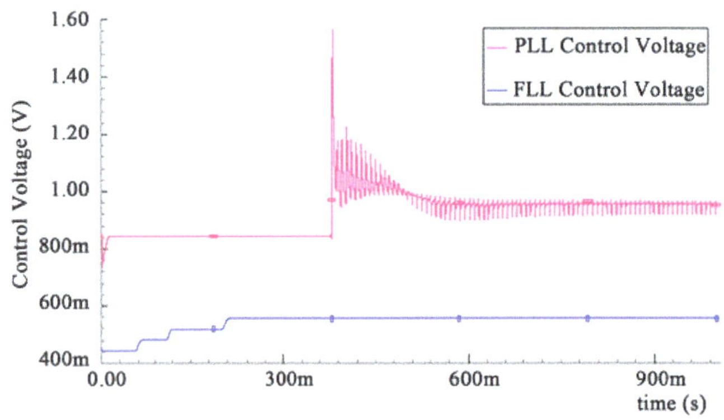

Figure 8. Capture transients of dual loop PLL at 2100 MHz output frequency.

Table 1. Jitter reduction as a function of K_{VCO} reduction (measured at a frequency of 2100 MHz).

K_{VCO} (MHz/V)	RMS jitter (pS)	K_{VCO} reduction factor M	Jitter reduction (compared with K_{VCO} = 3600 MHz/V case)
3600	2.27	1	1
1450	1.04	2.5	2.2
1000	0.94	3.6	2.4
630	0.49	5.7	4.6
540	0.43	6.6	5.3
467	0.30	7.7	7.6
340	0.22	10.5	10.3

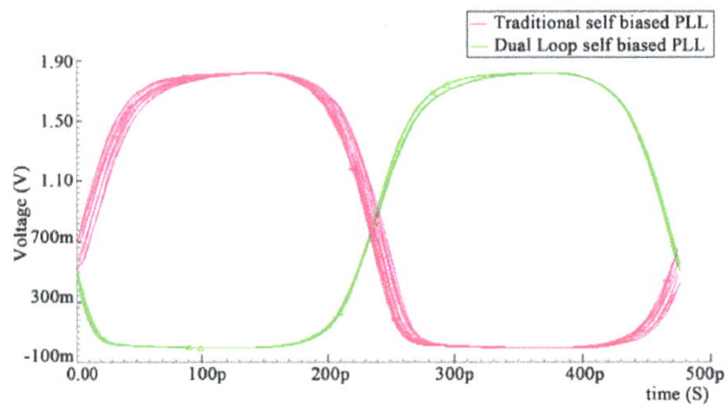

Figure 9. Jitter comparison of traditional self biased PLL with dual loop PLL.

Table 2. Performance comparison at extreme operating frequency.

	This work		Result reported in [7]	Result reported in [6]	Result reported in [8]
	Traditional self biased PLL	*Dual loop self biased PLL*			
Output frequency (MHz)	2720	2720	3100	2100	700
Reference (MHz)	170	170	108	65.6	1
Tuning range (GHz)	0.8 - 3.2	0.8 - 2.8	1.4 - 3.2	0.86 - 2.1	-
K_{VCO} (MHz/V)	3600	470	10	-	350
Lock time (frequency step)	0.16 µS (1.4 GHz)	<1 µS (710 MHz)	85 µS (940 MHz)	<3 µS (1.3 GHz)	-
RMS jitter (pS)	1.6 (0.0043UI)	0.8 (0.0022UI)	1.01 (0.0031UI)	1.37 (0.0028UI)	1.32 (0.0009UI)
Power (mW)	35.0	35.8	27.5	5.3	-
FOM (dB) (as in [7])	−220.7	−226.8	−225.5	−230	-
Supply voltage	1.8	1.8	1.2	1.8	-
Technology (nm CMOS)	180	180	65	180	-

In terms of jitter and Figure Of Merit (FOM) (as defined in [7]), it can be seen that dual loop self biased PLL outperforms the traditional self biased PLL and is comparable to the dual loop schemes of [6] - [8] . The acquisition time of the dual loop self biased PLL is found to be inferior compared to the traditional self biased PLL, but still far comparable to that reported in [6] [7] . Although this may not be a fair comparison since the present work is based on simulations only, the main objective here is to identify the key contributing factors and get estimates of the orders of improvements possible. Also, the work reported in [7] lacks supply noise immunity, process immunity and bandwidth adaptivity features that are inbuilt for the self biased PLLs. The jitter performance reported in [8] shows better jitter performance in comparison to the present but had not presented it settling time or power measure. The present work if made to operate with similar K_{VCO} would also have shown comparable jitter performance but with settling time greater than 1 µs.

CONCLUSION

This paper presents a systematic procedure for designing a low jitter self biased PLL based on simulation results. Noise generated from the individual functional blocks was analyzed, and a method was proposed to choose the circuit parameters for jitter reduction. The jitter improvement factor obtained by using a reduced K_{VCO} was determined experimentally through simulation. Two conclusions can be drawn from this simulation study. First is that, in the traditional self biased PLL, the minimum jitter is observed, by using device dimensions that reduce noise PSD from the charge pump and not from the VCO. This leads to the first conclusions that in wide bandwidth PLL, the significant noise contributor is the charge pump circuit. Second is that with K_{VCO} scaled, the PLL bandwidth becomes narrower; the VCO phase noise contribution to the overall phase noise has to increase; and the jitter improvement obtained can be mitigated. Even under this narrow bandwidth condition, a significant jitter performance improvement is observed. Hence it can be concluded that the dominant noise contributor is the charge pump systematic noise, and is observed to get scaled down by the same factor by which the K_{VCO} is reduced. The dual loop technique thus implemented with reduced K_{VCO} shows significant jitter reduction. Although comparison of the present work with silicon implementation results may not be entirely fair, in terms of jitter, lock time and FOM, it was shown that the systematically designed dual loop system performance is either better or comparable to the recently reported work on a ring oscillator based dual loop PLL [7] [8] .

REFERENCES

1. Maneatis, J.G. (1996) Low-Jitter Process-Independent DLL and PLL Based on Self-Biased Techniques. IEEE Journal of Solid-State Circuits, 31, 1723-1732. http://dx.doi.org/10.1109/JSSC.1996.542317
2. Allen, A., Desai, J., Verdico, F., Anderson, F., Mulvihill, D. and Krueger, D. (2009) Dynamic Frequency-Switching Clock System on a Quad-Core Itanium Processor. IEEE International Solid State Circuits Conference, San Francisco, 8-12 February 2009, 6263. http://dx.doi.org/10.1109/isscc.2009.4977308
3. Stackhouse, B., Bhimji, S., Bostak, C., Bradley, D., Cherkauer, B., Desai, J., Francom, E., Gowan, M., Gronowski, P., Krueger, D., Morganti, C. and Troyer, S. (2009) A 65 nm 2-Billion Transistor Quad-Core Itanium Processor. IEEE Journal of Solid-State Circuits, 44, 18 31. http://dx.doi.org/10.1109/JSSC.2008.2007150
4. Kurd, N., Mosalikanti, P., Neidengard, M., Douglas, J. and Kumar, R. (2009) Next Generation Intel? Core? Micro- Architecture (Nehalem) Clocking. IEEE Journal of Solid-State Circuits, 44, 1121-1129. http://dx.doi.org/10.1109/JSSC.2009.2014023
5. Duarte, D., Hsu, S., Wong, K., Huang, M. and Taylor, G. (2010) Interpolated VCO Design for Low Bandwidth, Low Jitter, Self Biased PLL in 45 nm CMOS. Custom Integrated Circuits Conference, San Jose, 19-22 September 2010, 1-4. http://dx.doi.org/10.1109/cicc.2010.5617473
6. Song, Y., Wang, Y., Jia, S. and Zhao, B.Y. (2008) A Novel Calibration Technique Applying to an Adaptive-Bandwidth PLL. International Conference on Solid State and Integrated Circuit Technology, Beijing, 20-23 October 2008, 1933- 1936. http://dx.doi.org/10.1109/icsict.2008.4734971
7. Sai, A., Kobayashi, Y., Saigusa, S., Watanabe, O. and Itakura, T. (2012) A Digitally Stabilized Type-III PLL Using Ring VCO with 1.01 psrms Integrated Jitter in 65 nm CMOS. IEEE International Solid-State Circuits Conference Digest of Technical Papers, San Francisco, 19-23 February 2012, 248-249.
8. Sun, H. and Moon, U.-K. (2015) MDLL/PLL Dual-Path Clock Generator. IEEE 58th International Midwest

Symposium on Circuits and Systems, Fort Collins, 2-5 August 2015, 1-4.

9. Ferriss, M., Plouchart, J.-O., Natarajan, A., Rylyakov, A., Parker, B., Tierno, J.A., Babakhani, A., Yaldiz, S., Valdes- Garcia, A., Sadhu, B. and Friedman, D.J. (2013) An Integral Path Self-Calibration Scheme for a Dual-Loop PLL. IEEE Journal of Solid-State Circuits, 48, 1-12. http://dx.doi.org/10.1109/JSSC.2013.2239114

10. Nonis, R., Da Dalt, N., Palestri, P. and Selmi, L. (2005) Modelling, Design and Characterization of a New Low-Jitter Analog Dual Tuning LC-VCO PLL Architecture. IEEE Journal of Solid-State Circuits, 40, 790-804. http://dx.doi.org/10.1109/JSSC.2005.848037

11. Van de Beek, R.C.H., Vaucher, C.S., Leenaerts, D.M.W., Klumperink, E.A.M. and Nauta, B. (2004) A 2.5-10-GHz Clock Multiplier Unit with 0.22-ps RMS Jitter in Standard 0.18-μm CMOS. IEEE Journal of Solid-State Circuits, 39, 1862-1872. http://dx.doi.org/10.1109/JSSC.2004.835833

12. Thambidurai, C. and Krishnapura, N. (2010) Spur Reduction in Wideband PLLs by Random Positioning of Charge Pump Current Pulses. IEEE International Symposium on Circuits and Systems, Paris, 30 May-2 June 2010, 3397-3400. http://dx.doi.org/10.1109/iscas.2010.5537876

13. Hajimiri, A., Limotyrakis, S. and Lee, T.H. (1999) Jitter and Phase Noise in Ring Oscillators. IEEE Journal of Solid- State Circuits, 34, 1303-1309. http://dx.doi.org/10.1109/4.766813

14. Dhurga Devi, J. and Ramakrishna, P.V. (2011) Performance Improvement for Maneatis PLL for Microprocessor Clock. 7th International Conference on Ph.D Research in Microelectronics & Electronics, Trento, 3-7 July 2011, 209-212.

CHAPTER 5

GENERATION OF SILICON-BASED HARDWARE PLC BY DIRECT CONVERSION OF THE LADDER DIAGRAM TO CIRCUIT DESIGN LANGUAGE

Daoshan Du Xiaodong Xu and Kazuo Yamazaki

Keywords: Programmable logic controller. Ladder diagram. Very high-speed circuit design hard description language. Field programmable gate array. Condensed simultaneity graph. Finite state machine

INTRODUCTION

Programmable logic controllers (PLCs) have been widely used for the logic control in the manufacturing system [1]. In industrial application, ladder diagram is the most popular programming language for PLC development. Traditionally, PLC includes a microprocessor, and the ladder diagram is sequentially executed inside this PLC microprocessor in a cyclic scan period. Based on this solution, PLC performance is limited by the cyclic scan period, which depends on the program length and the microprocessor's processing speed. In order to overcome these drawbacks with the programmable hardware solution, field programmable gate array (FPGA)-based PLC has been focused on by many researchers with its reconfigurable hardware structure and parallel execution advantage. Miyazawa [2] and Ikeshita et al. [3] developed a very rough manner in 1999 to convert the graphic ladder diagram into program description of very high-speed integrated circuit hardware description language (VHDL). Chen and Patyra [4] designed a VHDL model of the whole system directly from the original system requirements to build a controller. Abdel-Hamid et al. [5] and Kuusilinna et al. [6] developed an algorithm to convert finite state machine (FSM) into VHDL. Adamski, M. et al.[7–10] did an effective work in choosing Petri net model as a substitute of ladder diagram in manufacturing control. These studies show that reconfigurable hardware has advantages of simplicity of programming, size, and cost, while parallel execution of FPGA can improve the PLC performance dramatically. In FPGA design, parallel execution exists not only in one combinational logic operation but also in the multiple combinational logic operations. To the first case, in one combinational logic operation, there is only one output. But, all of the combinational logics which influence this output have been designed in a flat way as the existing circuit. The operation can be executed at the electrical speed. To the second case, in the multiple combinational logic operations, all of these combinational logic operations that influence every output have been designed in a flat way as the existing circuits. So they can occur in a parallel way. For the first case, it is very easy to be realized in VHDL. In this paper, the parallel execution to be discussed here specially means the second case. In addition, the current research stays at a very rough stage by only introducing case-based conversion from PLC's description language to VHDL. Also, most methods aim at obtaining the hardware description languages (HDL) or the register transistor level (RTL) architecture of FPGA from the original system requirement. As the majority of PLC programs in the manufacturing system are written by ladder diagram, it is essential to make use of the existing ladder diagram of current PLC system to conduct new PLC design. It is in a different way to implement PLC ladder diagram between inside microprocessor and inside FPGA. In microprocessor, the PLC instruction is sequentially executed line by line. Otherwise, in FPGA, all of probabilities of the PLC instructions have been considered and designed with circuit, which can be activated by sensitive signals. In order to implement PLC ladder diagram inside FPGA, the sequential operations in the original ladder diagram should be kept. At the same time, the capability of parallel execution can be used to get more efficient performance. This paper will discuss how to implement the PLC ladder diagram in the sequential execution way and the parallel execution way inside FPGA. Programming language

Citation: Daoshan Du, Xiaodong Xu, Kazuo Yamazaki, Generation of silicon-based hardware Plc by means of the direct conversion of the ladder diagram to circuit design language, doi:10.1007/s00170-009-2405-0.

for FPGA is VHDL. The left portion of this paper is organized as follows. Section 2 briefly introduces the FPGA-based PLC design, the research feasibility, and overall approach. Section 3 explains how to optimize the original ladder diagram with the sequential structure and parallel structure. The method to convert the optimized ladder diagram into VHDL is discussed in Section 4. A case study is shown to verify the proposed approach in Section 5. Finally, Section 6 concludes the paper with a brief description of the ongoing work and future research.

FPGA-BASED PLC DESIGN

FPGA technology has great advantages to conduct new PLC design by comparison with the traditional software based PLC solution. FPGA-based PLC scheme can improve PLC performance, reduce manufacturing cost, and enforce flexibility of the logic control of the manufacturing system. In order to approach these goals, the essential point is to convert the PLC program into gate-level digital circuit expressions so that the same control logic of PLC represented in its program can be exactly reflected in the FPGA-based solution. Since the internal structure of FPGA is reconfigurable with input/output block and configurable logic block, the required circuit can be built as long as a ladder diagram is converted into RTL architecture and downloaded to the FPGA chip. Such an implementation will perform the same functions as the original PLC ladder diagram, but not in the traditional sequential cyclic scan manner. The new solution can respond to the input signals with parallel execution at hardware processing speed, which dramatically improve the PLC speed. Moreover, it can be reconfigured any times as a new PLC program is converted and downloaded. As for the RTL architecture of FPGA implementation, it is a low-level description of internal logic circuit of FPGA. Several commercial software tools are available for FPGA development. The typical one is Xilinx ISE, which can establish RTL architecture by synthesizing a high-level program such as VHDL. Figure 1 shows the approach to implement a FPGA-based PLC from the ladder diagram. There are two steps to construct FPGA-based PLC: (a) convert ladder diagram into VHDL program with sequential and parallel design and (b) synthesize the VHDL design and implement it to FPGA device. With the special reconfigurable structure of the FPGA device, RTL architecture can be configured inside FPGA, which can execute the functions of the original ladder diagram inside FPGA with more efficiency and many other advantages such as capability of reconfiguration, low cost, etc. Usually, in the traditional PLC system, there are three obvious characteristics: (a) The PLC system is based on the microprocessor, which works as the central processing unit (CPU) to sequentially process the instructions which comprised the PLC program; (b) The PLC program mostly works as the software and is compiled as object code, which is executed by the microprocessor; and (c) The PLC works with a cyclic scan. During one period, the microprocessor finishes once scanning of the whole PLC program. In this kind of working way due to the limit of microprocessor, all of these operations are exactly executed in a sequential way inside the CPU. In fact, PLC system is a discrete event control system; some events must occur after other events while some events can occur in parallel. If the

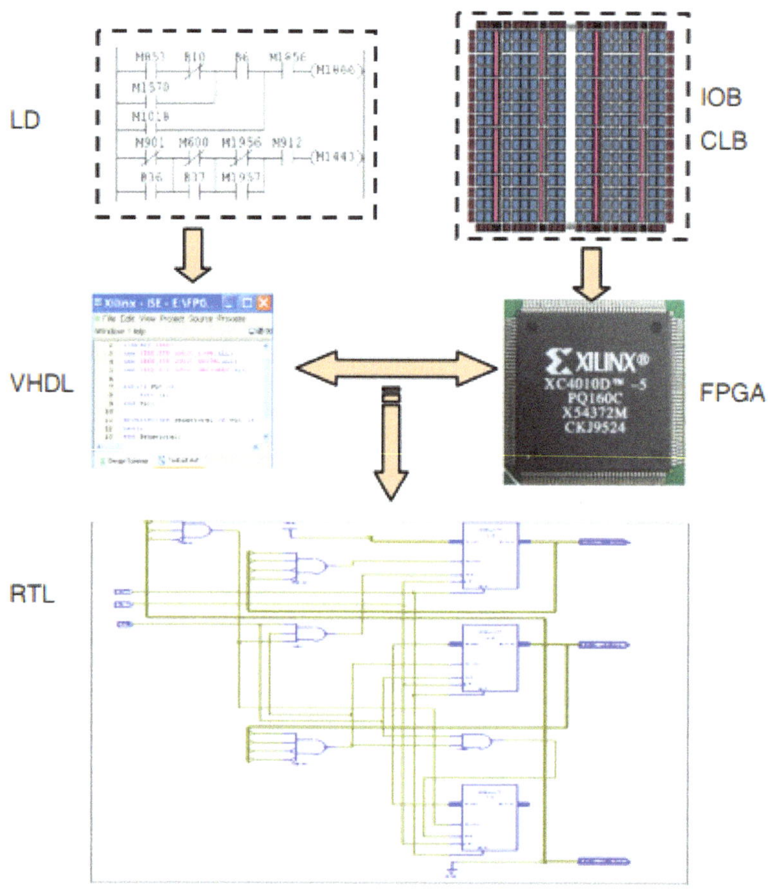

Figure. 1 Implementation of FPGA-based PLC from ladder diagram.

system is designed not only with the sequential structure as the traditional PLC program but also with parallel structure, it must be operated efficiently, and the performance can be dramatically improved. With the programmable hardware design technology such as PLD, FPGA, and complex programmable logic device, it becomes possible to design the discrete event control system based the hardware platform. FPGA-based PLC has been generally discussed by many previous researchers [2–13]. From practical view point, it is a very valuable solution to implement FPGA-based PLC by converting ladder diagram, which is the most popular programming language, to VHDL program, which is also the most popular hardware description language for FPGA design. The ladder diagram is organized in a sequential way, while the FPGA design can be organized in both the sequential and parallel way. In order to efficiently implement the original ladder diagram inside FPGA, it is necessary to analyze the ladder diagram and reorganize it with both the sequential and parallel structure. Therefore, how to convert a sequential ladder diagram into VHDL program with sequential and parallel design is the key of this research

OPTIMIZATION OF LADDER DIAGRAM PROGRAM WITH SEQUENTIAL AND PARALLEL STRUCTURE

In this paper, the condensed simultaneity graph (CSG) theory [14, 15] is used to realize the reorganization of the ladder diagram in both sequential and parallel way.

The condensed simultaneity graph theory

As a discrete event control system, the output of every rung in the ladder diagram can be described as a state variable, which is related to the status of one event switch having two possible values, "0" or "1." In ladder diagram, if the rung output depends upon the output of a previous rung or rungs, it is called dependent state. On the other hand, if the rung output is independent from the outputs of other rung or rungs and can occur at the same time with others in the ladder diagram, it is called independent state. Based on the CSG theory, two kinds of graph, the simultaneity graph and the dependency graph, can be used to describe those independent states and dependent states. Here, a simple example of ladder diagram is shown as follows (Fig. 2), which will be used to explain the

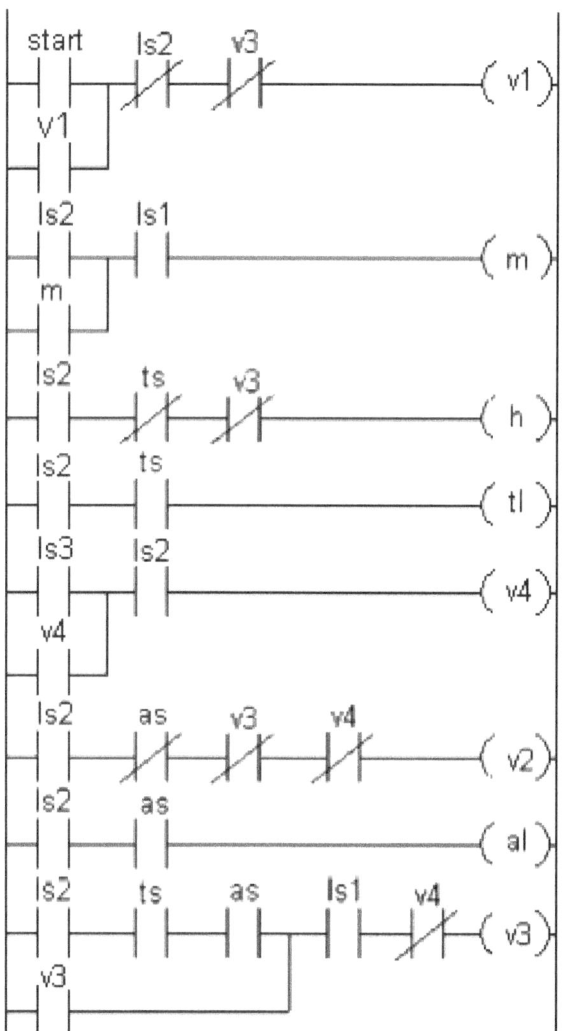

Figure. 2 An example of ladder diagram.

condensed simultaneity graph theory. The system has eight state variables: v1, m, h, tl, v4, v2, al, and v3 and six inputs: start, ls1, ls2, ls3, ts, and as.

The condensed simultaneity graph

In order to describe the independent states and the dependent states in the graph, two elements used are as follows: a node is used to represent one state, which is on behalf of one output of rung in the ladder diagram; an undirected or directed line is used to indicate that the relationship of the two states on the both ends of the line is independent or dependent.

The simultaneity graph (SG)

The first graph is the simultaneity graph (Fig. 3a). This graph uses undirected line to connect two states, which is contained if these two states can be on concurrently. This graph has a node for each rung output and an edge connecting the nodes corresponding to rung outputs that can be true simultaneously at the completion of one ladder diagram scan. The SG contains the sequential information of the process statements in VHDL program. The dependency information is very important to keep the necessary sequential operations in the original ladder diagram. If the graph is simple and there are not many states and edges, the adjacent matrix data structure is used to store the graph, which is easier to be processed. Or else, in order to save memory, adjacent list date structure is a better way to store the graph, which is a general method. Adjacent list is taken to describe the SG graph in this paper.

(1). The dependency graph (DG)

The second graph is the dependency graph (Fig. 3b). This graph uses directed line to connect two states if one state depends on the output of previous rung or rungs at the completion of one ladder diagram scan. The DG has a node for each state variable in the ladder diagram. Since the DG is a directed graph, each edge is represented by a directed pair; u is the tail and v the head of the edge and the edgeimplies that rung v depends on the output of rung u. The DG contains the concurrency information of the process statements in the VHDL model. The concurrency information is very important to implement concurrent operations inside FPGA for high efficient performance.

(2). The dependency graph (DG)

The second graph is the dependency graph (Fig. 3b). This graph uses directed line to connect two states if one state depends on the output of previous rung or rungs at the completion of one ladder diagram scan. The DG has a node for each state variable in the ladder diagram. Since the DG is a directed graph, each edge is represented by a directed pair; u is the tail and v the head of the edge and the edgeimplies that rung v depends on the output of rung u. The DG contains the concurrency information of the process statements in the VHDL model. The concurrency information is very important to implement concurrent operations inside FPGA for high efficient performance.

(3) The condensed simultaneity graph (DSG)

In order to integrate all of the concurrency information and the dependency information into one graph, the condensed simultaneity graph is introduced (Fig. 3c); an undirected graph and created by condensing the nodes of the simultaneity graph that are connected in the dependency graph into one node. The node interconnections are the same as in the original simultaneity graph. In this graph, more than one rung output can be associated with a node, and it contains information on the dependency in the SG and the concurrency in the DG.

(4) Implementation of the condensed simultaneity graph

According to the theory described in from (a) to (c), the CSG of the example can be built. The steps are as follows:

Step 1: Input the ladder diagram with instruction list expression, get through all of the rungs, count the rungs, and construct the list arrays for SG, DG, and CSG;

Step 2: Make SG according to (1) Fig. 3a;

Step 3: Make DG according to (2) Fig. 3b; and

Step 4: Build the CSG according to (3) and finally, get a completed list array to store this graph Fig. 3c.

Decomposition of the condensed simultaneity graph

The CSG includes the complete information on the dependency and the concurrency of the original ladder diagram. According to this information, the model of the ladder diagram with sequential structure and parallel

a

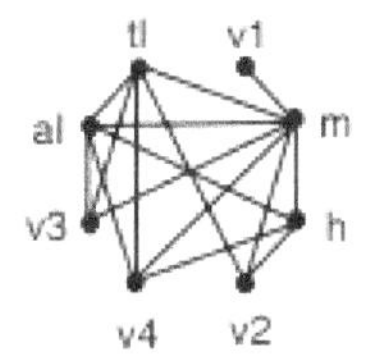

v1	→	m	/						
m	→	v1	h	tl	v4	v2	al	v3	/
h	→	m	v4	v2	al	/			
tl	→	m	v4	v2	al	v3	/		
v4	→	m	h	tl	al	/			
v2	→	m	h	tl	/				
al	→	m	h	tl	v4	v3	/		
v3	→	m	tl	al	/				

the simultaneity graph

b

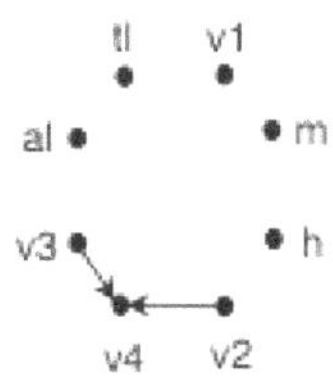

v1	→	/	
m	→	/	
h	→	/	
tl	→	/	
v4	→	/	
v2	→	v4	/
al	→	/	
v3	→	v4	/

the dependency graph

c

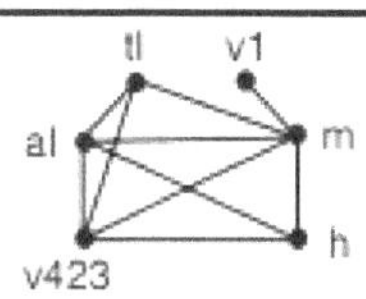

v1	→	m	/				
m	→	v1	h	tl	v423	al	/
h	→	m	v423	al	/		
tl	→	m	v423	al	/		
al	→	m	h	tl	v423	/	
v423	→	m	h	tl	al	/	

the condensed simultaneity graph

Figure. 3 The ladder diagram of the example is represented by condensed simultaneity graph theory. a The simultaneity graph. b The dependency graph. c The condensed simultaneity graph

(1) Connected component decomposition

Any condensed simultaneity graph G can be partitioned into its connected components. Given a graph G= (N, E), where N and E are sets of nodes and edges, the CCD of graph G produces a collection of subgraphs {G1; G2; ... Gi...}, such that each subgraph Gi= (Ni, Ei), and no node in Gi is connected to a node in another Gj, where i and j are used to represent the different subgraphs. Performing the CCD on the CSG produces one or more subgraphs to which new sequential structures are assigned. These newly created structures are produced in a sequential order. Since two structures cannot be activated concurrently, they are sequenced according to the order of state variables that appeared in the ladder diagram.

(2) Full connectivity decomposition

A connected graph can be partitioned into a collection of subgraphs via FCD. The FCD of graph G is denoted by FCD (G) and FCD (G) produces a collection of subgraphs {G1; G2; ...Gi...}, then every node in Gi is connected to every node in another Gj, where i and j are used to represent the different subgraphs. After accomplishing FCD, the interconnecting edges are removed since all the components in these two subgraphs are connected to each other. Accordingly, components in these two subgraphs can be activated in a parallel way and belong to the parallel structure.

(3) Implementation of CCD and FCD In order to process the CCD and FCD, data structure is constructed to store the state nodes in the CSG.

Table 1 lists the property specification of this state node structure

```
struct SStateNodeInfo
{
 int            nID;
 String         strName;
 int            nFirstChildID;
 int            nLeftSiblingID;
 int            nRightSiblingID;
 int            nAncestorID;
 bool           bCollection;
 int            nLevel;
 String         strBoolEquation;
 List           *plistLinkToState;
 SStateNodeInfo *pNext};
```

The structure of every state node should be established, and a structure list is constructed to manage all of these structures. In addition, a default structure is constructed for the original CSG as the head of the structure list: SStateNodeInfo Default Structure {0; NULL; 0; 0; 0; 0; false; 0; NULL; *plistLinkToFirstState; *pNext}; When implementing CCD or FCD, the sequential and concurrent information of every state can be changed by updating the corresponding properties in the structure.

The working flow to decompose the CSG is as follows:

(1) Read the CSG model by adjacent list. Update the property of listLinkToState in the structure.

(2) Go through the state nodes, check the property of listLinkToState, and select either CCD or FCD is needed.

(3) If CCD is needed, update the structure of this state node with sequential information. Or else, update the structure of this state node with concurrent information.

(4) Repeat Step 2, until the properties of all structures have been updated.

After decomposition, a structure list will be generated with the entire sequential and concurrent information. One structure list for an example of ladder diagram will be shown in the next section.

Apply CSG theory to optimize the ladder diagram

According to the CSG theory mentioned, it is possible to optimize the organization of the ladder diagram with sequential structure and parallel structure. This section applies the CSG to the ladder diagram shown in Fig. 2 to discuss the optimization. The whole working flow is shown in the Fig. 4.

Step 1: Create the CSG using the definition in the Section 3.1.1 (Fig. 4a).

Step 2: Apply FCD to node "m," the whole graph is divided into two parallel subgraphs as shown in Fig. 4b, {m} and {(v4, v2, v3), vl, h, tl, al}

Step 3: Apply CCD to node "v1." According to the Section 3.1.2 Part (2), the {v1} can be separated from the original graph of {(v4, v2, v3), v1, al, tl, h} as shown in Fig. 4c.

Two subgraphs are divided from the upper level graph. These two subgraphs are sequential. After CCD, they keep the original sequence as executed in the ladder diagram.

Step 4: Repeat Step 2 and apply FCD to node "al" and "v4, v2, v3." The subgraph {{v4, v2, v3}, al, tl, h} can be partitioned into three subgraphs: {al}, {(v4, v2, v3)}, and {tl, h}. The result of FCD to node "al" and "v4, v2, v3" is shown in Fig. 4c.

These three subgraphs are in parallel level and can be activated at the same time, so they are assigned to three parallel structures.

Step 5: Repeat Step 3 and apply CCD to node "tl" and "h." Subgraph {tl, h} is partitioned into two subgraphs, {tl} and {h}. These two subgraphs are sequential and keep the original sequence as executed in the ladder diagram. The result is shown in Fig. 4d.

Step 6: Since there are no nodes in CSG that can be decomposed, the connected nodes in DG are finally considered. It has been mentioned that DG shows information of rungs depending on the outputs of previous rungs during one scan of the ladder diagram.

Table 1 Property specification of state

Property name	Type	Description
nID	integer	Unique identifier assigned to this state node to distinguish it from others
strName	String	Label for this state
nFirstChildID	integer	ID for the first child of this state node
nLeftSiblingID	integer	ID for the left sibling of this state node
nRightSiblingID	integer	ID for the right sibling of this state node
nAncestorID	integer	ID for the parent of this state node
bCollection	bool	If this state node is a condensed node, it's true, else, it's false
nLevel	integer	Considering the parallel structure, the order number to execute this state
strBoolEquation	String	Boolean equation to assign this state variable
*plistLinkToState	list	List of the state connected to this state node in CSG, also the adjacent list of the state node to describe CSG.
*pNext	SStateNodeInfo	Pointer to the next structure

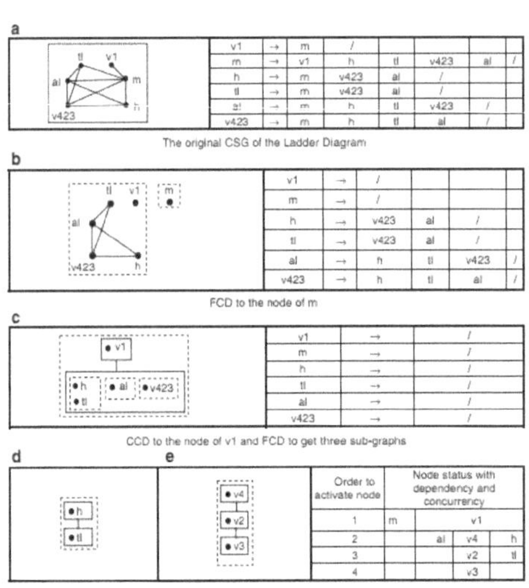

Figure. 4 a The original CSG of the ladder diagram. b FCD to the node of m. c CCD to the node of v1 and FCD to get three subgraphs. d Decomposition of nodes of h, tl. e Decomposition of nodes of v423l.

In this step, all of the states that are still not assigned will be lined up in the sequential way. The result is shown in Fig. 4e.

After the last step, a complete model with sequential and parallel structure describing the original ladder diagram has been established. The information on dependency and concurrency is shown in Table 2. This model can perform.

Table 2 Information on dependency and concurrency

{{0;	null;	0;	0;	0;	0;	false;	0;	null;	null;	*pNext};
{1;	"m";	0;	0;	2;	0;	false;	1;	"(ls2 # m) & ls1";	null;	*pNext};
{2;	"v1";	3;	1;	0;	0;	false;	1;	"(start # v1) & (! ls2) & (! v3)";	null;	*pNext};
{3;	"al";	0;	0;	4;	2;	false;	2;	"ls2 & as";	null;	*pNext};
{4;	"v4";	6;	3;	5;	2;	false;	2;	"(ls3 # v4) & ls2";	null;	*pNext};
{5;	"h";	7; 4;	0;	2;	false;	2;	"ls2 & (! ts) & (!v3)";	null;	*pNext};	
{6;	"v2";	8;	0;	0;	4;	false;	3;	"ls2 & (! as) & (! v3) & (! v4)";	null;	*pNext};
{7;	"tl";	0;	0;	0;	5;	false;	3;	"ls2 & ts";	null;	*pNext};
{8;	"v3";	0;	0;	0;	6;	false;	4;	"(ls2 & ts & as # v3) & ls1 & (! v4)";	null;	*pNext}};

the same function as the ladder diagram, not only in the sequential way but also in the parallel way. The structure list to represent the state nodes is also shown. In order to implement FPGA-based PLC with high efficient performance, the model of ladder diagram with the sequential and concurrent information needs to be realized inside FPGA. The VHDL is one of the most popular developing languages for FPGA. So, the next step is to convert this model to VHDL program and implement corresponding sequential and parallel structure in VHDL.

IMPLEMENTATION OF SEQUENTIAL AND PARALLEL OPERATIONS IN FPGA

In Section 3, the original ladder diagram has been analyzed and established a ladder diagram model with sequential and parallel structure. This section explains how to implement this ladder diagram model inside FPGA by converting ladder diagram to VHDL program. 4.1 Implementation of the Conversion from ladder diagram to VHDL program Usually, while PLC is working, ladder diagram is executed rung by rung from the top to the bottom. However, the FPGA design works in a parallel way once the FPGA device has been configured with the downloaded design. Thus, working principle is very different between the ladder diagram and VHDL program. For some components in ladder diagram, such as normal open and normal close, there are corresponding components in VHDL such as NOT, AND, and OR gates to match each other. For other components such as self-hold and interlock, there is no corresponding component. Therefore, a bridge is needed to eliminate this gap between these two programs. Several researchers used the modeling method like FSM or Petri net as the bridge [5–10, 15]. Actually, a complicated ladder diagram probably consists of thousands of states so that it may easily cause the FSM or Petri net model extreme complex and hard to handle. In this research, the Boolean equations are chosen as the bridge to connect ladder diagram to VHDL. Boolean algebra is a system of mathematics often used to manipulate.

Table 3 The operator used in Boolean expression

!	&	#	$
NOT	AND	OR	XOR

Table 4 Boolean equations for the example in Fig. 2

v1 :=(start # v1) & (! ls2) & (! v3)
m :=(ls2 # m) & ls1
h :=ls2 & (! ts) & (!v3)
tl :=ls2 & ts
v4 :=(ls3 # v4) & ls2
v2 :=ls2 & (! as) & (! v3) & (! v4)
al :=ls2 & as
v3 :=(ls2 & ts & as # v3) & ls1 & (! v4)

logic operation. It's very appropriate to describe logic circuit. As a logic description, Boolean expression almost has the same form as that in VHDL, so it is much easier to convert ladder diagram to Boolean expression than to other models. Table 3 lists the operator often used in Boolean expression. To the ladder diagram in Fig. 2, the equivalent Boolean equations are shown in Table 4.

IMPLEMENTATION OF THE SEQUENTIAL OPERATIONS IN VHDL

The sequential operation is critical to assure the correct performance. In the application system of the PLC, some actions must be done after the others, else, it will result serious damage or even the danger to life. This can be controlled over by activating the rungs of the ladder diagram in the correct order. Similarly, because VHDL program works based on process statements, if put, the operation of every dependent rung of the ladder diagram into process statements in VHDL, the sequential operation can be realized by using sensitive signals to sequentially activate different process statements In order to implement the same sequential operations as the original ladder diagram program, two steps are proposed in this paper: (a) generate sequential sensitive signals, which are used to activate different process states in order and (b) put the operation of every rung into the process statement in VHDL. These two steps can be implemented in VHDL with two types of process statements. The one is the finite state machine process statement. The other is the regular process statement. In the first step, finite state machine is proposed to generate the sensitive signals. PLC is a discrete event system. Finite state machine is often used to model the discrete event system. So, it is appropriate to control the sequential operation in PLC. In VHDL program, process statements are running in a parallel way. With the sensitive signals generated by finite state machine, process statements can be activated in a sequential way. Finite state machine can be implemented in VHDL as follows (Here, the limit of state status is taken as 8):

```
FSM_process: process(clock,reset)
begin
    if (reset='1') then
        current_process <= "000";
    elsif (clock'event and clock='1') then
        if (current_process ="111")then
            current_process <= "000";
        else
            current_process <= current_process + step_increment;
        end if;
    end if;
end process;
```

This is a process statement for finite state machine. Clock signal and reset signal are the sensitive signals for this process. The signal, current_process, is the sensitive signal which will be used to sequentially activate all of the regular process statements. The variable, step_increment, is used to adjust the order increment of regular process statements to be activated. Usually, it is assigned by "1," so the regular process statements will be activated one by one. If one of the regular process statements will execute a jump instruction, it will be assigned by a proper value, so the correct regular process statement will be activated in the next clock. In the second step, the Boolean equations such as those in Table 4 are mapped into regular process statements one by one. Here is an example as follows:

```
Regular_01: process ( current_process )
begin
    if current_ process = "000" then
        V1 <= (start or V2) and ( not V3);
    end if;
end process No_01;

Regular_11: process ( current_process )
begin
    if current_ process = "001" then
        V4 <= (ls3 or V4) and ls2;
    end if;
end process No_11;
```

Where, Regular_00 and Regular_10 mean the label of the regular process statement. One label is corresponding one rung in the ladder diagram program. The signal, current_process, is just the sensitive signal generated by the finite state machine process. In VHDL program, process statements are running in a parallel way. When the sensitive signal is changed, all of the regular process statements will be activated at the same time; but according to the value of the sensitive signal, only the proper regular process statements can be executed at this moment. With the finite state machine process statement periodically changing the sensitive signal, the sequential cyclic scan in the original ladder diagram can be realized in VHDL program.

Implement the parallel operations in FPGA Similarly, the parallel operation is critical to assure highefficient performance. In the practical application of the control system, some actions can be done at the same time. In the traditional PLC, these actions are executed one by one because of the limit of the microprocessor. However, inside FPGA, they can be implemented by the concurrent sensitive signals. If put, the operations of every independent rung in the ladder diagram program into process statements in VHDL, the parallel operation can be realized by using sensitive signals to activate different process statements at the same time. The independent rungs are mapped into the regular process statements shown in the following example. Both of these regular process statements will be activated by the sensitive signals, which are generated by the finite state machine process. But it is different with the sequential operations that the regular process statements will be executed at the same time because of the concurrent sensitive signal.

```
Regular_00: process ( current_process )
begin
    if current_ process = "000" then
        m <= (ls2 or m) and ls1;
    end if;
end process No_00;

Regular_01: process ( current_process )
begin
    if current_ process = "000" then
        V1 <= (start or V2) and ( not V3);
    end if;
end process No_01;
```

When the sensitive signal, current_ process, is assigned by "000," both of these regular process statements are activated and executed at this moment. Accordingly, it means two rungs related to these two process statements in the ladder diagram program are executed at the same time.

CASE STUDY

To the original ladder diagram program in Fig. 2, there are eight rungs. If this program is executed in sequential way as executed in microprocessor, eight instruction cycles are needed. In FPGA, it only needs four-clock cycles with both sequential operation and parallel operation. The performance becomes more efficient. At the same time, it keeps the same function as the original ladder diagram program. A converter has been developed to convert the ladder diagram model with information on dependency and concurrency to VHDL program. Once the VHDL program has been generated with the software tools such as the ISE Foundation from Xilinx and the ModelSIM from Mentor Graphics, the converted VHDL program can be debugged and simulated. In fact, the example of the ladder diagram mentioned in Fig. 2 is the control system for a well-known neutralization system in Fig. 5. The control rules are as follows [14, 15]:

(1) Initially, all the valves are closed, the mixer and heater are off, and the reaction tank is empty. When the start button is pressed, open valve v1 until level sensor ls2 is activated. This fills the tank with the solution to be neutralized.

(2) The following three processes proceed concurrently.

  (a) When the solution level rises above level sensor ls2, start mixer m. When the level drops below ls1, stop the mixer.

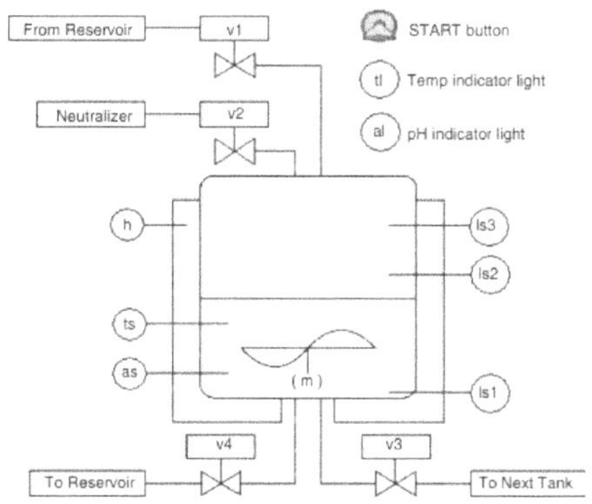

Figure. 5 Chemical vat neutralization system.

  (b) Whenever the temperature of the solution is below a preset point, energize heater h.

  (c) Whenever the pH of the solution is unbalanced, open valve v2 to add the neutralizer.

(3) If the tank becomes full, as indicated by the activation of level sensor ls3, close v2 to stop the in flow of the neutralizer. Next, open valve v4 to reduce the level of the solution to the point indicated by ls2. Then close v4 and proceed with step (c).

(4) When both the temperature and pH of the solution are correct, de-energize the heater and close v2. Then open valve v3 to drain the tank. When the tank is empty, as indicated by the deactivation of ls1, close v3 and proceed with step (1).

(5) Two indicator lights, tl and a1, are regulated. Light tl should turn on whenever the solution level is above ls2 and the temperature has reached the preset value. The same rule is used to light al and pH.

With the ModelSIM, the VHDL program can be simulated and the result is shown in Fig. 6.

From the Fig. 6, the following results can be verified:

- In the first four-clock cycle, when the start button is pressed, if both ls1 and ls2 are on, the mixer and the heater will turn on, v2 will be opened to add the neutralizer.
- In the second four-clock cycle, if ls3 is on, the solution reaches level 3. v4 will be opened and v2 will be closed. It means open the valve 4 for reducing the level of the solution and at the same time close the valve to stop adding the neutralizer.

In the third four-clock cycle, if the level of the solution goes down under level2, ls2 and ls3 are off, valve v1 will be reopened to add solution. Before the level of solution reaches level 2, the heater turns off. And the valve 4 should be closed. In the fourth four-clock cycle, if the level of the solution return back to level 2, close the valve 1, turn on the heater, open the valve 2 for neutralizing. In the fifth four-clock cycle, once the temperature and pH are satisfied (ts and as turn to high), tl and al turn on. The valve 2 is closed and the heater is off. Open the valve 3 to drain the tank. The results of simulation verify that the converted VHDL program can perform the same function as the ladder diagram does. Synthesize this VHDL program and download it to the FPGA, the control system will be implemented inside FPGA.

CONCLUSIONS

In order to overcome the performance limitation of the traditional microprocessor-based PLC and dramatically improve the PLC performance, in this paper, FPGA-based PLC is proposed. With FPGA device, it is possible to implement the ladder diagram in programmable hardware solution. In order to efficiently implement the ladder diagram inside FPGA, CSG optimization method is introduced to reorganize ladder diagram with sequential and parallel structure. Two important components in VHDL design, finite state machine scheme and sensitive signal, make sure to implement the original ladder diagram in VHDL design not only in a sequential way but also in a parallel way. The authors developed the software to convert the ladder diagram to VHDL design. Using this converter, a case

Figure. 6 The logic sequence of the example implemented inside FPGA.

study was made to verify that the proposed approach to implement FPGA-based PLC is feasible, and the PLC performance is improved dramatically. Furthermore, as a programmable hardware solution, FPGA device is reconfigurable, which make it easier for the application system to be modified and maintained. The design method is more flexible.

ACKNOWLEDGMENT

The authors wish to exprcss their sincere appreciation for the generous support from Mori Seiki Co. Ltd., which made this work possible.

REFERENCES

1. Auslander DM, Pawlowski C, Ridgely J (1998) Reconciling Programmable Logic Controllers with Mechatronics Control Software, Proceedings of the 1998 IEEE international Conference on Control Applications, pp 415–420
2. Ikeshita M, Takeda Y, Murakoshi H, Funakubo N, Miyazawa I (1999) Application of FPGA to high-speed programmable controller - development of the conversion program from SFC to Verilog IEEE Symposium on Emerging Technologies and Factory Automation, IEEE, Piscataway, NJ, USA, ETFA.v2:1386–1390
3. Miyazawa I, Nagao T, Fukagawa M, Itoh Y, Mizuya T, Sekiguchi T (1999) Implementation of ladder diagram for programmable controller using FPGA. IEEE Symposium on Emerging Technologies and Factory Automation, IEEE, Piscataway, NJ, USA, ETFA. v2:1381–1385
4. Chen J, Patyra MJ (1994) VHDL modeling of a multivariable fuzzy logic controller hardware system IEEE International Conference on Fuzzy Systems. IEEE, Piscataway, NJ, USA, v1:129–132
5. Abdel-Hamid AT, Zaki M, Tahar S (2004) A tool converting finite state machine to VHDL, Canadian Conference on Electrical and Computer Engineering Canadian Conference on Electrical and Computer Engineering; Technology Driving Innovation, 4:1907–1910
6. Kuusilinna K, Lahtinen V, Hamalainen T, Saarinen J (2001) Finite state machine encoding for VHDL synthesis. Comput Digi Tech 148:23–30
7. Adamski M, Monteiro JL (2000) From interpreted Petri net specification to reprogrammable logic controller design. IEEE International Symposium on Industrial Electronics 1:13–19
8. Adamski M (1998) SFC, Petri nets and application specific logic controllers, Proceedings of the IEEE International Conference on Systems, Man and Cybernetics. IEEE, Piscataway, NJ, USA, 1:728–733
9. Uzam M, Jones AH (1998) Discrete event control system design using automation Petri nets and their ladder diagram implementation. Int J Adv Manuf Technol 14(n10):716–728
10. Lee JS, Hsu PL (2005) An improved evaluation of ladder logic diagrams and Petri nets for the sequence controller design in manufacturing systems. Int J Adv Manuf Technol 24(n3–4):279–287
11. Welch JT (1992) Translating unrestricted relay ladder logic into Boolean form. Computers in Industry 20:45–61
12. Shanta S, Dipali S (2005) A new generation of PLC—an FPGA based PLC. Proceedings of the SICE Annual Conference, SICE 2005 Annual Conference in Okayama—Proceedings. pp 2367–2370
13. Yadong L, Kazuo Y, Makoto F, Masahiko M (2005) Model-driven programmable logic controller design and FPGA-based hardware implementation. Proceedings of the ASME International Design Engineering Technical Conferences and Computers and Information in Engineering Conference—DETC2005, 4:81–88
14. Falcione A, Krogh BH (1992) Design recovery for relay ladder logic, the first IEEE conference on control applications. Dayton, OH, pp 90–98
15. Lee JI, Chun SW, Kang J (2002) Virtual prototyping of PLCbased embedded system using object model of target and behavior model by converting RLL-to-state chart directly. J Systems Archit 48:17–35

CHAPTER 6

FUTURE CHIP MULTIPROCESSORS

Kyriakos Stavrou and Pedro Trancoso

INTRODUCTION

The doubling of microprocessor performance every 18 months has been the result of two factors: more transistors per chip and superlinear scaling of the processor clock with technology generation [1]. However, technology scaling together with frequency and complexity increase result in a significant increase of the power density. This trend, which is becoming a key-limiting factor to the performance of current state-of-the-art microprocessors [2–5], is likely to continue in future generations as well [4, 6]. The higher power density leads to increased heat dissipation and consequently higher operating temperature [7, 8]. To handle higher operating temperatures, chip manufactures have been using more efficient and more expensive cooling solutions [6, 9]. While such solutions were adequate in the past, these packages are now becoming prohibitively expensive, as the relationship between cooling capabilities and cooling costs is not linear [4, 6]. To reduce packaging cost, current processors are usually designed to sustain the thermal requirement of typical workloads and utilize dynamic thermal management (DTM) techniques when temperature exceeds the design-set point [4, 10]. When the operating temperature reaches a prede- fined threshold, the DTM techniques reduce the processor's power consumption in order to allow it to cool down [4, 6, 7, 11–13]. An example of such a DTM mechanism is to reduce the consumed power through duty-cycle-based throttling. While it is very effective achieving its goal, each DTM event comes with a significant performance penalty [4, 7]. Moreover, the reliability of electronic devices and therefore of microprocessors depends exponentially on the operation temperature [4, 5, 14–17]. Viswanath et al. [5] note that even small differences in operating temperature, in the order of 10°C–15°C, can result in a 2x difference in the lifespan of the devices. Finally, higher temperature leads to power and energy inefficiencies mainly due to the exponential dependence of leakage power on temperature [4, 6, 7, 13]. As in future generations, leakage current is expected to consume about 50% of the total power [1, 3] this issue will become more serious. Additionally, the higher the operating temperature is, the more aggressive the cooling solution must be (e.g., higher fan speeds) which will lead to further increase in power consumption [11, 12]. The chip multiprocessors (CMP) architecture has been proposed by Olukotun et al. [2] as a solution able to extend the performance improvement rate without further complexity increase. The benefits resulting from this architecture are proved by the large number of commercial products that adopted it, such as IBM's Power 5 [18], SUN's Niagara [19], Intel's Pentium-D [20], and AMD's Athlon 64 X2 [21]. Recently, CMPs have been successfully used for multimedia applications as they prove able to offer significant speedup for these types of workload [22–24]. At the same time, embedded devices have an increasing demand for multiprocessor solutions. Goodacre [25] states that 3 G handsets may use parallel processing at a number of distinct levels, such as when making a video call in conjunction with other background applications. Therefore, the CMP architecture will be soon used in the embedded systems. The trend for future CMPs is to increase the number of on-chip cores [26]. This integration is likely to reduce the per-core cooling ability and increase the negative effects of temperature-induced problems [27]. Additionally, the characteristics of the CMP, that is, multiple cores packed together, enable execution

Citation: Kyriakos Stavrou and Pedro Trancoso, Future Chip Multiprocessors, doi:10.1155/2007/48926.

scenarios that can cause excessive thermal stress and significant performance penalties. To address these problems, we propose thermal-aware scheduling. Specifically, when scheduling a process for execution, the operating system determines on which core the process will run based on the thermal state of each core, that is, its temperature and cooling efficiency. Thermal-aware scheduling is a mechanism that aims to avoid situations such as creation of large hotspots and thermal violations, which may result in performance degradation. Additionally, the proposed scheme offers opportunities for performance improvements arising not only from the reduction of the number of DTM events but also from enabling per-core frequency increase, which benefits significantly single-threaded applications [10, 28]. Thermal-aware scheduling can be implemented purely at the operating system level by adding the proper functionality into the scheduler of the OS kernel. The contributions of this paper are the identification of the thermal issues that arise from the technological evolution of the CMP chips, as well as the proposal and evaluation of a thermal-aware scheduling algorithm with two optimizations: thermal threshold and neighborhood awareness. To evaluate the proposed techniques, we used the TSIC simulator [29]. The experimental results for future CMP chip configurations showed that simple thermal-aware scheduling algorithms may result in significant performance degradation as the temperature of the cores often reach the maximum allowed value, consequently triggering DTM events. The addition of a thermal threshold results in a significant reduction of DTM events and consequently in better performance. By making the algorithm aware of the neighboring core thermal characteristics (neighborhood aware), the scheduler is able to take better decisions and therefore provide a more stable performance comparing to the other two algorithms.

The rest of this paper is organized as follows. Section 2 discusses the relevant related work, Section 3 presents the most important temperature-induced problems and analyzes the effect they are likely to have on future chip multiprocessors. Section 4 presents the proposed thermal-aware scheduling algorithms. Section 5 describes the experimental setup and Section 6 the experimental results. Finally, Section 7 presents the conclusions to the work.

RELATED WORK

As temperature increase is directly related to the consumed power, techniques that aim to decrease the power consumption achieve temperature reduction as well. Different techniques, however, target power consumption at different levels. Circuit-level techniques mainly optimize the physical, transistor, and layout design [30, 31]. A common technique uses different transistor types for different units of the chip. The architectural-level techniques take advantage of the application characteristics to enable on-chip units to consume less power. Examples of such techniques include hardware reconfiguration and adaptation [32], clock gating and modification of the execution process, such as speculation control [33]. At the application level, power reduction is mainly achieved during the compilation process using specially developed compilers. What these compilers try to do is to apply power-aware optimizations during the application's optimization phase such as strength reduction and partial redundancy elimination. Another solution proposed to deal with the thermal issues is thermal-aware floor planning [34]. The rationale behind this technique is placing hot parts of the chip in locations having more efficient cooling while avoiding the placement of such parts adjacent to each other. To handle situations of excessive heat dissipation, special dynamic thermal management (DTM) techniques have been developed. Skadron et al. in [4] present and evaluate the most important DTM techniques, dynamic voltage and frequency scaling (DVFS), units toggling and execution migration. DVFS decreases the power consumed by the microprocessor's chip by decreasing its operating voltage and frequency. As power consumption is known to have a cubic relationship with the operating frequency [35], scaling it down leads to decreased power consumption and consequently decreased heat dissipation. Although very effective in achieving its goal, DVFS introduces significant performance penalty, which is related to the lower performance due to the decreased frequency and the overhead of the reconfiguration event. Toggling execution units [4], such as fetch engine toggling, targets power consumption decrease indirectly. Specifically, such techniques try to decrease the number of instructions on-the-fly in order to limit the consumed power and consequently allow the chip to cool. The performance penalty comes from the underutilization of the available resources.

Execution migration [13] is another technique targeting thermal issues and maybe the only one from those mentioned above, that does it directly and not through reducing power consumption. When a unit gets too hot, execution is migrated to another unit that is able to perform the same operation. For this migration to be possible, replicated and idle units must exist. Executing a workload in a thermal-aware manner has been proposed by Mooref et al. [12] for large data-centers. Specifically, the placement of applications is such that servers executing intensive applications are in positions favored by the cold-air flow from the air conditioners. Thermal-aware scheduling follows the same principles but applies this technique to CMPs. Donald and Martonosi [36] present a throughout analysis of thermal management techniques for multicore architectures. They classify the techniques they use in terms of core throttling policy, which is applied locally to a core or to the processor as a whole, and process migration policies. The authors concluded that there is significant room for improvement.

CMP THERMAL ISSUES

The increasing number of transistors that technology advancements provide, will allow future chip multiprocessors to include a larger number of cores [26]. At the same time, as technology feature size shrinks, the chip's area will decrease. This section examines the effect these evolution trends will have on the temperature of the CMP chip. We start by presenting the heat transfer model that applies to CMPs and then discuss the two evolution scenarios: smaller chips and more cores on the same chip

Heat transfer model in CMPs

Cooling in electronic chips is achieved through heat transfer to the package and consequently to the ambient, mainly through the vertical path (Figure 1(a)). At the same time, there is heat transfer between the several units of the chip and from the units to the ambient through the lateral path. In chip multiprocessors, there is heat exchange not only between the units within a core but also across the cores that coexist on the chip (Figure 1(b)). As such, the heat produced by each core affects not only its own temperature but also the temperature of all other cores. The single chip microprocessor of Figure 1(a), can emit heat to the ambient from all its 6 cross-sectional areas whereas each core of the 4-core CMP (Figure 1(b)) can emit heat from only 4. The other two cross-sectional areas neighbor to other cores and cooling through that direction is feasible only if the neighboring core is cooler. Even if the temperature of the neighboring core is equal to that of the ambient, such heat exchange will be poor when compared to direct heat dissipation to the ambient due to the low thermal resistivity of silicon [4]. Furthermore, as the number of on-chip cores increases, there will be cores with only 2 "free" edges (cross-sectional areas at the edge of the chip), further reducing the per-core cooling ability (Figure 1(c)). Finally, if the chip's area does not change proportionally, the per-core "free" cross-sectional area will reduce harming again the cooling efficiency. All the above lead us to conclude that CMPs are likely to suffer from higher temperature stress compared to single chip microprocessor architectures.

CMP evolution trends

Trend 1: decreasing the chip size

As mentioned earlier, technology improvements and feature size shrink will allow the trend of decreasing chip's size to continue. This chip's area decrease results in higher operating temperature as the ability of the chip to cool by vertically dissipating heat to the ambient is directly related to its area (Section 3.1). As such, the smaller the chip size is, the less efficient this cooling mechanism is. The most important consequence of higher operating temperature is the significant performance penalty caused by the increase of DTM events. Further details about this trend are presented in Section 6.1.

Trend 2: increasing the number of cores

As the number of on-chip core increases, so does the throughput offered by the CMP. However, if the size of the chip does not scale, the per-core area will decrease. As shown previously in Section 3.2, this has a negative effect on the operating temperature and consequently on the performance of the multiprocessor. A detailed study about the effect of increasing the number of on-chip cores will be presented in Section 6.1 together with the experimental results.

Reliability Adding more cores to the chip improves the fault tolerance by enabling the operation of the multiprocessor with the remainder cores. Specifically, a CMP with 16 cores can be made to operate with 15 cores if one fails. More cores on the chip, however, will decrease the chipwide reliability in two ways. The first is justified by the characteristics of failure mechanisms. According to the sum-offailure-rates (SOFR) model [37, 38], the failure rate of a CMP can be modeled as a function of the failure rate of its basic core (λ^{BC}) as shown by (1). In this equation, n is the number of on-chip cores, all of which are assumed to have the same failure rate (λ^{BC} i = λ^{BC} ∀i). Even if we neglect failures due to the interconnects, the CMP chip has n-times greater failure rate compared to its Basic Core,

$$\lambda^{CMP} = \sum_{i=1}^{n} (\lambda_i^{BC}) + \lambda^{Interconnects} = n \cdot \lambda^{BC} + \lambda^{Interconnects}. \tag{1}$$

The second way, more cores on the chip affect chip wide reliability is related to the fact that higher temperatures exponentially decrease the lifetime of electronic devices [4, 5, 14–17]. As we have shown in Section 3.2, large-scale

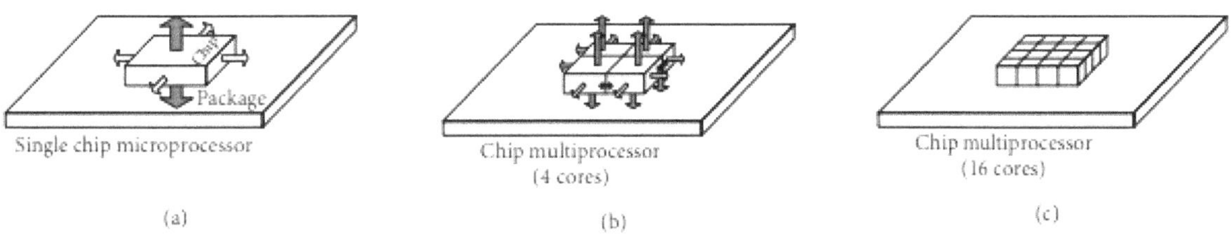

Figure 1: Cooling mechanisms in single chip microprocessors and in chip multiprocessors.

CMPs will suffer from larger thermal stress, accelerating these temperature-related failure mechanisms. It is also necessary to mention that other factors that affect the reliability are the Spatial (different cores having different temperatures at the same time point) and temporal (differences in the temperature of a core over the time) temperature diversities.

Thermal-aware floorplanning

Thermal-aware floorplanning is an effective widely used technique for moderating temperature-related problems [17, 34, 39, 40]. The rationale behind it is placing hot parts of the chip in locations having more efficient cooling while avoiding the placement of such parts adjacent to each other. However, thermal-aware floorplanning is likely to be less efficient when applied to CMPs as core-wide optimal decisions will not necessarily be optimal when several cores are packed on the same chip. Referring to Figure 2(d), although cores A and F are identical, their thermally optimal floorplan is likely to be different due to the thermally different positions they have on the CMP. These differences in the optimal floorplan are likely to increase as the number of on-chip cores increases due to the fact that the number of thermally different locations increase with the number of on-chip cores. Specifically, as Figures 2(a) to 2(d) show, for a CMP with n2 cores, there will be $(\lceil n/2 \rceil \cdot (\lceil n/2 \rceil + 1))/2$ different possible locations. A CMP with the majority of its cores being different in terms of their floorplan would require a tremendous design and verification effort making the optimal design prohibitively expensive.

The ideal operation scenario

In the ideal case, each core has a constant temperature since the processor was powered-on and therefore no temporal temperature diversities exist. Additionally, this temperature is the same among all cores eliminating spatial temperature diversities.

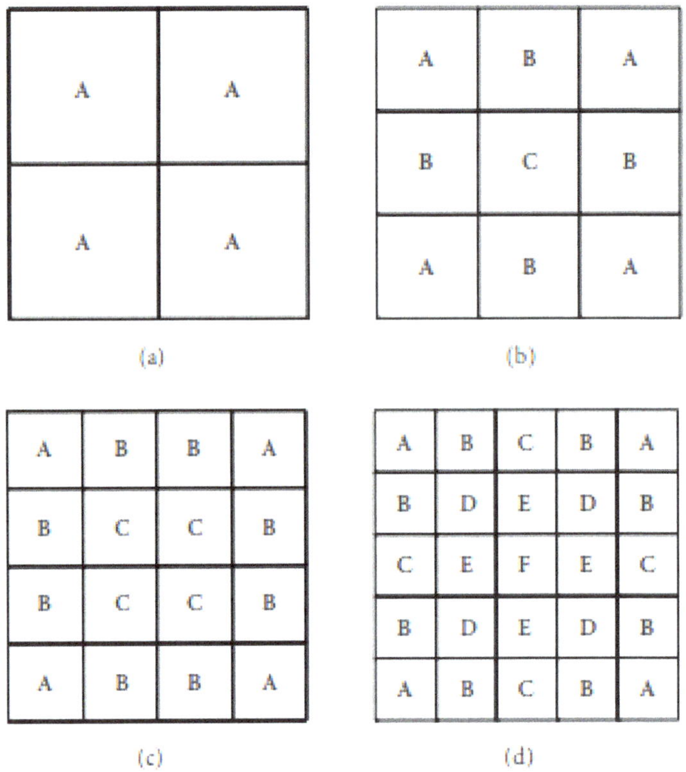

Figure 2: The thermally different locations on the chip increase with the number of cores. For a CMP with n^2 identical square cores, there will be $(\lceil n/2\rceil \cdot (\lceil n/2\rceil + 1))/2$ different locations

The decrease of spatial and temporal temperature diversities will have a positive effect on chip's reliability. Of course, this common operating temperature should be as low as possible for lower power consumption, less need for cooling, increased reliability, and increased performance. Finally, the utilization of each core, that is, the fraction of time a core is nonidle should be the same in order to avoid cases where a core has "consumed its lifetime" whereas others have been active for very short. Equal usage should also take into account the thermal stress caused to each core by the applications it executes. Specifically, the situation where a core has mainly being executing temperature intensive applications whereas others have mainly been executing moderate or low stress applications is unwanted. Equal usage among cores will result in improving the chip-wide reliability.

Highly unwanted scenarios

Several application-execution scenarios that can lead to highly unwanted cases, such as, large performance penalties

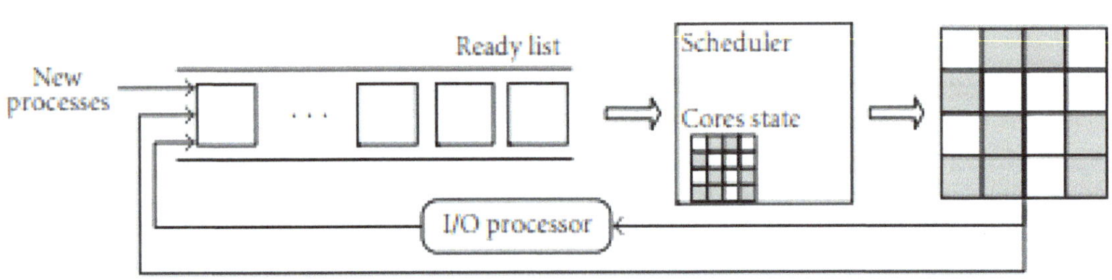

Figure 3: Basic scheduling scheme in operating systems. Cores state array, shown as part of the scheduler, tracks the state of each core as busy or idle.

or high thermal stress are discussed in this section. These scenarios do not necessarily describe the worse case, but are presented to show that temperature unaware scheduling can lead to situations far from the ideal with consequences opposite to those presented above. Simple thermal-aware scheduling heuristics are shown to prevent such cases.

Scenario 1: large performance loss

As mentioned earlier, the most direct way the processor's temperature can affect its performance is due to more frequent activation of DTM events, which occur each time the temperature of the core exceeds a predefined threshold. The higher the initial temperature of the core is, the easier it is to reach this predefined threshold is. For the temperature of a core to rise, its own heat generation (local) must be larger than the heat it can dissipate to the ambient and to the neighboring cores. However, a core can only dissipate heat to its neighbors if they are cooler. The local heat generation is mainly determined by the application running on the core which may be classified as "hot," "moderate", and "cool" [4, 10, 34] depending on the heat it generates. Therefore, the worse case for large loss of performance is to execute a hot process on a hot core that resides in a hot neighborhood. Let us assume that the CMP's thermal snapshot (the current temperature of its cores) is the one depicted in Figure 4(a), and that a hot process is to be scheduled for execution. Four cores are idle and thus candidate for executing the new process: C3, D4, E3, and E4. Although C3 is the coolest core, it is the choice that will cause the largest performance loss. C3 has reduced cooling ability due to being surrounded by hot neighbors (C2, C4, B3, and D3) and due to not having free edges, that is, edges of the chip. As such, its temperature will reach the threshold soon and consequently activate a DTM event, leading to a performance penalty. A thermal-aware scheduler could identify the inappropriateness of C3 and notice that although E4 is not the coolest idle core of the chip, it has two advantages: it resides in a rather cool area and neighbors to the edge of the chip both of which enhance its cooling ability. It would prefer E4 compared to E3 as E4 has two idle neighbors and compared to D4 as it is cooler and has more efficient cooling.

Scenario 2: hotspot creation

The "best" way to create a hotspot, that is, an area on the chip with very high thermal stress is to force very high temperature on adjacent cores.

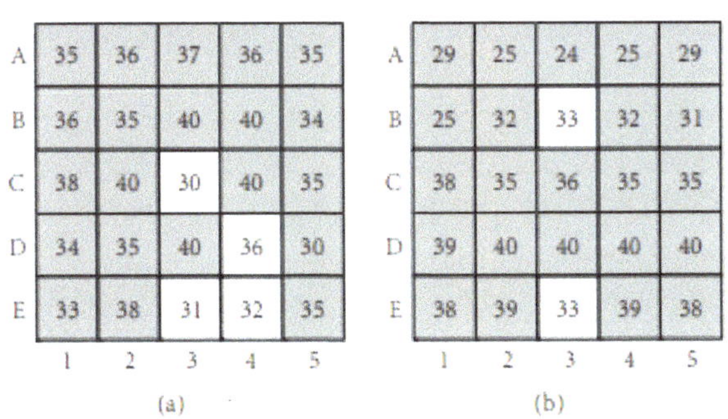

Figure 4: Thermal snapshots of the CMP. Busy cores are shown as shaded. Numbers correspond to core's temperature (∘C) above the ambient.

This could be the result of running hot applications on the cores and at the same time reducing their cooling ability. Such a case would occur if a hot application was executed on core E3 of the CMP depicted in Figure 4(b). This would decrease the cooling ability of its already very hot neighbors (E2, E4, and D3). Furthermore, given that E3 is executing a hot application and that it does not have any cooler neighbor, it is likely to suffer from high temperature, soon leading to the creation of a large hotspot at the bottom of the chip. A thermal-aware scheduler would take into account the impact such a scheduling decision would have, not only on the core under evaluation but also on the other cores of the chip, thus avoiding such a scenario.

Scenario 3: high spatial diversity

The largest spatial diversities over the chip appear when the temperature of adjacent cores differs considerably. Chess like scheduling (Figure 5) is the worse case scenario for spatial diversities as between each pair of busy and probably hot cores an idle, thus cooler, one exists. A thermal-aware scheduler would recognize this situation, as it is aware of the temperature of each core, and moderate the spatial diversities.

Scenario 4: high temporal diversity

A core will suffer from high temporal diversities when the workload it executes during consecutive intervals has opposite thermal behavior. Let us assume that the workload consists of 2 hot and 2 moderate applications.

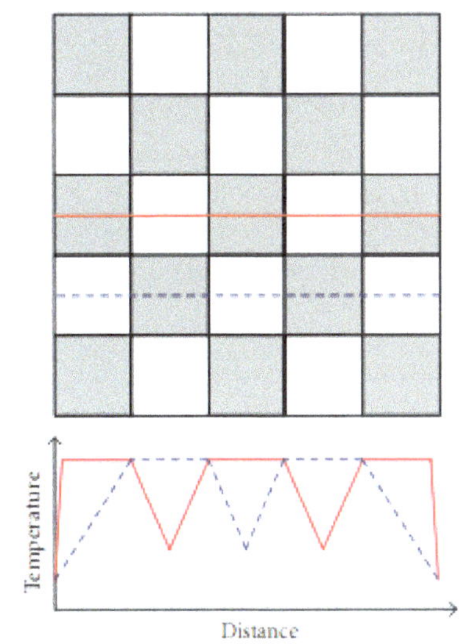

Figure 5: Chess-like scheduling and its effect on spatial temperature diversity. The chart shows the trend temperature is likely to follow over the lines shown on the CMP.

A scenario that would cause the worse case temporal diversities is the one depicted in Figure 6(a). In this scenario, process execution intervals are followed by an idle interval. Execution starts from the two hot processes and continues with the moderate one maximizing the temporal temperature diversity. A thermal-aware scheduler that has information about the thermal type of the workload can efficiently avoid such diversities (Figures 6(b) and 6(c)).

Thermal-aware scheduling on chip multiprocessors

Thermal-Aware Scheduling (TAS) [27] is a mechanism that aims to moderate or even eliminate the thermal-induced problems of CMPs presented in the previous section. Specifically, when scheduling a process for execution, TAS selects one of the available cores based on the core's "thermal state," that is, its temperature and cooling efficiency. TAS aims at improving the performance and thermal profile of the CMP, by reducing its temperature and consequently avoiding thermal violation events.

TAS implementation or a real OS

Implementing the proposed scheme at the operating system level enables commodity CMPs to benefit from TAS without any need for microarchitectural changes. The need for scheduling is inherent in multiprocessors operating systems and therefore, adding thermal awareness to it, by enhancing its kernel, will cause only negligible overhead for schedulers of reasonable complexity. The only requirement is an architecturally visible temperature sensor for each core, something rather trivial given that the Power 5 processor [18] already embeds 24 such sensors. Modern operating systems already provide functionality for accessing these sensors through the advanced configuration and power interface (ACPI) [41]. The overhead for accessing these sensors is minimal and so we have not considered it in our experimental results.

Thermal-aware schedulers

In general, a thermal-aware scheduler, in addition to the core's availability takes into account its temperature and other information regarding its cooling efficiency. Although knowing the thermal type of the workload to be executed can increase the efficiency of TAS, schedulers that operate without this knowledge, as those presented below, are shown by our experimental results to provide significant benefits. Our study is currently limited to simple, stateless scheduling algorithms which are presented next.

Coolest

The new process is assigned to the Coolest idle core. This is the simplest thermal-aware algorithm and the easiest to implement.

Neighborhood

This algorithm calculates for each available core a cost function (equation (2)) and selects the core that minimizes it. This cost function takes into consideration the following:

(i) temperature of the candidate core (T_c),

(ii) average temperature of directly neighboring cores (T_{DA}),

(iii) average temperature of diagonally neighboring cores (T_{dA}),

(iv) number of nonbusy directly neighboring cores (NBDA), (v) the number of "free" edges of the candidate core (N_{fe}).

$$\text{Cost} = a_1 \cdot T_c + a_2 \cdot \overline{T}_{DA} + a_3 \cdot \overline{T}_{dA} + a_4 \cdot NB_{DA} + a_5 \cdot N_{fe}. \quad (2)$$

Threshold neighborhood

The Threshold Neighborhood algorithm uses the same cost function as the Neighborhood algorithm, but schedules a process for execution only if a good enough core exist. This good enough threshold is a parameter of the algorithm. A core is considered appropriate if its cost function is lower than thi threshold (in contrast, when the neighborhood algorithm is used, a process is scheduled no matter the value of the cost function). This algorithm is nongreedy as it avoids scheduling a process for execution on a core that is available but in a thermally adverse state. Although one would expect that the resulting underutilization of the cores could lead to performance degradation, the experimental results showed that with careful tuning, performance is improved due to the reduction of the number of DTM events.

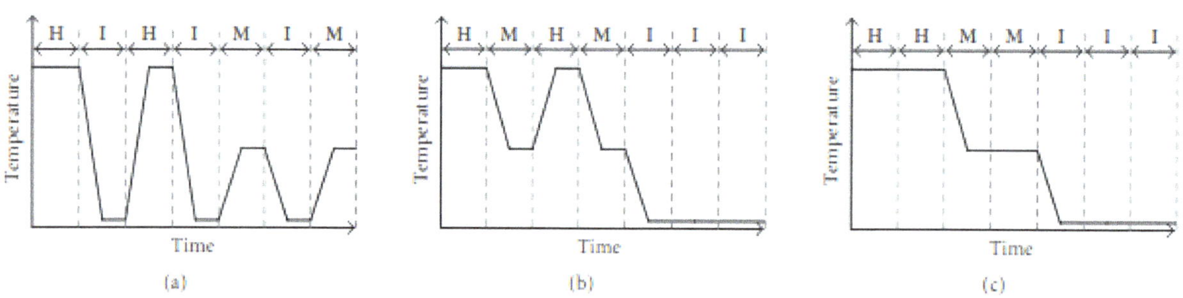

Figure 6: Temporal temperature diversity. H stands for 'hot" process, M for a process of moderate thermal stress, and I for an idle interval. The charts show the trend temperature is likely to follow over the time. (a) The worse case temporal diversity scenario. (b) A scenario with moderate temporal diversity. (c) The scenario that minimizes temporal diversity

MST heuristic

The maximum scheduling temperature (MST) heuristic, is not an algorithm itself but an option that can be used in combination with any of the previously mentioned algorithms. Specifically, MST prohibits scheduling a process for execution on idle cores when their temperature is higher than a predefined threshold (MSTT).

EXPERIMENTAL SETUP

To analyze the effect of thermal problems on the evolution of the CMP architecture and to quantify the potential of TAS in solving these issues, we conducted several experiments using a specially developed simulator.

The simulated environment

At any given point in time, the operating system's ready list contains processes ready to be executed. At the same time, each core of the CMP may be either busy executing a process or idle. If idle cores exist, the operating system, using a scheduling algorithm selects one such core and schedules on it a process from the ready list. During the execution of the simulation, new processes are inserted into the ready list and wait for their execution. When a process completes its execution, it is removed from the execution core, which is thereafter deemed as idle. The heat produced during the operation of the CMP and the characteristics of the chip define the temperature of each core. For the simulated environment, the DTM mechanism used is that of process migration. As such, when the temperature of a core reaches a predefined threshold (45°C above the ambient), the process it executes is "migrated" to another core. Each such migration event comes with a penalty (migration penalty—DTM-P), which models the overheads and performance loss it causes (e.g., invocation of the operating system and cold caches effect).

The simulator

The simulator used is the Thermal Scheduling SImulator for Chip Multiprocessors (TSIC) [29], which has been developed specially to study thermal-aware scheduling on chip multiprocessors. TSIC models CMPs with different number of cores whereas it enables studies exploring several other parameters, such as the maximum allowed chip temperature, chip utilization, chip size, migration events, and scheduling algorithms.

Process model

The workload to be executed is the primary input for the simulator. It consists of a number of power traces, each one modeling one process. Each point in a power trace represents the average power consumption of that process during the corresponding execution interval. Note that all intervals have the same length in time. As the power consumption of a process varies during its execution, a power trace is likely to consist of different power consumption values for each point. The lifetime of a process, that is, the total number of simulation intervals that it needs to complete its execution, is defined as the number of points in that power trace. TSIC loads the workload to be executed in a workload list and dynamically schedules each process to the available cores. When the temperature of a core reaches a critical point (DTM-threshold), the process running on it must be either migrated to another core or suspended to allow the core to cool. Such an event is called thermal violation event. If no cores are available, that is, they are all busy or do not satisfy the criteria for the MST heuristic of Threshold Neighborhood algorithm, the process is moved back to the workload list and will be rescheduled when a core becomes available.

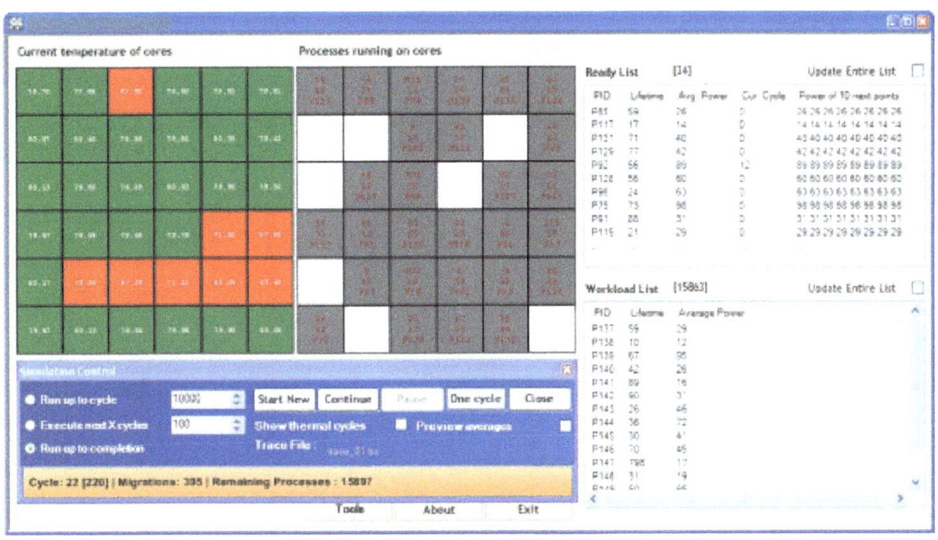

Figure 7: The main window of Thermal Scheduling SImulator for Chip Multiprocessors (TSIC).

Each time a process is to be assigned for execution, a scheduling algorithm is invoked to select a core, among the available ones, to which the process will be assigned for execution. For the experiments presented in this paper, the workload used consists of 2500 synthetic randomly produced processes with average lifetime equal to 100 simulation intervals (1 millisecond per interval) and average power consumption equal to 10 W. The rationale behind using a short average lifetime is to model the OS's context-switch operation. Specifically, each simulated process is to be considered as the part of a real-world process during two consecutive context switches.

The chip multiprocessor

TSIC uses a rather simplistic model for the chip's floorplan of the CMP. As depicted in Figure 7, each core is considered to cover a square area whereas the number of cores on the chip is always equal to n^2 where n is the number of cores in each dimension. In current TSIC implementation, cores are assumed to be areas of uniform power consumption. The area of the simulated chip is equal to 256 mm^2 (the default of the Hotspot simulator [4]).

Thermal model

TSIC uses the thermal model of Hotspot [4] which has been ported into the simulator. The floorplan is defined by the number of cores and the size of the chip.

Metrics

During the execution of the workload, TSIC calculates the total number of intervals required for its execution (Cycles), the number of migrations (Migrations) as well as several temperature-related statistics listed below.

(i) Average Temperature: the Average Temperature represents the average temperature of all the cores of the chip during the whole simulation period. The Average Temperature is given by (3), where $T^t_{i,j}$ is the temperature of core i, j during simulation interval t, S_T is the total number of simulation intervals, and n is the number of cores,

Average Temperature = $$\overline{T} = \sum_{t=0}^{S_T} \left[\frac{\sum_{i=0}^{n} \sum_{j=0}^{n} (T^t_{i,j})}{n \cdot S_T} \right]. \quad (3)$$

(ii) Average Spatial Diversity: the Spatial Diversity shows the variation in the temperature among the cores at a given time. The Average Spatial Diversity (equation (4)) is the average of the Spatial Diversity during the simulation period. A value equal to zero means that all cores of the chip have the same temperature at the same time, but possibly different temperature at different points in time. The larger this value is, the grater the variability is. In the Average Spatial Diversity equation, $T^t_{i,j}$ is the temperature of core i, j during simulation interval t, $\overline{T^t} = 1/n^2 \cdot \sum_{i=0}^{n} \sum_{j=0}^{n} T^t_{i,j}$ is the average chip temperature during simulation interval t, ST is the total number of simulation intervals, and n is the number of cores,

Average Spatial Diversity = $$\sum_{t=0}^{S_T} \left[\frac{\sum_{i=0}^{n} \sum_{j=0}^{n} |T^t_{i,j} - \overline{T^t}|}{n \cdot S_T} \right] \quad (4)$$

(iii) Average Temporal Diversity: the Average Temporal Diversity is a metric of the variation of the average chip temperature, across all cores, and is defined by (5). In the Average Temporal Diversity equation $T^t_{i,j}$ is the temperature of core i, j during simulation interval t, $\overline{T^t} = 1/n^2 \cdot \sum_{i=0}^{n} \sum_{j=0}^{n} T^t_{i,j}$ is the average chip temperature during simulation interval t, T is the average chip temperature as defined by (3), S_T is the total number of simulation intevals, and n is the number of cores,

Average Temporal Diversity = $$\sum_{i=0}^{S_T} \left[\frac{\sum_{j=0}^{S_T} |\overline{T^t} - \overline{T}|}{n \cdot S_T} \right]. \quad (5)$$

(iv) Efficiency: efficiency is a metric of the actual performance the multiprocessor achieves in the presence of thermal problems compared to the potential offered by the CMP. Effi- ciency is defined by (6) as the ratio between the time required for the execution of the workload (Workload Execution Time) and the time it would require if no thermal violation events existed (Potential Execution Time, (7)). The maximum value for the Efficiency metric is 1 and represents full utilization of the available resources,

$$\text{Efficiency} = \frac{\text{Potential Execution Time}}{\text{Workload Execution Time}}, \quad (6)$$

$$\text{Potential Execution Time} = \sum_{n=1}^{\#\text{processes}} \frac{\text{Lifetime}(\text{Process}_n)}{\text{Number of Cores}}. \quad (7)$$

Scheduling algorithms

For the experimental results presented in Section 6, all threshold values for the scheduling algorithms, the ai factors in (2), the MST-T, and the "Threshold Neighborhood," have been statically determined through experimentation. Although adaptation of these threshold values could be done dynamically, this would result in an overhead for the scheduler of the operating system. We are however currently studying these issues.

RESULTS

Thermal behavior and its implications for future CMPs

In this section we present the thermal behavior and its impact on the performance for future CMP configurations which are based on the technology evolution. This leads to chips of decreasing area and/or more cores per chip. For the results presented, we assumed that the CMPs are running an operating system that supports a minimal overhead thermal scheduling algorithm such as Coolest (baseline algorithm for this study). Consequently these results are also an indication of the applicability of simple thermal scheduling policies.

Trend 1: decreasing the chip size

As mentioned earlier, technology improvements and feature size shrink will allow the trend of decreasing the chip size to continue. Figure 8(a) depicts the effect of this chip size decrease while keeping the consumed power constant for a CMP with 16 cores. The results clearly show the negative effect of chip's area decrease on the average temperature and

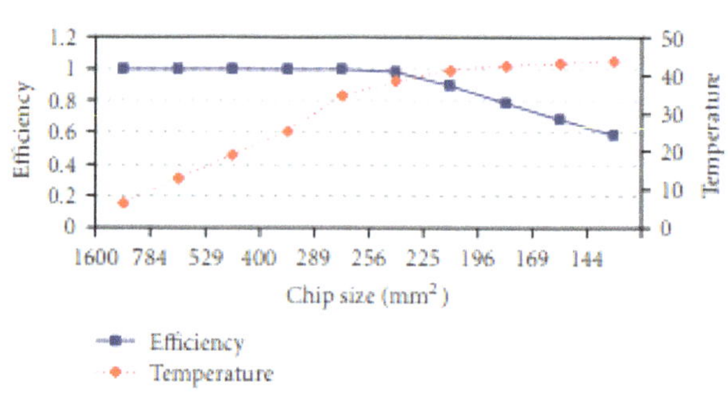

(a)

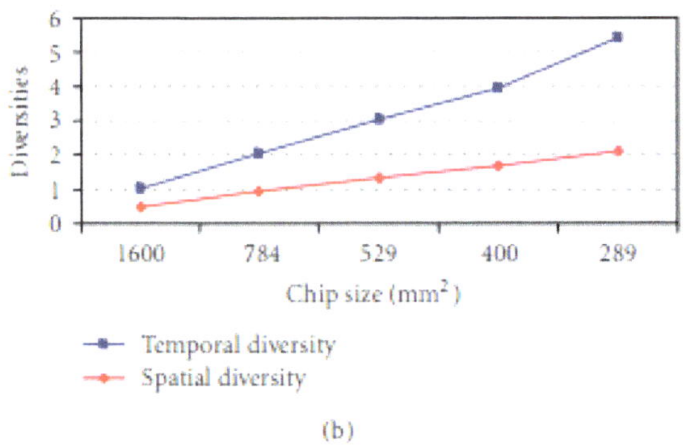

Figure 8: (a) Efficiency and temperature (◦C above the ambient) and (b) spatial and temporal diversities for different chip sizes.

the efficiency of the multiprocessor. This is explained by the fact that the ability of the chip to cool by vertically dissipating heat to the ambient is directly related to its area (Section 3.1). Lower cooling ability leads to higher temperature, which in turn leads to increased number of migrations, and consequently to significant performance loss. The reason for which the temperature only asymptotically approximates 45◦C is related to the protection mechanism used (process migration) which is triggered at 45◦C. Notice that the area of typical chips today does not exceed 256 mm2, which is the point beyond which it is possible to observe considerable performance degradation. A migration penalty (DTM-P) of one interval is used for these experiments. This value is small compared to what would apply in a real world system and consequently these charts present an optimistic scenario. Another unwanted effect is related to the spatial and temporal diversities, which also become worse for smaller chips (Figure 8(b)) and is justified mainly by the higher operating temperatures. Notice that in this chart we limit the chip size range to that for which no migrations exist in order to exclude from the trend line the effect of migrations.

Trend 2: increasing the number of cores

As explained in Section 3.2, due to thermal limitations, the throughput potential offered by the increased number of cores cannot be exploited unless the size of the CMP is scaled proportionally. Figure 9 depicts the efficiency and

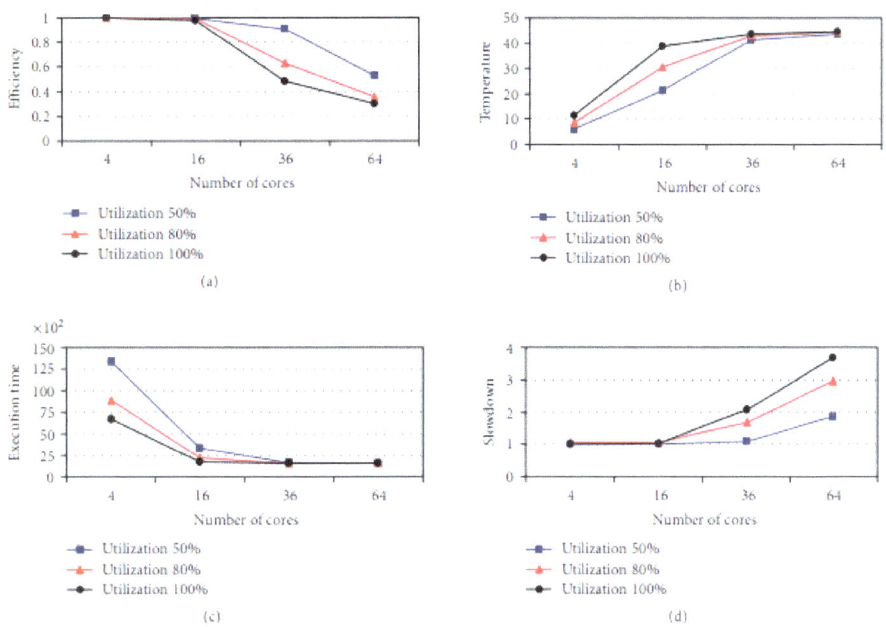

Figure 9: (a) Efficiency (b) temperature (◦C above the ambient) (c) workload execution time (in terms of simulation intervals) (d) slowdown orienting from temperature issues for CMPs with different number of cores and different utilization points.

temperature for CMPs with different number of cores (4, 16, 36, and 64) for three different utilization points (50%, 80%, and 100%). Utilization shows the average fraction of cores that are active at any time point and models the execution stress of the multiprocessor. The efficiency of the different CMP configurations studied is depicted in Figure 9(a). The decrease in efficiency with the increase in the number of on-chip cores is justified by the decrease in the per-core area and consequently of the vertical cooling capability. The increased utilization also decreases the cooling capabilities of cores but this is related to the lateral heat transfer path. Specifically, if a neighboring core is busy, and thus most likely hot, cooling through that direction is less efficient. In the worse scenario, a core will receive heat from its neighbors and instead of cooling, it will get hotter. Both factors have a negative effect on temperature (Figure 9(b)) and consequently in the number of migration events, which is the main reason for performance loss. It is relevant to notice that for the 36- and 64-core CMPs the average temperature is limited by the maximum allowed threshold, which has been set to 45◦C for these experiments. The workload execution time for the different CMP con- figurations studied is depicted in Figure 9(c)). For the 4-core CMP, higher utilization leads to a near proportional speedup, which is significantly smaller for the 16-core CMP and almost diminishes for multiprocessors with more cores. This indicates the constrain thermal issues pose on the scalability offered by the CMP's architecture. It is relevant to notice that for the 100% utilization point, the 64-core chip has almost the same performance as the 16-core CMP. This behavior is justified by the large number of migration events suffered by the large scale CMPs. Figure 9(d) displays the slowdown of each configuration due to temperature related issues taking the utilization into consideration, that is, if a configuration with utilization 50% executes the workload in 2X cycles where the same configuration with 100% utilization executes it in X cycles, the former is considered to have zero slowdown. The results emphasize the limitations posed by temperature issues on fully utilizing the available resources. Notice that these limitations worsen as the available resources increase. Finally, Figure 10 depicts the spatial and temporal diversities of the CMP configurations studied, when utilization is equal to 100%. Both diversities are shown to worsen when more cores coexist on the chip. This is not only due to the higher temperature but also due to variability caused by the larger number of on-chip cores.

Optimization 1: thermal threshold

The results from the previous section showed a significant drop in performance as the maximum operating temperature

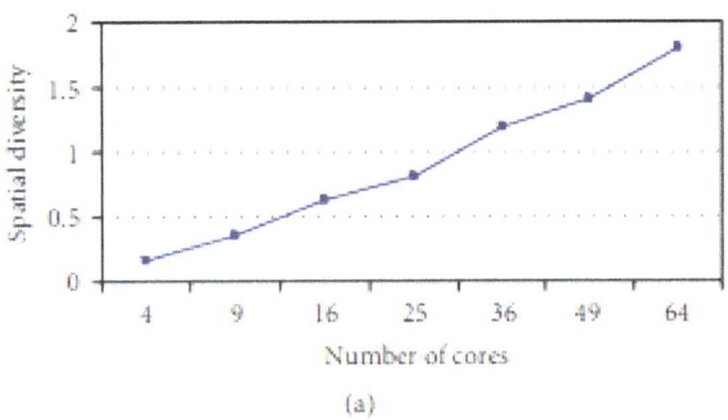

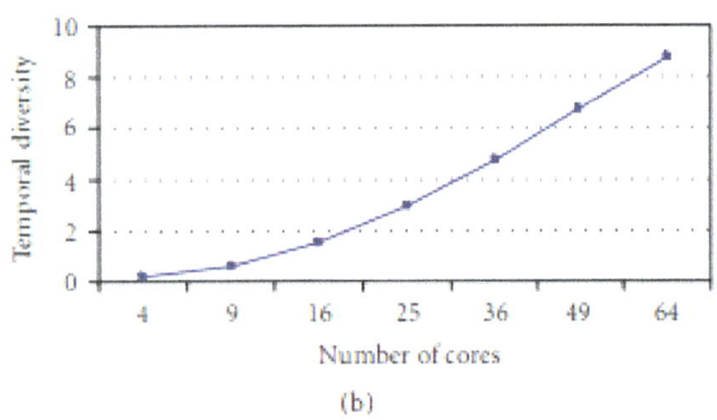

Figure 10: (a) Spatial and (b) temporal diversity as the number of on-chip cores increases.

is reached. To avoid this performance degradation, we propose to enhance the basic Thermal-aware scheduling policy (Coolest) by using a threshold on the core's temperature. This is what we named the Coolest + MST scheduling scheme, that is, a process is executed on the coolest available core only if a core with temperature lower than N°C exists. In our case, we use 40°C as the threshold value (MST-T), that is, five degrees lower than the maximum allowed temperature. The goal for these experiments is to show how, Coolest + MST, is able to improve the performance by reducing the number of migrations. In addition, we set the DTM-Penalty to zero, which is the reason why we will not present performance results. Table 1 presents the number of migrations and the average temperature for the execution scenarios mentioned before. As can be seen from the results, the Coolest + MST heuristic is able to significantly decrease the number of migration events. The potential of this algorithm increases with the number of cores. This is a first-class indication that performance improvement can be achieved. At the same time, this TAS scheme decreases the average chip temperature by approximately 2°C for the 16-core and 2.5°C for the 25-core CMP. Figure 11 depicts the number of migrations and temperature of CMPs with different number of cores as the MST-Threshold (MST-T) ranges from 40°C to 45°C. Note that when MST-T is equal to the DTM-threshold (DTM-T) (45°C), scheduling is the same as what would apply without MST (Coolest). As depicted in Figure 11(a) for both the 16-core and 25- core CMP, the number of migrations increases with the MSTT. This is due to the fact that cores with very high temperature are allowed to be used. The same trend stands for the average temperature of the chip (Figure 11(b)), which justifies what is observed for migrations.

Table 1: Number of migrations and average temperature (°C) above the ambient for CMPs of different number of cores for two scheduling schemes.

	Number of migrations		Temperature (°C)	
Number of cores	Coolest	Coolest + MST	Coolest	Coolest + MST
4	0	0	11.38	11.38
9	0	0	23.63	23.63
16	1193	6	38.74	36.88
25	109 733	1222	42.58	39.94

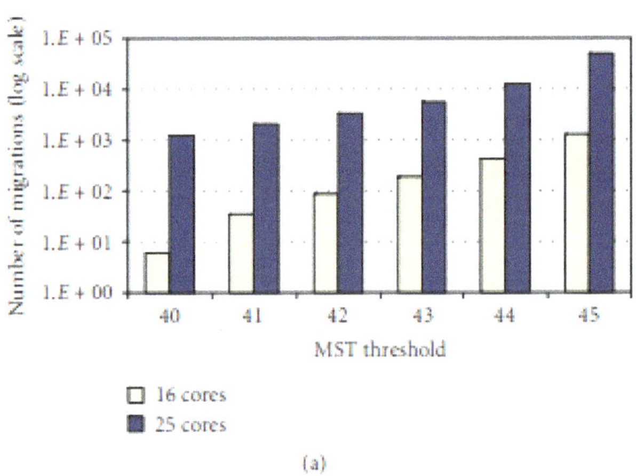

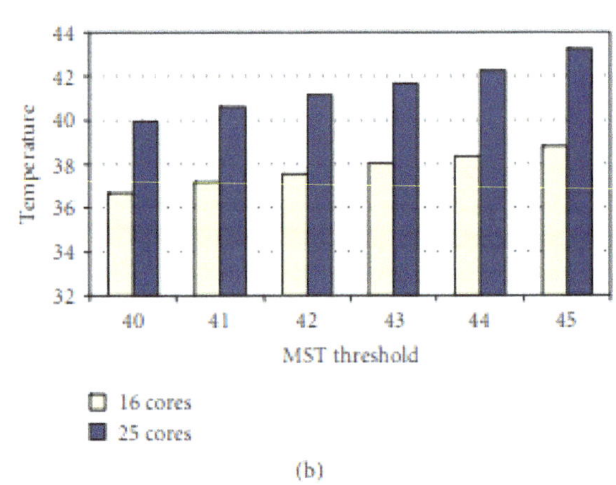

Figure 11: (a) Number of migrations graphed in a logarithmic scale and (b) average chip temperature (°C above the ambient) versus MST-T.

No performance results are presented for this experiment as the DTM-Penalty value has been set to be equal to 1 interval only. As such, its impact on performance is minimal. However, as mentioned earlier, in a real-world system the DTM-Penalty will be significantly larger. When DTM events are penalized, execution using the Coolest policy may not complete. If the scheduling algorithm is greedy in that it tries to fully utilize the available resources no matter their thermal state, a scenario described by a vicious-circle of continued process-migrations is possible. Such a scenario appears when cores with very high temperature are used and at the same time, the average temperature of the chip is close to DTM-T. For example, this scenario happens when executing the experimental workload on a 36- core CMP with Coolest.

Optimization 2: neighborhood awareness

The results from the previous section showed that adding a Threshold to the simple thermal-aware scheduling (Coolest) policy can significantly decrease the number of migration events. Nevertheless, the Coolest + MST algorithm uses local information to make the scheduling decisions, that is, it considers only the temperature of the candidate cores. In this section, we present the results for an algorithm that takes into consideration not only the temperature of the candidate cores but also the temperatures of all the surrounding or neighboring cores. We presented previously, in Section 4.4, two algorithms that use this information, the Neighborhood and the Threshold Neighborhood. The results presented in this section are for the Threshold Neighborhood as its performance is much better compared to the simple Neighborhood. As we have not yet completely tuned the Threshold Neighborhood algorithm, we present results only for a single setup of a CMP with 16 cores and 100% utilization. Figure 12(a) depicts the number of migrations for the two algorithms under evaluation (Threshold Neighborhood and Coolest + MST), for different DTM-Penalties. The number of migration events suffered by the Threshold Neighborhood algorithm is always smaller compared to those of the Coolest + MST. For the Coolest + MST algorithm, this number of migrations increases with the DTM-Penalty as the additional time required for the execution of the workload worsens the already problematic thermal state of the CMP. On the other hand, for the Threshold Neighborhood algorithm the number of migrations decrease with the DTMPenalty. This shows the ability of the algorithm to adapt to the different DTM-Penalty values. Specifically, as this penalty increases, the algorithm becomes more strict when evaluating the thermal appropriateness of a core. It must be noted here that the parameters of the Threshold Neighborhood algorithm are not the same for all situations. As the migration penalty increases, the configuration that performs better is the one that has smaller weight for the temperature of the candidate core and larger weight for the number of nonbusy directly adjacent cores (Sec (2)). However, when the migration penalty is small, a conservative selection for execution cores is not desired as the effect of migrations is less important. Figure 12(b) depicts the execution time of the experimental workload for the different scenarios studied. The performance of the Coolest + MST algorithm worsens as the DTM Penalty increases mainly due to two reasons. The first is related to the increase of migration events whereas the sec-

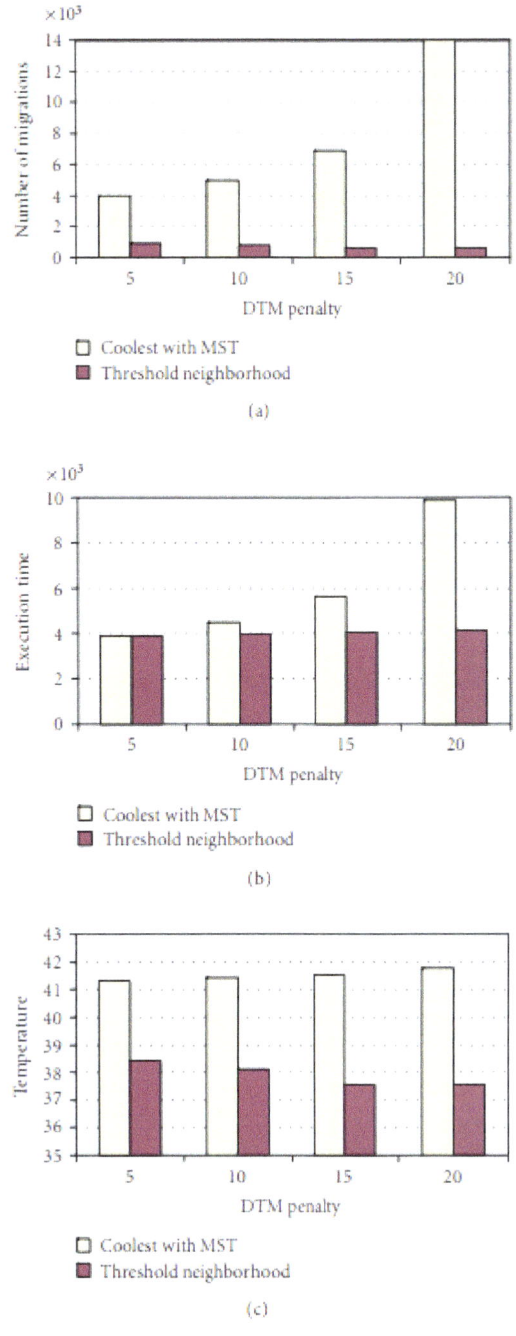

Figure 12: (a) Number of migrations, (b) temperature (°C above the ambient), and (c) execution time (number of simulation intervals) for different DTM-Penalties and scheduling schemes.

ond to the fact that each migration has a larger cost. In contrast, the performance of Threshold Neighborhood algorithm is almost constant. This is due to the ability of the algorithm to decrease the number of migrations it suffers, as their cost increases with the migration penalty.

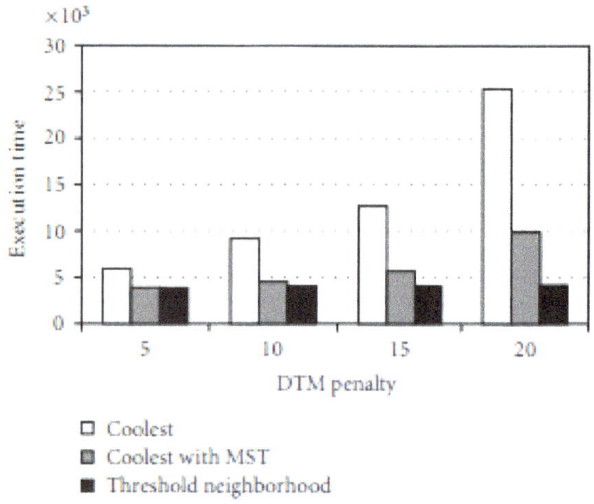

Figure 13: Execution time for different CMP scheduling schemes and DTM-Penalties.

Finally, Figure 12(c) depicts the average temperature of the chip for the different configurations studied. It is obvious that the Threshold Neighborhood algorithm manages not only to increase performance but also to decrease the temperature of the chip. This was expected, as only when the chip has better temperature characteristics, migration events can be controlled. This exploration clearly shows that trying to fully utilize the available resources without taking into consideration the thermal issues may significantly affect performance. Another conclusion is that it is often beneficial to be conservative on utilizing the on-chip resources as this will allow better cooling, will decrease the number of migrations, and consequently enhance performance.

Summary

In the previous sections, we showed the performance improvements that may be achieved by two optimizations to the basic TAS algorithm (Coolest). On the one hand, Coolest + MST uses a threshold to reduce the number of migrations. On the other hand, Threshold Neighborhood uses local and information about the surrounding cores to take the scheduling decisions. In addition to reducing the number of migrations, this algorithm has also the potential to achieve a better chip-wide thermal behavior. A simple comparison between the three TAS algorithms is presented in Figure 13. This Figure depicts the execution time for the three algorithms for different DTM penalty values on a 16-core CMP. The results in Figure 13 show that Coolest is intolerant to the increase of the DTM penalty, resulting in large performance loss. This is due to the larger number of migrations compared with the other algorithms. The Coolest + MST performs well for smaller values of DTM penalty. Nevertheless, it is possible to observe that the execution time almost doubled for Coolest + MST when the penalty increased from 15 to 20. In contrast with the previous two algorithms, for Threshold Neighborhood the execution time is not affected by the increase in the DTM penalty resulting in almost no performance degradation. As such, we are led to conclude that the Threshold Neighborhood is the most stable TAS algorithm that achieves the best overall results.

REFERENCES

1. V. Agarwal, M. S. Hrishikesh, S. W. Keckler, and D. Burger, "Clock rate versus IPC: the end of the road for conventional microarchitectures," in Proceedings of the 27th Annual International Symposium on Computer Architecture (ISCA '00), pp. 248–259, ACM Press, Vancouver, BC, Canada, June 2000.
2. K. Olukotun, B. A. Nayfeh, L. Hammond, K. Wilson, and K. Chang, "The case for a single-chip multiprocessor," in Proceedings of the 7th International Conference on Architectural Support for Programming Languages and Operating Systems (ASPLOS '96), pp. 2–11, ACM Press, Cambridge, Mass, USA, October 1996.
3. T. Mudge, "Power: a first-class architectural design constraint," Computer, vol. 34, no. 4, pp. 52–58, 2001.
4. K. Skadron, M. R. Stan, W. Huang, S. Velusamy, K. Sankaranarayanan, and D. Tarjan, "Temperature-aware microarchitecture: extended discussion and results," Tech. Rep. TRCS-2003-08, University of Virginia, Charlottesville, Va, USA, 2003.
5. R. Viswanath, V. Wakharkar, A. Watwe, and V. Lebonheur, "Thermal performance challenges from silicon to systems," Intel Technology Journal, vol. 4, no. 3, p. 16, 2000.
6. S. H. Gunther, F. Binns, D. M. Carmean, and J. C. Hall, "Managing the impact of increasing microprocessor power consumption," Intel Technology Journal, vol. 5, no. 1, p. 9, 2001.
7. W. Liao and L. He, "Coupled power and thermal simulation with active cooling," in Proceedings of the 3rd International Workshop on Power-Aware Computer Systems (PACS '03), vol. 3164 of Lecture Notes in Computer Science, pp. 148–163, San Diego, Calif, USA, December 2003.
8. A. Dhodapkar, C. H. Lim, G. Cai, and W. R. Daasch, "TEM2 P2 EST: a thermal enabled multi-model power/performance ESTimator," in Proceedings of the 1st International Workshop on Power-Aware Computer Systems-Revised Papers (PACS '00), pp. 112–125, Cambridge, Mass, USA, November 2000.
9. R. Mahajan, K. Brown, and V. Atluri, "The evolution of microprocessor packaging," Intel Technology Journal, p. 10, 2001.
10. Y. Li, D. Brooks, Z. Hu, and K. Skadron, "Performance, energy, and thermal considerations for SMT and CMP architectures," in Proceedings of the 11th International Symposium on HighPerformance Computer Architecture (HPCA '05), pp. 71–82, San Francisco, Calif, USA, February 2005.
11. D. Genossar and N. Shamir, "Intel pentium M processor: power estimation, budgeting, optimization and validation," Intel Technology Journal, vol. 7, no. 2, pp. 44–49, 2003.
12. J. Mooref, J. Chasef, P. Ranganathanf, and R. Sharmaf, "Making scheduling "cool": temperature-aware workload placement in data centers," in Proceedings of the USENIX Annual Technical Conference, pp. 61–75, Anaheim, Calif, USA, April 2005.

13. S. Heo, K. Barr, and K. Asanovic, "Reducing power density ′ through activity migration," in Proceedings of the International Symposium on Low Power Electronics and Design (ISLPED '03), pp. 217–222, Seoul, South Korea, August 2003.

14. Z. Lu, W. Huang, S. Ghosh, J. Lach, M. Stan, and K. Skadron, "Analysis of temporal and spatial temperature gradients for IC reliability," Tech. Rep. CS-2004-08, University of Virginia, Charlottesville, Va, USA, March 2004.

15. Z. Lu, J. Lach, M. Stan, and K. Skadron, "Banking chip lifetime: opportunities and implementation," in Proceedings of the 1st Workshop on High Performance Computing Reliability Issues (HPCRI '05), San Francisco, Calif, USA, February 2005.

16. C. Bailey, "Modelling the effect of temperature on product reliability," in Proceedings of 19th Annual IEEE Semiconductor Thermal Measurement and Management Symposium (SEMITHERM '03), pp. 324–331, San Jose, Calif, USA, March 2003.

17. R. Mukherjee, S. O. Memik, and G. Memik, "Temperatureaware resource allocation and binding in high-level synthesis," in Proceedings of the 42nd Design Automation Conference (DAC '05), pp. 196–201, ACM Press, Anaheim, Calif, USA, June 2005.

18. R. Kalla, B. Sinharoy, and J. M. Tendler, "IBM Power5 chip: a dual-core multithreaded processor," IEEE Micro, vol. 24, no. 2, pp. 40–47, 2004.

19. P. Kongetira, K. Aingaran, and K. Olukotun, "Niagara: a 32- way multithreaded Sparc processor," IEEE Micro, vol. 25, no. 2, pp. 21–29, 2005.

20. Intel, "Intel Pentium D Processor Product Information," 2006, http://www.intel.com/products/processor/ pentium d/. [21] AMD, "AMD Athlon 64X2 Dual-Core Processor," 2006, http://www.amd.com/.

21. R. Sasanka, S. V. Adve, Y.-K. Chen, and E. Debes, "The energy efficiency of CMP vs. SMT for multimedia workloads," in Proceedings of the 18th Annual International Conference on Supercomputing (ICS '04), pp. 196–206, Saint-Malo, France, June-July 2004.

22. E. Debes, "Recent changes and future trends in general purpose processor architectures to support image and video applications," in Proceedings of IEEE International Conference on Image Processing (ICIP '03), vol. 3, pp. 85–88, Barcelona, Spain, September 2003.

23. F. Mombers and D. Mlynek, "Multithreaded multimedia processor merging on-chip multiprocessors and distributed vector pipelines," in Proceedings of IEEE International Symposium on Circuits and Systems (ISCAS '99), vol. 4, pp. 287–290, Orlando, Fla, USA, May-June 1999.

24. J. Goodacre, "Understanding the Options for Embedded Multiprocessing," 2003, TechOnLine: http://www. techonline.com/ community/tech group/soc/tech paper/29359.

25. Intel, "Intel Pentium D Processor Product Information," 2006, http://www.intel.com/products/processor/pentium d/.

26. K. Stavrou and P. Trancoso, "Thermal-aware scheduling: a solution for future chip multiprocessors thermal problems," in Proceedings of the 9th EUROMICRO Conference on Digital System Design: Architectures, Methods and Tools (DSD '06), pp. 123–126, Dubrovnik, Croatia, August-September 2006.

27. E. Grochowski, R. Ronen, J. Shen, and H. Wang, "Best of both latency and throughput," in Proceedings of IEEE International Conference on Computer Design: VLSI in Computers and Processors (ICCD '04), pp. 236–243, San Jose, Calif, USA, October 2004.

28. K. Stavrou and P. Trancoso, "TSIC: thermal scheduling simulator for chip multiprocessors," in Proceedings of the 10th Panhellenic Conference on Informatics (PCI '05), vol. 3746 of Lecture Notes in Computer Science, pp. 589–599, Volos, Greece, November 2005.

29. P. I. Penzes, M. Nystr ′ om, and A. J. Martin, "Transistor siz- ¨ ing of energy-delay-efficient circuits," in Proceedings of the 8th ACM/IEEE International Workshop on Timing Issues in the Specification and Synthesis of Digital Systems (TAU '02), pp. 126–133, Monterey, Calif, USA, December 2002.

30. J. Ebergen, J. Gainsley, and P. Cunningham, "Transistor sizing: how to control the speed and energy consumption of a circuit," in Proceedings of the 10th International Symposium on Advanced Research in Asynchronous Circuits and Systems (ASYNC '04), vol. 10, pp. 51–61, Crete, Greece, April 2004.

31. R. Balasubramonian, D. Albonesi, A. Buyuktosunoglu, and S. Dwarkadas, "Memory hierarchy reconfiguration for energy and performance in general-purpose processor architectures," in Proceedings of the 33rd Annual IEEE/ ACM International Symposium on Microarchitecture (MICRO '00), pp. 245–257, ACM Press, Monterey, Calif, USA, December 2000.

32. S. Manne, A. Klauser, and D. Grunwald, "Pipeline gating: speculation control for energy reduction," in Proceedings of the 25th Annual International Symposium on Computer Architecture (ISCA '98), pp. 132–141, Barcelona, Spain, June-July 1998.

33. Y. Han, I. Koren, and C. A. Moritz, "Temperature aware floorplanning," in Proceedings of the 2nd Workshop on TemperatureAware Computer Systems (TACS '05), Madison, Wis, USA, June 2005.

34. R. McGowen, C. A. Poirier, C. Bostak, et al., "Power and temperature control on a 90-nm Itanium family processor," IEEE Journal of Solid-State Circuits, vol. 41, no. 1, pp. 229–237, 2006.

35. J. Donald and M. Martonosi, "Techniques for multicore thermal management: classification and new exploration," in Proceedings of the 33rd International Symposium on Computer Architecture (ISCA '06), pp. 78–88, Boston, Mass, USA, June 2006.

36. NIST/SEMATECH, Assessing Product Reliability, Chapter 8, e-Handbook of Statistical Methods, http:// www.itl.nist.gov/ div898/handbook/.

37. J. Srinivasan, S. V. Adve, P. Bose, J. Rivers, and C.-K. Hu, "RAMP: a model for reliability aware microprocessor

design," IBM Research Report RC23048, p. 26, IBM, White Plains, NY, USA, 2003.

38. W.-L. Hung, Y. Xie, N. Vijaykrishnan, M. Kandemir, and M. J. Irwin, "Thermal-aware task allocation and scheduling for embedded systems," in Proceedings of the Conference on Design, Automation and Test in Europe (DATE '05), vol. 2, pp. 898–899, IEEE Computer Society, Munich, Germany, March 2005.

39. P. Rosinger, B. Al-Hashimi, and K. Chakrabarty, "Rapid generation of thermal-safe test schedules," in Proceedings of the Conference on Design, Automation and Test in Europe (DATE '05), vol. 2, pp. 840–845, IEEE Computer Society, Munich, Germany, March 2005.

40. ACPI, "Advanced Configuration and Power Interface," 2006, http://www.acpi.info/ .

CHAPTER 7 DESIGN AND ARCHITECTURE OF MIXED-SIGNAL EMBEDDED SYSTEMS

Sergio Saponara, Pierluigi Nuzzo, Paolo D'Abramo, and Luca Fanucci

INTRODUCTION

The continuous evolution of VLSI technologies, especially CMOS and its variants (HV MOS, BCD, RF CMOS, etc.), has enabled the integration of complex functionalities in a single, heterogeneous embedded system. Digital subsystems, such as microprocessors, memories, or communication interfaces, can be integrated onto the same substrate (System-on-Chip) or the same package (System-in-Package) together with RF blocks, analog acquisition and processing circuits, analog power drivers, and even micromechanical parts for sensors and actuators.

Overcoming the classic dichotomy between analog and digital domains, such new generation of mixed-signal embedded systems is fueling the development of more efficient and performance solutions in several technology areas, to name just a few, distributed control-sensing-actuation units to increase safety, comfort, and engine efficiency in vehicles, software defined and cognitive radios for multimode multimedia communication, wireless sensor networks for ambient intelligence in home, industrial scenarios, or for environment control.

The opportunity given by mixed-signal embedded systems comes, however, with lots of challenges. Assembly and integration of heterogeneous blocks onto the same chip or the same package is already a big technological challenge. However, these systems also bring with them several design challenges. At different levels of the design flow, innovative methods, computer-aided design (CAD) tools, as well as architecture and circuit techniques need to be developed to allow successful design of mixed-signal integrated systems. Codesign of the different subsystems, hardware and software but also RF/analog with digital and power, has become instrumental to achieve high performance at reasonable costs. However, system-level specification, simulation, and optimal hardware/software, analog/digital partitioning are still open research fields. Moreover, whenever the largest payoffs in design originate from system level optimization, implementing hardware building blocks with increased flexibility and programmability becomes a wanting to enable intellectual property (IP) block sharing and reuse. Finally, efficient and reliable verification and testing strategies need to be devised, by leveraging a combination of simulation, formal methods, and rapid prototyping.

This special issue, including 7 papers from both Europe and America, will give to the readers an overview of the ongoing research in academia and industry in the field of design methodologies and architectures for mixed-signal systems. In particular, the papers address several case studies in automotive, consumer electronics, and wireless communication, demonstrating how the realization of mixed-signal embedded systems can either enable new services and applications or improve performance, efficiency, and cost of the existing ones.

The first two papers deal with design methodologies. The first paper "A platform-based methodology for system-level mixed-signal design," by the University of California at Berkeley and Marvell Inc., presents a system-level design methodology for electronic circuits, utilizing the platform-based design (PBD) paradigm as the natural framework for mixed-domain design formalization. In PBD, a meet-in-

---

Citation: Sergio Saponara, Pierluigi Nuzzo, Paolo D'Abramo, and Luca Fanucci, Design and Architectures of Mixed-Signal Embedded Systems, doi:10.1155/2010/641261.

the-middle approach allows systematic exploration of the design space through a series of top-down mapping of system constraints onto component feasibility models in a platform library, which is based on bottom-up characterization of the design components. A robust optimization scheme is proposed to account for modeling errors as well as process, voltage, and temperature variations. In this framework, new designs can be assembled from the precharacterized library components, giving the highest priority to design reuse, correct assembly, and efficient design flow from specifications to implementation. The effectiveness of the PBD methodology is demonstrated on the design of a pipeline A/D converter and two receiver front-ends, for UMTS and UWB communications, in submicron CMOS technologies.

The second paper "On mixed abstraction, languages and simulation approach to refinement with systemC AMS" by the University of Vienna proposes a methodology based on SystemC-AMS to be applied to complex mixed-signal embedded systems in performing early design decision trade-off exploration, architecture experimentation, and, particularly, model refinement and critical behavior analysis. An RTL model of a data encryption standard is demonstrated as an example.

The third paper "Systematic development methodology for mixed-mode behavioral models of in-vehicle embedded electronic systems," by the Universities of Genoa and Santa Catarina (Florianopolis, Brazil) and On Semicondutors, presents a systematic development methodology for mixed-mode behavioral models of automotive embedded systems, with particular reference to in-vehicle networks. The methodology allows achieving accurate models, which provide reliable system simulations. Two automotive case studies are presented concerning a Flexray physical layer transceiver and a CAN bus transceiver-integrated circuit. Thanks to the increased simulation speed and flexibility with respect to transistor level modeling techniques, fast communication network design, modification, and verification are permitted.

Automotive applications are also the focus of the following two papers, proposing innovative mixed-signal architectures for multisensor signal conditioning and piezoelectric actuator driving. The paper "A mixed-signal embedded platform for automotive sensor conditioning," from the University of Pisa and SensorDynamics AG, presents the Intelligent Sensor InterFace (ISIF) system, a mixed signal system able to fast identify, trim, and verify sensor interfaces. The system, developed according to a platform-based design approach, consists in a set of optimized high-performance analog, digital, and software IP modules that can interface to several kinds of sensors. These IPs can be easily defined for fast prototyping of the interface circuit of a given sensor. Final ASIC implementation of the desired conditioning circuitry can then be easily derived with reduced risk and short development time. Several case examples of sensor conditioning (rate-gyro, magnetic angular position, low-g accelerometer) are presented to demonstrate the effectiveness and flexibility of this system.

In the paper "Model and design of a power driver for piezoelectric satack actuators," from the Polytechnic of Turin, a power driver has been developed to control piezoelectric stack actuators used in automotive application. An FEM model of the actuator has been implemented starting from experimental characterization of the stack and mechanical and piezoelectric parameters. Experimental results are reported to show a correct piezoelectric actuator driving method and the possibility to obtain a sensorless positioning control.

Architectures for consumer electronics, both audio and wireless communications are presented in the last two papers. In "Mixed-signal architectures for high efficiency and low distortion digital audio processing and power amplification," from the University of Pisa, the algorithmic and architectural design of digital input power audio amplifiers is addressed. A modeling platform, based on a meet-in-the-middle approach between top-down and bottom-up strategies, allows a fast but still accurate exploration of the mixed-signal design space. Different amplifier architectures are configured and compared to find optimal trade-offs among different cost-functions: low distortion, high efficiency, low circuit complexity, and low sensitivity to parameter changes. A novel amplifier architecture is derived; its prototype targeting power levels of tens of Watts implements digital processing IP macrocells on a single low-complexity FPGA while off-chip components are used only for the power output stage (LC filter and power MOS bridge), and no heatsink is required. Discussions towards the full-silicon integration of the mixed-signal amplifier in embedded devices, by using a BCD technology and targeting power levels of a few Watts, are also reported.

We conclude with a paper on Phase-Locked Loop (PLL) design. PLLs are an essential component in many wireline and wireless applications for clocking and frequency synthesis. Moreover, PLL is among those building blocks that mostly took advantage of CMOS scaling and the integration of digital logic with analog circuits. The paper "AD-PLL for WiMAX with digital bang-bang TDC and glitch correction logic", from the Polytechnic of Milan, describes the design of an All-Digital Phase-Locked Loop (ADPLL) in 90 nm CMOS for WiMAX wireless applications. In the proposed architecture a time-to-digital converter (TDC) is implemented as a delay-locked loop (DLL) to be insensitive to process spreads and it uses a lead-lag phase detector and a digital loop filter to further take advantage of the digital approach. A glitch-corrector logic is also described to remove spurs due to time skew between counter and TDC, thus allowing to meet the phase noise specifications.

We hope that this special issue will provide the readers with a good understanding of ongoing research in both academia and industry and of the challenges on the road ahead.

CHAPTER 8

DESIGN AND IMPLEMENTATION OF SPACE VECTOR PWM INVERTER BASED ON A LOW COST MICROCONTROLLER

Mahmoud M. Gaballah

Keywords: Space vector pulse width modulation (SVPWM), Uninterruptible power supplies (UPS's), Voltage source inverter (VSI)

INTRODUCTION

Voltage source inverter (VSI) synthesizes AC voltage and frequency from a constant DC voltage using PWM techniques. Nowadays, VSI is used in large applications such as variable speed drives (VSDs), uninterruptible power supplies (UPS's), frequency converters, and active filters [1,2]. PWM techniques have been studied extensively during the last few decades. A large variety of methods, differing in concept and performance, have been developed to achieve one or more of the following objectives: wide linear modulation range, fewer switching losses, less total harmonic distortion (THD), easy implementation and less computation time [3,4]. Several modulation strategies differing in concept and performance have been developed. With the development of microprocessors, space vector modulation has become one of the most important PWM methods for three-phase converters [5]. It uses the space vector concept to compute the duty cycle of the switches. It is simply the digital implementation of PWM modulators. An aptitude for easy digital implementation and wide linear modulation range for line to line voltages are the noticeable features of space vector modulation. Many methods have been developed to implement the space vector pulse width modulation (SVPWM) for driving VSI's. Generally, the SVPWM implementation involves sector identification, switching time calculation, switching vector determination, and optimum-switching-sequence selection for the inverter voltage vectors [6,7]. Sector identification can be done by coordinate transformation as introduced in [8–10] or by repeated comparison of the three-phase reference voltages as introduced in [6–11]. The lookup tables can be used for determining the switching vectors in best switching sequence as introduced in [12]. Calculating the duration of the switching vectors can be simplified by mapping the sector of the multilevel inverter to a corresponding sector of the two-level inverter as introduced in [13,14]. The objective of this paper is to introduce a simplified SVPWM technique in which the inverter leg switching times are directly obtained from the instantaneous sampled reference phase voltages, and the inverter switching vectors are generated automatically. This method is much simpler and more executable than conventional means without lookup tables or complex logical judgments. In addition, an objective of this paper is to introduce a practical SVPWM inverter design based on a low cost microcontroller. The practical design is modeled using the MATLAB SIMULINK software package, and experimentally implemented on the low cost microchip PIC microcontroller 18F4431 platform.

Three-Phase Inverter The major purpose of the PWM inverter is to generate a variable-voltage variable-frequency (VVVF) three-phase voltage from a DC voltage. Two-level VSI consists of six power semiconductor switches with antiparallel diodes. In the widely used pulse width modulation (PWM) methods, the inverter output voltage approximates the reference value through high frequency switching for the six power semiconductor switches. The circuit model of a typical two-level inverter is as shown in Fig. 1. S1–S6 are the six power switches that shape the output, these are controlled by the signal to terminals a, à, b, b, c, and ` ` c.

Citation: Mahmoud M. Gaballah, Design and Implementation of Space Vector PWM Inverter Based on a Low Cost Microcontroller, DOI 10.1007/s13369-012-0464-2

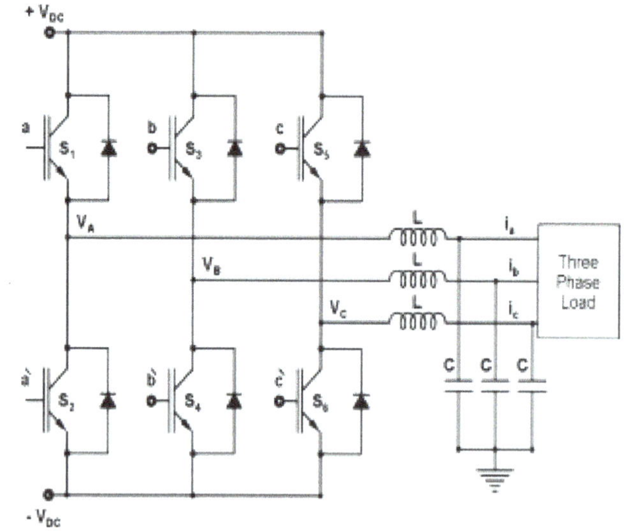

Figure. 1 Two-level voltage source inverter.

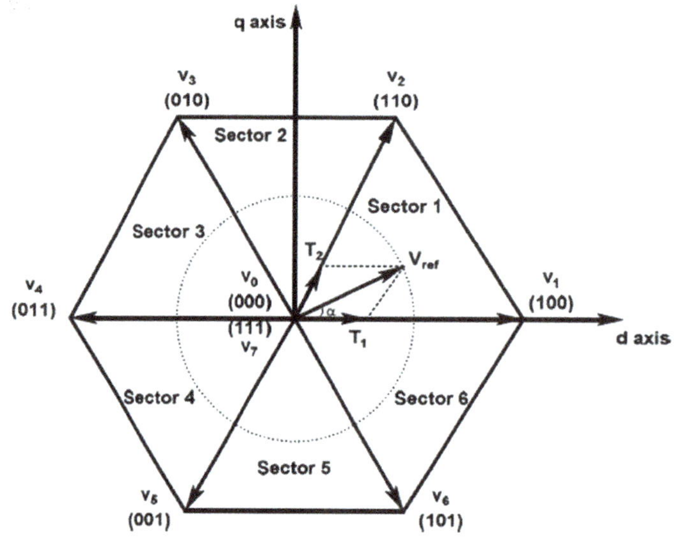

Figure. 2 Output voltage space of the two-level inverter in dq coordinates.

It is assumed that S1 and S2, S3, and S4 as well as S5 and S6 are switched in a complementary way. There are only eight possible switching vectors. Six out of these eight vectors produce a non-zero voltage and are known as non-zero switching states: the remaining two vectors produce zero output voltage and known as zero switching states. The output voltages of the inverter are composed by these eight switch states. The six active vectors divide the space vector plane into six equal sized sectors of 60° with equal magnitude which forms an origin centered hexagon, and two zero space vectors found at the origin as shown in Fig. 2. The hexagon is the maximum boundary of the space vector, and the circle is the trajectory of the regular sinusoidal outputs in linear modulation. Table 1 lists all of the possible switching vectors and the respective line to line/line to neutral voltages.

Table 1 Possible switching vectors, phase voltage, and output line to line voltage

Voltage vectors	Switching vectors			Pole voltage			Line voltage		
	a	b	c	V_{a0}	V_{b0}	V_{c0}	V_{ab}	V_{bc}	V_{ca}
V_0	0	0	0	0	0	0	0	0	0
V_1	1	0	0	2/3	−1/3	−1/3	1	0	−1
V_2	1	1	0	1/3	1/3	−2/3	1	0	−1
V_3	0	1	0	−1/3	2/3	−1/3	−1	1	0
V_4	0	1	1	−2/3	1/3	1/3	−1	0	1
V_5	0	0	1	−1/3	−1/3	2/3	0	−1	1
V_6	1	0	1	1/3	−2/3	1/3	1	−1	0
V_7	1	1	1	0	0	0	0	0	0

To obtain a sinusoidal waveform from the VSI, a voltage reference V_{ref} is provided in terms of a revolving space vector. The magnitude and the frequency of the fundamental component are specified by the magnitude and frequency, respectively, of the reference vector. The reference vector is sampled once in every sub-cycle. The inverter is maintained in different states for appropriate durations such that an average voltage vector equal to the sampled reference vector is generated over a given sub-cycle.

Conventional SVPWM Algorithm

In the space vector approach, the inverter states used are the two zero states, and the two active states. These voltage vectors are the closest to the commanded voltage vector. The SVM algorithm has four switching rules: (a) the trajectory of V_{ref} should be a circle, (b) only one switching by state transition, (c) not more than three switching in one sampling period, and (d) the final state of one sample must be the initial state of the next sample. These rules help in limiting the number of switching actions and therefore, there is a decrease in the switching losses. In addition, they maintain symmetry in switching waveforms at the VSI output to achieve the lower THD. For example, for a commanded vector in sector 1 as shown in Fig. 3, switching states 0, 1, 2, and 7 can be used. If we assume that during the sampling interval T_s, the reference voltage $\bar{V}_{ref}$ remains steady. For implementing the conventional SVPWM using the SVM rules $\bar{V}_{ref}$ can be expressed as follows:

$$\bar{V}_{ref} = \left(\frac{T_1}{T_s} \times \bar{V}_1\right) + \left(\frac{T_2}{T_s} \times \bar{V}_2\right) + \left(\frac{T_{0/7}}{T_s} \times \bar{V}_{0/7}\right) \quad (1)$$

Equation 1 means the inverter is in active state 1 for a period T_1 and in active state 2 for a period T_2. For the remaining time of the sampling interval period T_s there is no voltage applied. This can be achieved by applying inactive state 0 or

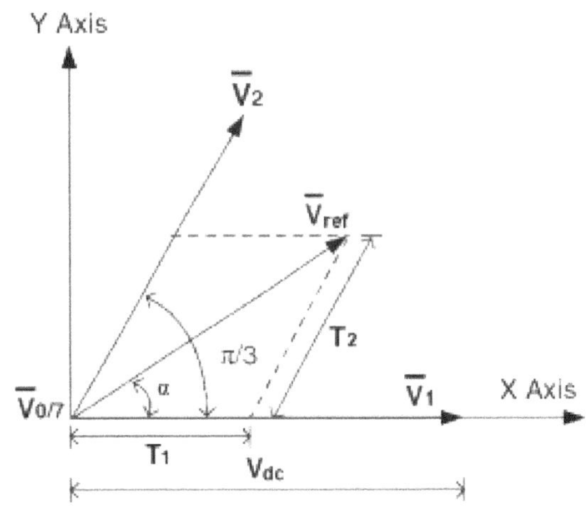

Figure. 3 Reference vector in sector 1.

7 for the remaining time T_0 or T_7. To generate this vector in an average sense, the durations for which the active state 1, the active state 2, and the two zero states together must be applied which are given by T_1, T_2 and T_Z, respectively, obtained as:

$$T_1 = \bar{V}_{ref} \sin(60° - \alpha) \quad (2)$$

$$T_2 = \frac{\bar{V}_{ref} \sin(\alpha)}{\sin(60°)} \quad (3)$$

$$T_Z = T_s - T_1 - T_2 \quad (4)$$

where α is the angle of rotating vector $\bar{V}_{ref}$.

The division of the duration T_Z between the two zero vectors T0 and T7 is a degree of freedom in the space vector approach. This division of T_Z in a sub-cycle is equivalent to adding a common mode component to the three-phase average pole voltages. The typical VSI switching waveforms in sector 1, as defined in Eq. 1 are as given in Fig. 4. Realization of conventional SVPWM involves the following steps:

(1) Coordinate transformation for the reference vector $\bar{V}_{ref}$ from rotating reference frame to stationary reference frame.

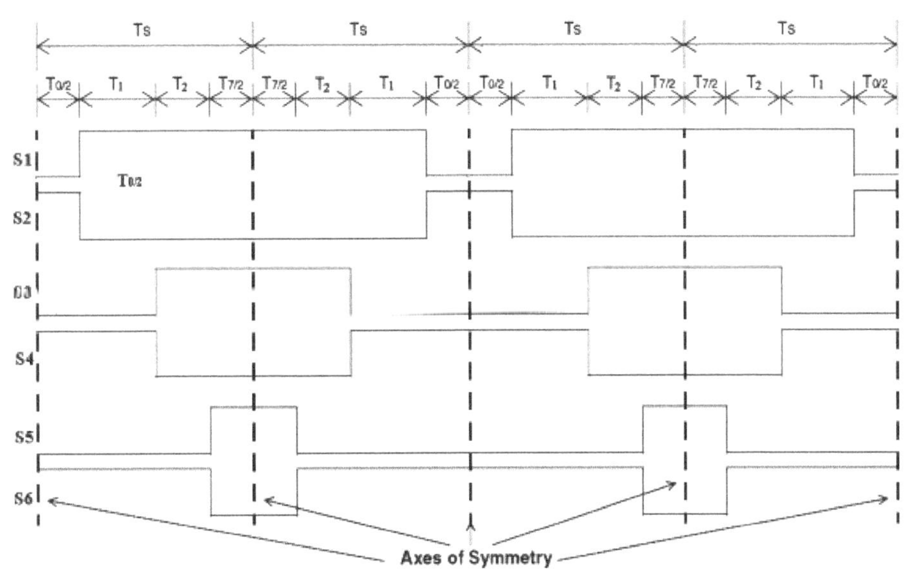

Figure. 4 Typical VSI switching waveforms in sector 1.

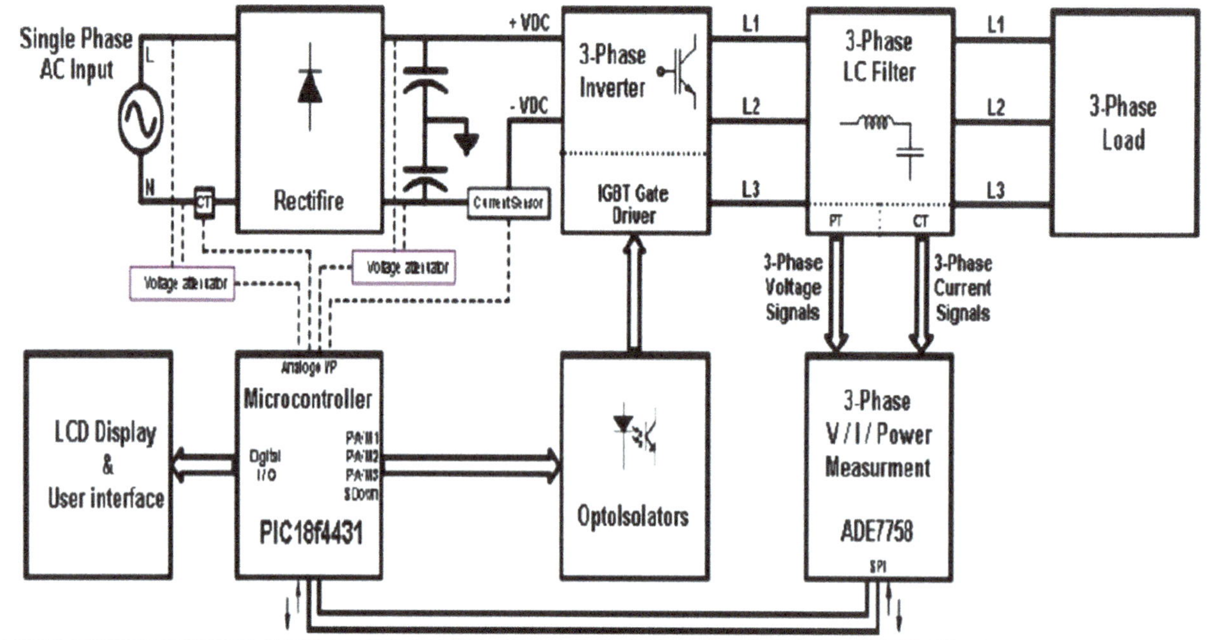

Figure. 5 Overall block diagram of the experiment.

(2) Determine time durations T_1, T_2, and T_0.

(3) Determine the switching time of each transistor (S1–S6).

SVPWM Using a Reduced Computation Method

This method is based on the principle of equivalence of SVPWM with sinusoidal PWM (SPWM) and can generate the SVPWM signals directly from the instantaneous reference phase voltages. In the sinusoidal PWM scheme for a two-level inverter, each reference phase voltage is compared with the triangular carrier, and the individual pole voltages are generated independent of each other [3]. To obtain the maximum possible peak amplitude of the fundamental phase voltage in linear modulation a common mode voltage V_{offset} is added to the reference phase voltages [15,16], where the magnitude of V_{offset} is given by Eq. (5)

$$V_{offset} = \frac{-(V_{max} - V_{min})}{2} \quad (5)$$

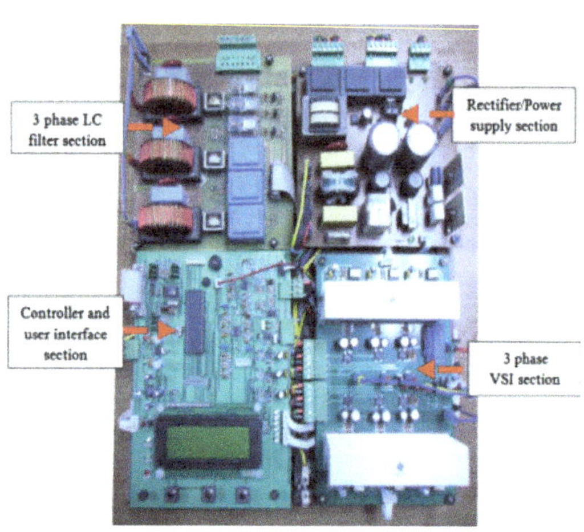

Figure. 6 Real image for the experiment

In Eq. (5), V_{max} is the maximum magnitude of the three sampled reference phase voltages, while V_{min} is the minimum magnitude of the three sampled reference phase voltages. In a sampling interval the addition of the common mode voltage V_{offset} results in the active inverter switching vectors being centered in a sampling interval, making the SPWM technique equivalent to the SVPWM technique [3]. Equation (5) is based on the fact that in a sampling interval, the reference phase which has lowest magnitude (termed the min-phase) crosses the triangular carrier first and causes the first transition in the inverter switching state, while the reference phase which has the maximum magnitude (termed the max-phase) crosses the carrier last and causes the last switching transition in the inverter switching states in a two-level SVPWM scheme [15,16]. Thus, the switching periods of the active vectors can be determined from the (max-phase and minphase) sampled reference phase voltage amplitudes in a two level inverter scheme [17]. The idea behind this SVPWM technique is to determine the sampled reference phase, from the three sampled reference phases, which crosses the triangular first (first-cross) and the reference phase which crosses the triangular carrier last (third-cross). This SVPWM technique presents a simple way to determine the time instants at which the three reference phases cross the triangular carriers using only the instantaneous reference phase amplitudes. These time instants are sorted to find the offset voltage [18,19]. This voltage is then added to the reference phase voltages, so the middle inverter switching vectors are centered (during a sampling interval), as in the conventional two-level SPWM scheme [20]. Implementing this SVPWM method for driving a two-level VSI involves three steps:

A: Read the sampled reference phase amplitudes of VAN, VBN and VCN for the present sampling interval and then calculate the time equivalents of phase voltages, i.e. T_{as}, T_{bs} and T_{cs} as:

$$T_{as} = V_{AN} \times \frac{T_s}{V_{DC}} \quad (6)$$

$$T_{bs} = V_{BN} \times \frac{T_s}{V_{DC}} \tag{7}$$

$$T_{cs} = V_{CN} \times \frac{T_s}{V_{DC}} \tag{8}$$

where T_s is the sampling time period and VDC is the DC link voltage across inverter terminals. B: Find T_{offset} as:

$$T_{offset} = 0.5T_s - 0.5(T_{max} - T_{min}) \tag{9}$$

where T_{max} and T_{min} are the maximum and minimum of T_{as}, Tbs and T_{cs}.

C: Find T_{ga}, T_{gb} and T_{gc} as:

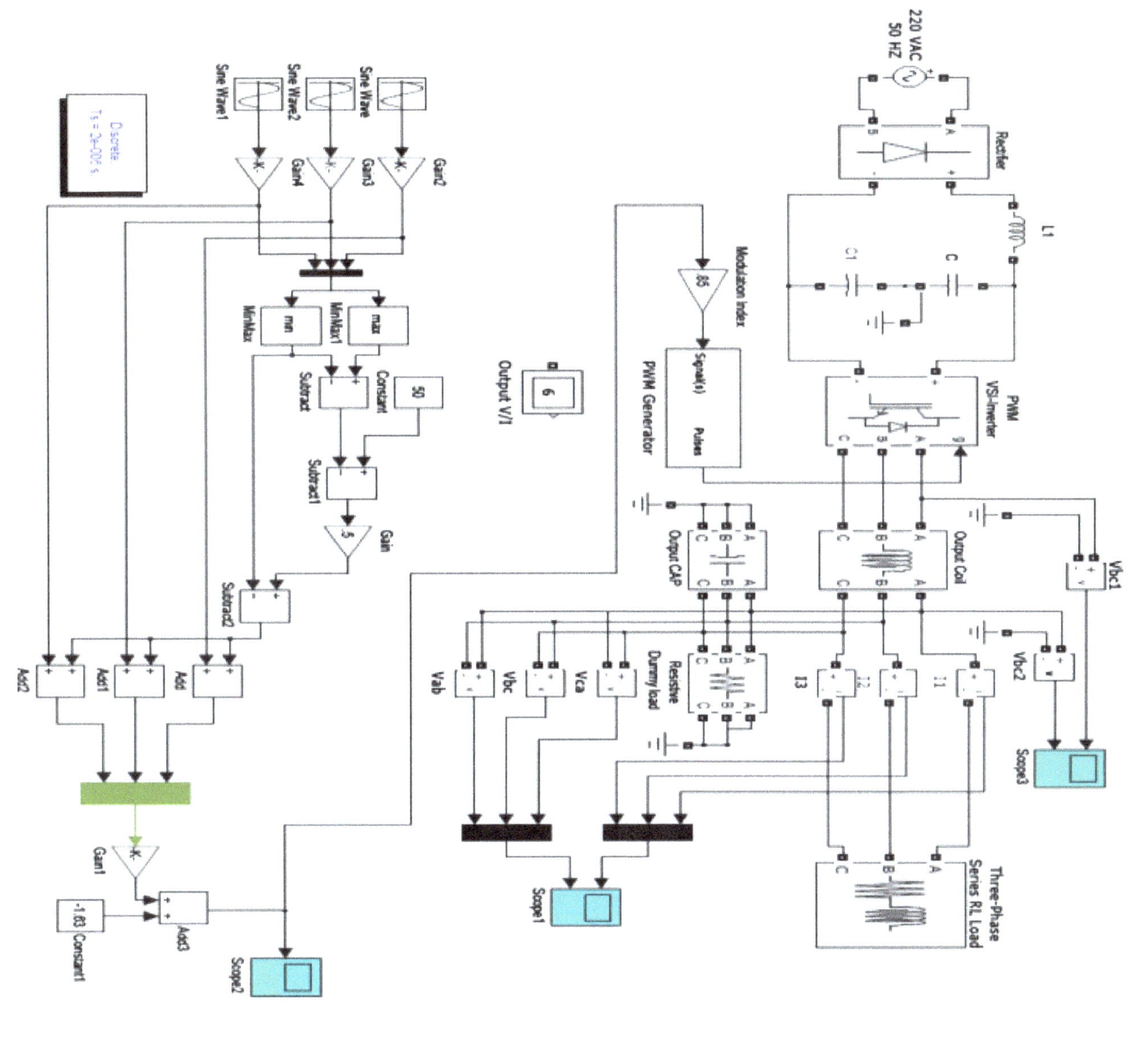

Figure. 7 MATLAB SIMULINK model for the system

$$T_{ga} = T_{as} + T_{offset} \tag{10}$$

$$T_{gb} = T_{bs} + T_{offset} \tag{11}$$

$$T_{gc} = T_{cs} + T_{offset} \tag{12}$$

where T_{ga}, T_{gb} and T_{gc} are the gating signals during which the top switches in a leg are turned on. Equations (6)–(12) show that the centering of the middle inverter switching vectors of the SVPWM in this method reduces the computation time required to determine the switching times for inverter legs, making the algorithm suitable for real-time implementation. Furthermore, the complicated calculations for inverter switching vector times and lookup tables for selecting the inverter switching vector are avoided in this scheme. A comparison between the conventional SVPWM algorithm and the SVPWM based on equivalence with SPWM is introduced below:

<table>
<tr><th>Conventional SVPWM</th><th>SVPWM Based on Equivalence with SPWM</th></tr>
<tr><td>Generates the gating drive signals using switching rules as listed below:
<table>
<tr><th>Sector</th><th>T_{ga}</th><th>T_{gb}</th><th>T_{gc}</th></tr>
<tr><td>1</td><td>$T_0/2$</td><td>$T_0/2 + T_1$</td><td>$T_S - T_0/2$</td></tr>
<tr><td>2</td><td>$T_0/2 + T_2$</td><td>$T_0/2$</td><td>$T_S - T_0/2$</td></tr>
<tr><td>3</td><td>$T_S - T_0/2$</td><td>$T_0/2$</td><td>$T_0/2 + T_1$</td></tr>
<tr><td>4</td><td>$T_S - T_0/2$</td><td>$T_0/2 + T2$</td><td>$T_0/2$</td></tr>
<tr><td>5</td><td>$T_0/2 + T_1$</td><td>$T_S - T_0/2$</td><td>$T_0/2$</td></tr>
<tr><td>6</td><td>$T_0/2$</td><td>$T_S - T_0/2$</td><td>$T_0/2 + T_2$</td></tr>
</table>
</td><td>The gating drive signals are generated directly from the instantaneous samples of the reference phase amplitudes.</td></tr>
<tr><td>Needs sector identification to define the switching rules in each sector.</td><td>It needs no sector identifications.</td></tr>
<tr><td>Needs Vref and α angle information.</td><td>Vref and α angle information is not needed.</td></tr>
<tr><td>Uses look-up tables for switching times calculations.</td><td>Do not use any look-up tables for switching times calculations.</td></tr>
<tr><th colspan="2">Digital Implementation</th></tr>
</table>

<table>
<tr><td>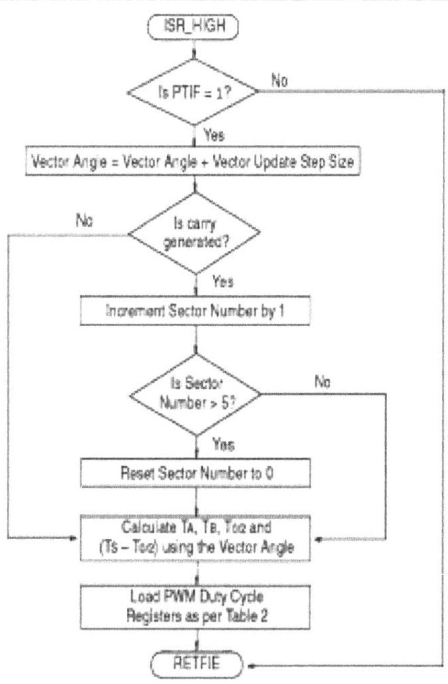
</td><td>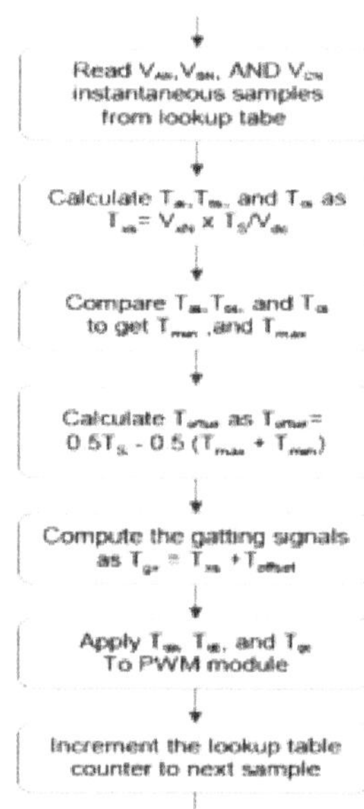
</td></tr>
<tr><td>When the microcontroller is running at 20 MHZ: this routine consumes processing time = 25.3uSec. This means:
<ul><li>High CPU usage</li><li>High execution time</li><li>Big code size</li></ul></td><td>When the microcontroller is running at 20 MHZ: This routine consumes processing time = 15.2 uSec. This means:
<ul><li>low CPU usage</li><li>low execution time</li><li>small code size</li></ul></td></tr>
</table>

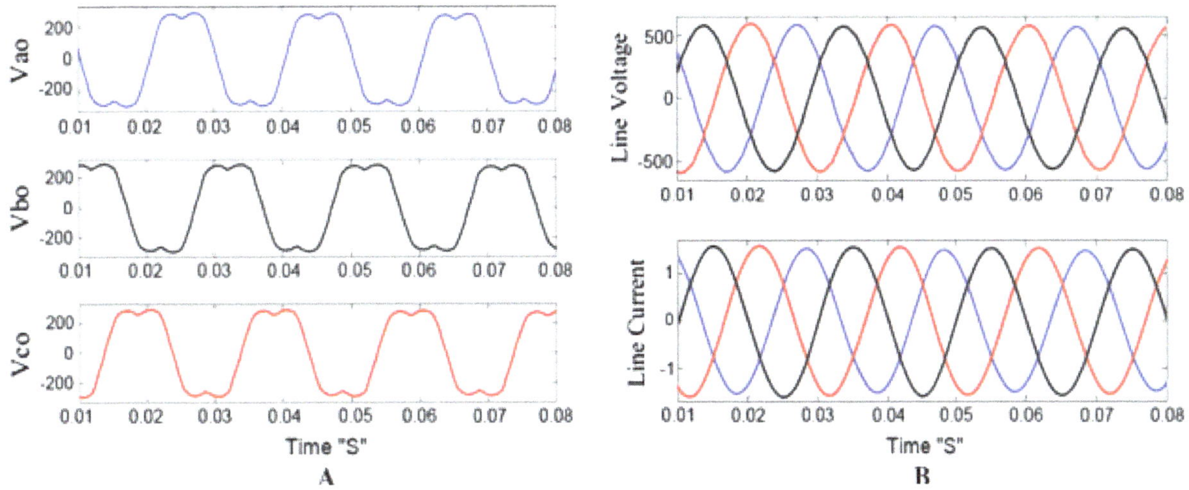

Figure. 8 The simulation results at modulation index = 0.1. a Output pole voltage waveforms, b line voltage/current waveforms across load terminals

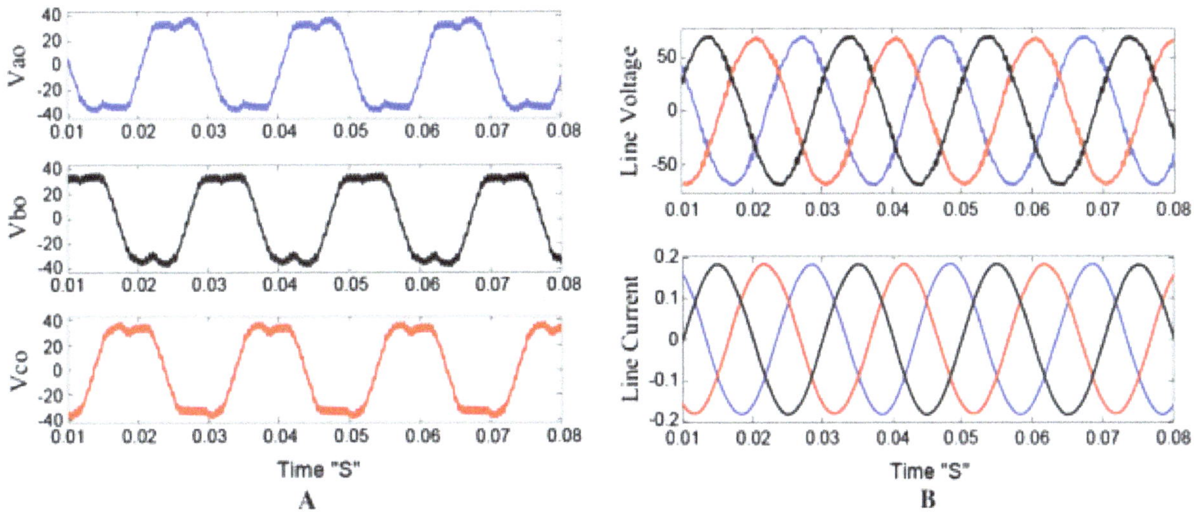

Figure. 9 The simulation results at modulation index = 0.85. a Output pole voltage waveforms, b line voltage/current waveforms across load terminals.

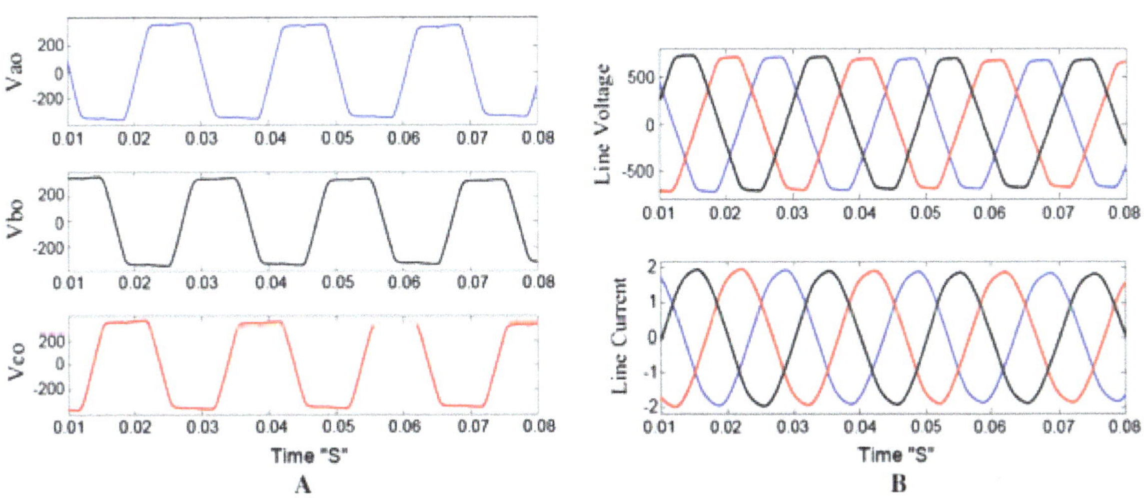

Figure. 10 Shows the simulation results at modulation index = 1.15. a Output pole voltage waveforms, b line voltage/current waveforms across load terminals

EXPERIMENTAL SETUP

Hardware: The practical work is as shown in Fig. 5. The input stage consists of a dual half bridge diode rectifier, which provides the DC bus voltage±310 V from the 220 VAC input and controlled by enable signal from a microcontroller (MCU). The output is provided by a three-phase two-level VSI followed by a three-phase LC filter and an output enable relay. The inverter converts the DC bus voltage back to a sinusoidal voltage using the SVPWM technique. The output inverter is fully controlled by the MCU and generates a pure sinusoidal waveform, free of any disturbance. The system consists of four printed circuit board (PCB), as shown in Fig. 6.

(1) A rectifier/power supply PCB which provides the DC link voltage via the rectifier circuit and provides all DC power supply voltages for the various control circuits such as the main control circuits and the inverter driving circuits. (2) A three-phase two-level inverter PCB, which converts the DC voltage of the DC bus to a three-phase sinusoidal voltage with the required amplitude and frequency via six power IGBTs derived from the control circuit. (3) A three-phase LC filter PCB which removes the undesired component within the output waveforms coming from inverter and maintains the fundamental waveform with pure sinusoidal voltage. (4) A controller/user interface PCB, which manages all control algorithms and measurements of the system, also it contains a user interface that includes a 4 × 20 characters LCD with three input switches and three indication LEDs.

Software: The firmware is developed for a microchip PIC18F4431 microcontroller using ASSEMBLY language, Microchip MPLAB IDE ver. 8.3 compiler, and a PICKIT 3 programmer/debugger kit. The reference sine waveform is generated using the built-in PWM module within the microcontroller. The sine reference is stored in a lookup table. The table values are periodically taken from the table and then multiplied by the required amplitude. The resulting value gives the duty cycle of the PWM output. The pointer to the table is incremented by a value which corresponds to the desired output frequency. All the values over one period give a sinusoidal modulated square wave output. If such a signal passes through an LC filter, a pure sine-wave voltage is generated on the inverter output. At every instant of reading sample from the sine table, the program will compute the minimum and maximum phase amplitudes then compute the offset time and then get/apply the new duty which will drive the VSI switches.

Simulation Results

The practical system introduced in Sect. 5 is modeled/simulated using MATLAB SIMULINK software package. The MATLAB SIMULINK model is as shown in Fig. 7.

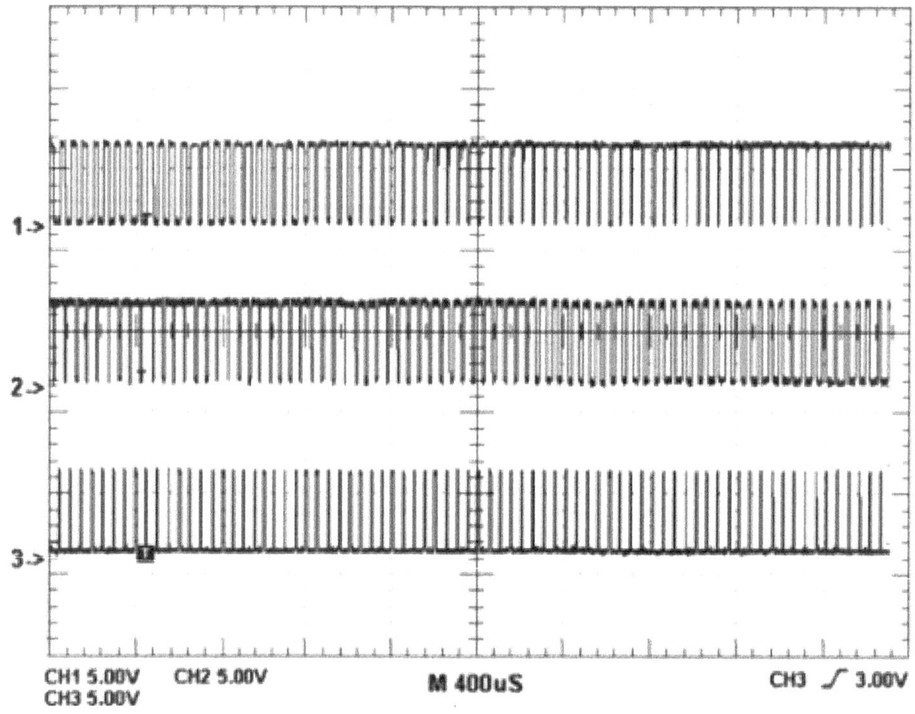

Figure. 11 Experimental SV-PWM gating signals. CH1 = T_{ga},CH2 = T_{gb}, and the CH3 = T_{gc}, X axis 400 μs/div, Y axis 5 V/div.

The simulation is performed under the following conditions: input voltage = 220 VAC, V_{DC} = 620, output voltage fundamental harmonic f = 50 Hz, switching frequency f_{sw} = 20 kHz, and variable modulation index "0.1, 0.85 and 1.15", also the output of the system is connected to 1.5 KVA 0.7 P.F Inductive load. The simulation is done for three load modulation indexes. Figure 8 shows the simulation results for 1.5 KVA inductive load with 0.7 power factor at modulation index = 0.1, Fig. 9 shows the simulation results at modulation index = 0.85, and Fig. 10 shows the simulation results at modulation index = 1.15. From these simulation results we can say that the SVPWM signal generation using the algorithm in Sect. 4 can work in the under-modulation region with some small harmonics in pole and line voltages as shown in Fig. 8, also it can work in the over-modulation region with some small distortions in the output line voltage at the largest modulation index of SVPWM as shown in Fig. 10, and finally it works good in the linear modulation region as shown in Fig. 9.

Experimental Results

The algorithm in Sect. 4 is implemented on a microchip PIC18F4431 microcontroller platform and the experimental results are presented for a two-level VSI with output LC filter as shown in Fig. 1. The modulation index is varied from the under-modulation region to the over-modulation region. The experiment is done under the following conditions: DC link = 400 V is used for the inverter, output voltage fundamental harmonic f = 50 Hz, switching frequency f_{sw} = 20 kHz, and a three-phase 1.5 KVA/0.7 power factor load. The experimental results are presented in Figs. 11, 12

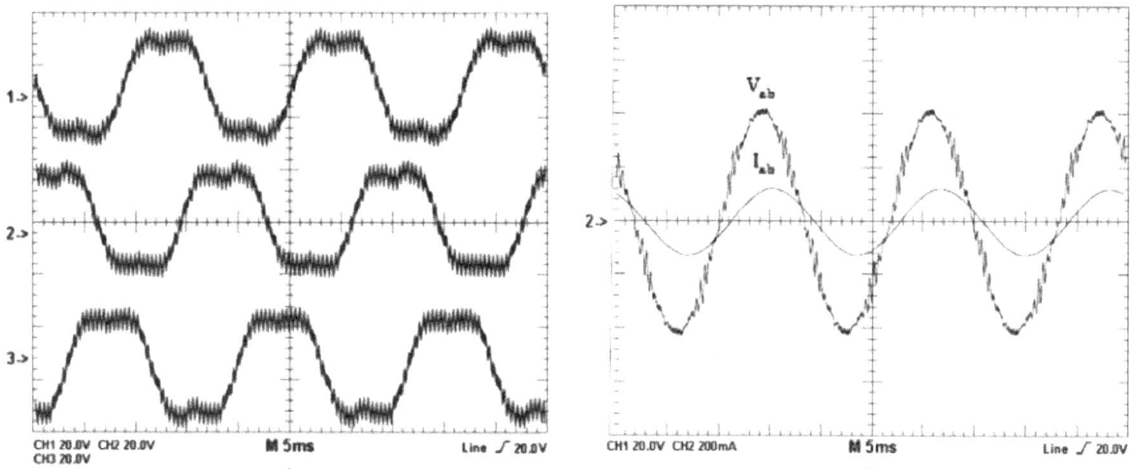

A B

Figure. 12 Experimental results for modulation index = 0.1, a pole voltage CH1 = V_{ao}/CH2 = V_{bo}/CH_3 = V_{co} waveforms, B line voltage CH1 = V_{ab}, line current CH2 = I_{ab} waveforms

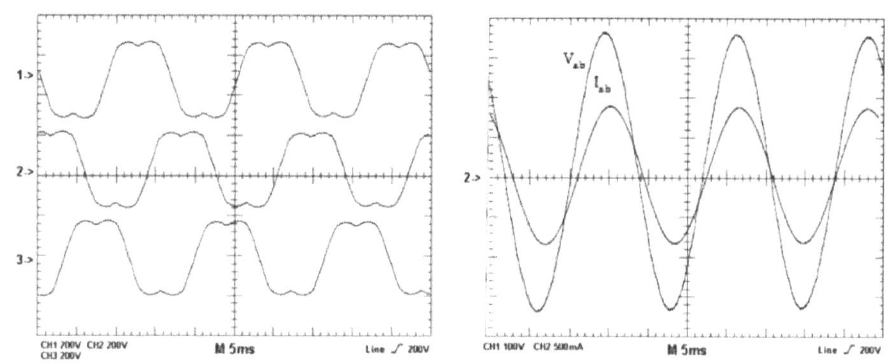

A B

Figure. 13 Experimental results for modulation index = 0.85, a pole voltage CH1 = V_{ao}/CH2 = V_{bo}/CH3 = V_{co} waveforms, b line voltage CH1 = V_{ab}, line current CH2 = I_{ab} waveforms

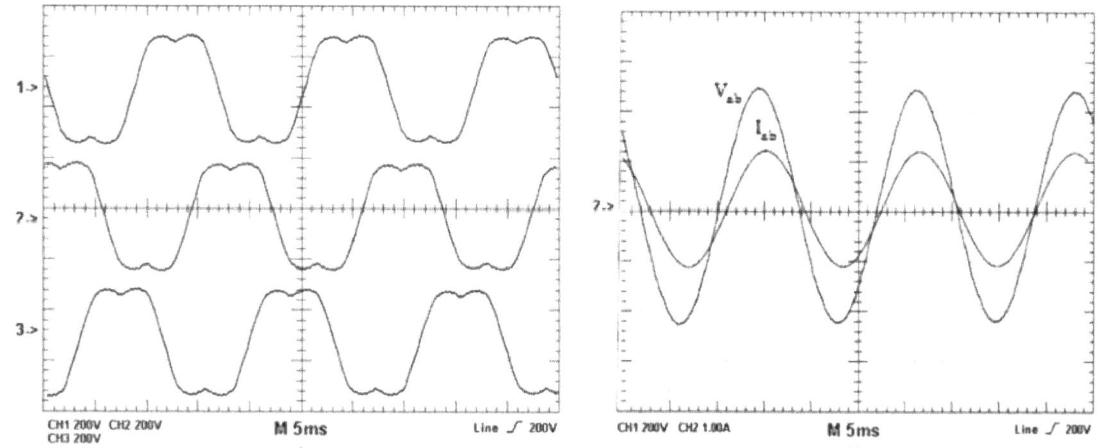

A B

Figure. 14 Experimental results for Modulation index = 1.00, a pole voltage CH_1 = V_{ao}/CH_2 = V_{bo}/CH_3 = V_{co} waveforms, b line voltage CH_1 = V_{ab}, line current CH2 = I_{ab} waveforms

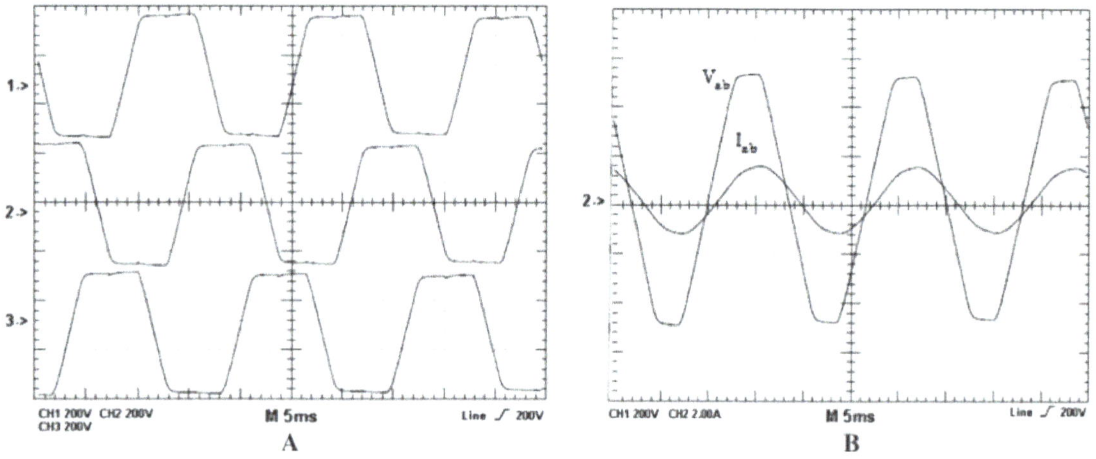

Figure. 15 Experimental results for modulation index = 1.15, a pole voltage $CH_1 = V_{ao}/CH_2 = V_{bo}/CH3 = V_{co}$ waveforms, b line voltage $CH_1 = V_{ab}$, line current $CH_2 = I_{ab}$ waveforms

Table 2 Summary of measured parameters

Parameter	Load	
	Linear load	Non-linear
Efficiency	93 %	92 %
Output voltage THD without load	0.4 %	
Output voltage THD	1 %	5 %
Power factor	0.99	
Output frequency	50 Hz ± 0.5 %	

13, 14 and 15. Figure 11 shows the generated SVPWM signal used to drive the inverter switches SW1, SW3 and SW5, respectively. Figures 12, 13, 14 and 15 show the pole voltage/line voltage/line current waveforms at modulation index 0.1, 0.85, 1.00, and 1.15, respectively. The experimental results for the under-modulation region with a modulation index 0.1 are as shown in Fig. 12. Figure 12a shows the pole voltage of three phases, Fig. 12b shows the line voltage and line current for phase A; it may be noted that the appearance of harmonics in voltage waveforms is due to low modulation index. The experimental results for the modulation region with the modulation indexes 0.85 and 1.00 are as shown in Figs. 13 and 14. The line voltage and line current signals seem to be good with low harmonic and no distortion. Finally the experimental result for the over-modulation region with a modulation index 1.15 is as shown in Fig. 15. Figure 15a shows the pole voltage of three phases; it may be noted that these waveforms tend to be square waves due to over-modulation. Figure 15b shows the line voltage and line current for phase A, the small distortion within line voltage is due to large modulation index. All measured parameters for the inverter performance are summarized in Table 2.

The simulation and experimental waveforms are identical. It demonstrates that the simplified implementation of SVPWM is feasible and effectual in driving three-phase twolevel inverter, and it is much faster "about 1.66" and more executable than conventional means without lookup tables or complex logical judgments.

CONCLUSIONS

A SVPWM technique based on a reduced computation method was presented. The SVPWM scheme can drive the inverter gating signals from the sampled amplitudes of the reference phase voltages. The switching vectors for the inverter are derived using a simple digital logic which does not involve any complex computations and hence reduces the implementation time. A practical design and a real implementation for a SVPWM inverter, including MATLAB SIMULINK model, and simulation results for different modulation indexes were also presented. The SVPWM scheme was implemented on a microchip PIC microcontroller 18F4431 platform and the experimental results were presented for a two-level VSI with a three-phase LC filter. The practical results show a good performance with less computation time "about 1.66 faster" and easy software implementation for the presented SVPWM scheme rather than conventional SVPWM.

REFERENCES

1. Channegowda, P.; John, V.: Filter optimization for grid interactive voltage source inverter. IEEE Trans. Ind. Electron. 157, 4106–4114 (2010)

2. Yang, S.; Lei, Q.; Peng, F.Z.; Qian, Z.: Robust control scheme for grid interactive voltage source inverter. IEEE Trans. Ind. Electron. 158, 202–212 (2011)

3. Holtz, J.: Pulse-width-modulation—a survey. IEEE Trans. Ind. Electron. 39, 410–419 (1992)

4. Bowesand, S.R.; Lai, Y.S.: The relationship between space-vector modulation and regular-sampled PWM. IEEE Trans. Ind. Electron. 44, 670–679 (1997)

5. Van Der Broeck, H.W.; Skudenly, H.-C.; Stanke, G.V.: Analysis and realization of a pulse width modulator based on voltage space vector. IEEE Trans. Ind. Electron. 24, 142–150 (1988)

6. Celanovic, N.; Boroyevich, D.: A fast space-vector modulation algorithm for multilevel three-phase converters. IEEE Trans. Ind. Electron. 37(2), 637–641 (2001)

7. Tekwani, N.; Kanchan, S.; Gopakumar, K.: A dual five-level induction motor drive with common-mode voltage elimination and DClink capacitor voltage balancing using only the switching-state redundancy—Part I. IEEE Trans. Ind. Electron. 54(5), 2600–2608 (2007)

8. Seixas, P.F.; Severo Mendes, M.A.; Donoso Garcia, P.; Lima, A.M.N.: A space-vector PWM method for three-level voltage source inverters, vol. 1. In: Proceedings of IEEE APEC, pp. 549– 555 (2000)

9. Yao, W.; Hu, H.; Lu, Z.: Comparisons of space-vector modulation and carrier-based modulation of multilevel inverter. IEEE Trans. Ind. Electron. 23(1), 45–51 (2008)

10. Somasekhar, V.T.; Gopakumar, K.; Baiju, M.R.; Mohapatra, K.K.; Umanand, L.: A multilevel inverter system for an induction motor with open-end windings. IEEE Trans. Ind. Electron. 52(3), 824– 836 (2005)

11. Shivakumar, E.G.; Gopakumar, K.; Sinha, S.K.; Pittet, A.; Ranganathan, V.T.: Space-vector PWM control of dual inverter fed open-end-winding induction motor drive. In: Proceedings of the IEEE APEC, pp. 399–405 (2001)

12. Somasekhar, V.T.; Srinivas, S.; Kumar, K.K.: Effect of zero-vector placement in a dual-inverter fed open-end-winding induction motor drive with a decoupled space-vector PWM strategy. IEEE Trans. Ind. Electron. 55(6), 2497–2505 (2008)

13. Shivakumar, E.G.; Gopakumar, K.; Sinha, S.K.; Pittet, A.; Ranganathan, V.T.: Space-vector PWM control of dual inverter fed open-end-winding induction motor drive. In: Proceedings of the IEEE APEC, pp. 399–405 (2001)

14. Gopinath, A.; Mohamed A.S.A.; Baiju, M.R.: Fractal based space vector PWM for multilevel inverters—a novel approach. IEEE Trans. Ind. Electron. 56(4), 1230–1237 (2009)

15. Holmes, D.G.: The general relationship between regular sampled pulse width modulation and space vector modulation for hard switched converters. In: Conf. Rec. IEEE Industry Applications Society (IAS) Annual Meeting, pp. 1002–1009 (1992)

16. Baiju, M.R., Mohapatra, K.K., Somasekhar, V.T., Gopakumar, K., Umanand, L.: A five-level inverter voltage space phasor generation for an open-end winding induction motor drives. IEE Proc. Electr. Power Appl. 150(5), 531–538 (2003)

17. Kim, J.; Sul, S.: A novel voltage modulation technique of the Space Vector PWM'. In: Proc. Ind. Power Electronics Conf., Yokohama, Japan, pp. 742–747 (1995)

18. Baiju, M.R.; Gopakumar, K.; Somasekhar, V.T.; Mohapatra, K.K.; Umanand, L.: A space-vector-based PWM method using only the instantaneous amplitudes of reference phase voltages for threelevel inverters. EPE J. 13(2), 35–45 (2003)

19. Kanchan, R.S.; Baiju, M.R.; Mohapatra, K.K.; Ouseph, P.P.; Gopakumar, K.: Space-vector PWM signal generation for multilevel inverters using only the sampled amplitudes of reference phase voltages. Proc. Ind. Elect. Eng. Elect. Power Appl. 152(2), 297–309 (2005)

20. van der Broeck, H.W.; Skudelny, H.Ch.; Stanke, G.: Analysis and realization of a pulse width modulator based on voltage space vectors. IEEE Trans. Ind. Electron. 24, 142–150 (1988)

CHAPTER 9

A COMPARISON OF FUZZY LOGIC AND PID CONTROLLER FOR A SINGLE-AXIS SOLAR TRACKING SYSTEM

Emre Kiyak and Gokhan Gol

Keywords: Solar tracking system, Fuzzy controller, PID controller, Energy efficiency

BACKGROUND

Renewable energy is a type of energy that is derived from ongoing natural processes and energy of natural processes converted into available forms. Renewable energy sources can be listed as sunlight, wind, flowing water, biological processes, and geothermal. The use of renewable energy sources is growing rapidly because of the fact that fossil fuels are limited, being rapidly depleted, pollute the environment and cause climate change. In addition, another factor that increases the use of renewable energy sources is that they can be installed everywhere, and can be developed using various technologies.

The most common sources of renewable energy are solar and wind energy. The siting of wind turbines must ensure maximum exposure to wind in order to achieve wind energy, but not every province in Turkey can provide such conditions. Also, wind turbines must be built far from residential areas, due to the fact that the sound of wind turbines greatly inconveniences people. In addition to this, the rural setting of wind turbines is generally on the migration route of birds. Research shows that birds often change their migration routes or that birds die by hitting operational turbines due to the siting of such turbines (Akkaya et al. 2002). Accordingly, it is observed that solar farms, as a source of renewable energy, are more widespread than wind turbines.

Due to the fact this method of obtaining electrical energy using solar energy is easier, more practical, less harmful, and at lower cost than other renewable energy sources, it is rapidly becoming widespread. However, the efficiency of solar cells is only around 20 %, so it prevents the conversion of solar energy into electrical energy at full efficiency (Green et al. 2000). Therefore, solar tracking systems have emerged, in order to obtain the greatest efficiency from sunlight.

A solar tracking system is a monitoring system which aims for solar panels to operate by tracking the sun at full efficiency during the day to allow for the sun rays arrive perpendicular to the panels. It can be seen that the efficiency achieved from the solar panels increases from between 25 and 55 % (Irina and Cătălin 2010).

In recent years, there have been many studies concerning the development of solar tracking systems. Studies are currently being conducted to get sun rays perpendicular to panels using a variety of control techniques (Shugar et al. 1996; Roth et al. 2004; Colak et al. 2005). According to the study, applications on solar energy have increased rapidly in recent years, and new materials and methods are being developed for this energy source.

In another study, a SCADA system is developed in order to allow for optimal sun tracking and real-time control using a programmable logic controller (PLC) and step motor in a two degrees of freedom (DOF) tracking system. In this study, power generation increases, compared to other PV systems, and non-tracking systems because real-time control and monitoring is provided (Figueiredo and Costa 2008).

It can be seen that the tracking system is made very stable by the development of a microprocessor-based solar tracking system. A microprocessor-based solar tracking controller was

Citation: Emre Kiyak and Gokhan Gol, A comparison of fuzzy logic and PID controller for a single-axis solar tracking system, DOI 10.1186/s40807-016-0023-7.

designed and manufactured in 1990, in New Delhi, India. The controller has several features which makes it versatile for tracking and system control/monitoring applications (Saxena and Dutta 1990).

Environmental conditions effect power production in solar energy, but our devices work on a constant voltage. If light intensity and temperature do not change, a maximum power point (MPP) will occur at a constant voltage. However, if the environmental conditions change over time, the voltage in the MPP will also change. In this case, for better performance, a more complicated controller is required with parameter changes according to atmospheric conditional changes (Nopporn et al. 2005). The output power of a solar panel depends on the amount of light on the panel (Li et al. 2005). The designed Fuzzy logic controller technique can find peak power by doing wide range of illumination and temperature variations (Ghassami et al. 2013).

A fuzzy logic-based two-axis solar tracking system increases efficiency by 33.416 % compared to a non-tracking system. In this study, a stepper motor is used for the direction control and an Arduino Uno is used for the microcontroller. In addition, the proposed fuzzy logic controller has been implemented and tested using MATLAB for this study (Bawa and Patil 2013).

Fuzzy logic is used in many engineering applications, because it is considered by designers to be the simplest solution available for a specific problem, for instance, in household electrical appliances, auto electronics applications, and industrial automation systems (Peri and Simon 2005). With complicated processes where control is hard, it becomes necessary to use a fuzzy logic controller (Takagi and Sugeno 1985). Fuzzy logic controller is adaptive and nonlinear nature, which provides it robust enactment under load, supply voltage disturbances, and parameter variation (Punithaa et al. 2013).

So far Fuzzy logic control-based solar trackers of different configurations have been implemented on FPGA and PIC microcontrollers, but the control logic for this research was implemented on a simple microcontroller board Arduino Uno (Hamed and Mohammed 2012; Khaehintung et al. 2007).

In the solution there is a dual-axis solar tracking system, based on solar maps, which can predict the exact apparent position of the sun, by latitude location, thereby avoiding the need to use sensors or guidance systems (Abdallah and Nijmeh 2004).

In another study, a solar tracking algorithm is designed and implemented on a solar tracking experimental platform, using a tri-positional control strategy. It makes use of measured values for radiation from appropriate sensors and assures command of the platform's two positioning motors. The implementation technique reduces the cost of the tracking method and makes it cost-effective technology (Arghira and Iliescu 2013).

A solar cooling system is important for improving of energy efficiency. This study's aim is to improve energy efficiency of a solar cooling system by an innovative combination of optimized solar cooling, storage techniques, and an absorption chiller, with highly developed techniques for control using known tool; TRNSYS and MATLAB with Simulink (Visek et al. 2014).

A single-axis sun tracking system with two sensors was designed. The data acquisition, control, and monitoring of the mechanical movement of the photovoltaic module were implemented based on a programmable logic-controlling unit (Al-Mohamad 2004).

In this study, the design and application for a single-axis solar tracking system is developed based on both fuzzy logic and PID controller in a real system. The necessary control circuits are designed. The control circuit is built on an Atmega 328 microcontroller and necessary software is installed into the control unit according to MATLAB simulations. In this study, it has been shown that fuzzy controllers are more efficient than PID controllers in single-axis solar tracking system. This result is reached using real-time measurement data obtained in the scope of the study.

Solar Tracking System

The purpose of the solar tracking system is to allow sun rays to reverberate to a solar panel with full angle in the solar system which has low efficiency and to generate energy at full efficiency. Therefore, sun rays placed at different points on the system are controlled using an intensity sensor, and the solar panel is thereby moved toward the point gathering most sun rays by a DC motor. The materials used for the solar tracking application are as follows:

- Atmega 328 microcontroller
- DC motor
- DC motor drive circuit
- Solar panel
- Mechanical systems for single-axis control
- Feed circuit
- Sensors

Atmega 328 microcontroller

An Atmega 328 on an Arduino UNO is a microcontroller, with 14 digital input/output, and 6 analog inputs. Six of the digital pins can be used for pulse width modulation (PWM). Input/output pins can provide up to a maximum 40 mA of current. There are 32 GB flash memories for storing written code, a 2 kB static random access memory (SRAM) used for creating a work area, and a 1 kB erasable programmable read only memory (EEPROM) for users on the microcontroller.

Certain additional elements, such as an external crystal and capacitors are necessary in order that the Atmega 328 microcontroller can be used as a control unit in a circuit. A 16 MHz crystal, and a two 22 pF capacitor for operation of the microcontroller are also required (Fig. 1).

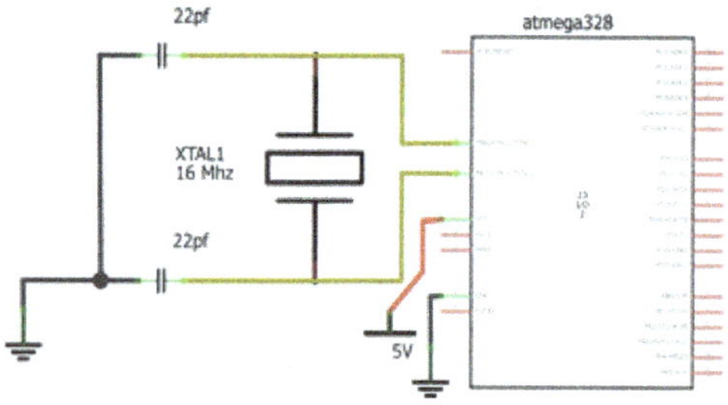

Figure. 1 Atmega 328 circuit connection diagram.

DC motor

An electrical device which can direct current electric energy and convert it to mechanical energy is called a DC motor. In this study, a low speed, high torque direct current motor is used for moving the solar panels. The motor used (titan motor) operates at a low speed of 5 revolutions per minute (rpm) and can produce 32 kg torque per cm. It uses a low amount of current with 400 mA at maximum load capacity and 24 mA at idle.

DC motor drive circuit

DC motors are motors which start to rotate when provided with energy from a battery directly. The motor rotation direction changes when the positive and the negative terminals of the battery are applied to a motor intake interchange. Due the fact that motor direction cannot be changed during use, it is necessary to create driver circuits. There are numerous direct current motor driver circuits. However, in this study, direction and speed control of the motor is performed using the popular H-bridge method. The operating principles of the H-bridge, as illustrated in Fig. 2, are based on the change of logic signal which is applied to the base of the transistors used. As shown in the circuit, if the base terminals of 1 and 4 transistors are given logic 1, and the base terminals of 2 and 3 transistors are given logic 0, the motor rotates to the left. Therefore, a related truth table is drawn according to this principle.

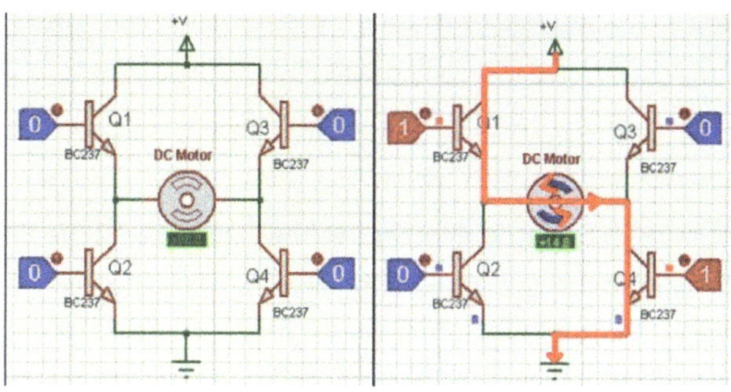

Figure. 2 H-bridge operation diagram.

The H-bridge is made using BC237 transistors. However, when this circuit is designed, it can be seen that the transistors overheat and that the motor cannot be controlled properly. This circuit, designed in order to understand how driver circuits work, is not used for motor control in this study. In this study, as shown in Fig. 3, the motor speed and the direction control circuit are mounted on a pcb board.

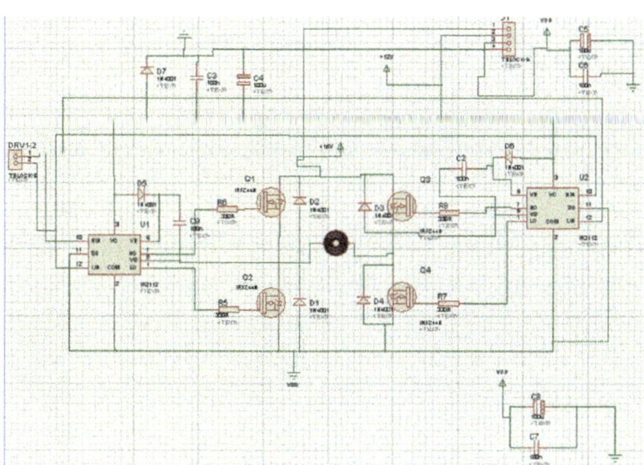

Figure. 3 DC motor driver circuit.

As can be seen from the circuit, there are two inputs to enter the control signal coming from the microcontroller to the driver circuit. According to Table 1, the motor direction control can be performed by sending a logic signal sent from this input.

Table 1 Driver circuit truth table

Input 1	Input 2	Motor direction
ON	OFF	Right
OFF	ON	Left
OFF	OFF	Stop
ON	ON	Un-solicited status

DC motor speed control is performed using the PWM method. In this method, the voltage applied to the motor signal cable is a fixed frequency and square wave. That is, the speed of the motor is set by changing the pulse width of logic 0 and 1 sent to the driver. Due to the fact that the motor used in this study runs at a 600 Hz frequency, the period is 1666 µs. The speed of the motor is adjusted depending on pulse width modulation. If the motor is desired to be rotated at full speed, a 255 value is sent to the motor with the help of the microcontroller using an analog Write (255) command. The 255 value is due to the 8-bit output ADC of the microcontroller. In Fig. 4, pulse width modulation is seen according to a duty cycle.

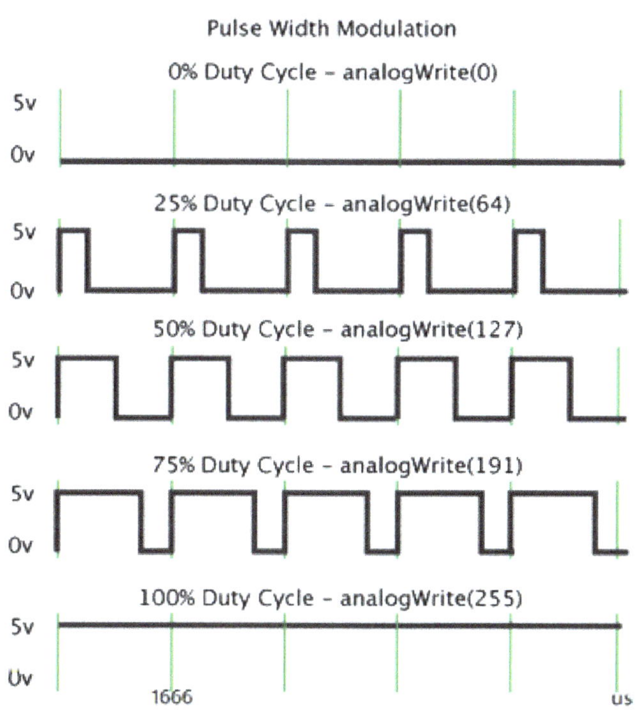

Figure. 4 DC motor speed control.

Solar panel

A solar panel is an energy source containing many solar cells enabling the absorption of solar energy. Solar cell/photovoltaic cells convert solar energy into electrical energy using the photovoltaic effect characteristic of semiconductor devices. In this study, 1 solar panel of 15 V 20 W capacity is used.

Feed circuit

Two different voltage levels were required for this study. 5 V is required for the control circuit and 12 V is required for the motor drive circuit. Voltage from the solar panel is not applied to the circuits directly, and the battery is charged by the battery charging circuit. The charging circuit arranged for a 12 V battery is shown in Fig. 5.

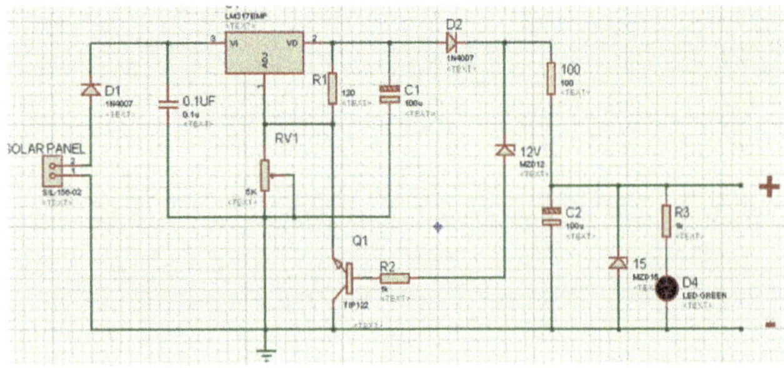

Figure. 5 Battery charging circuit.

The 5 V required for the control unit was made by designing a buck converter circuit. A buck converter is a voltage step-down and current step-up converter. The converter circuit was made using an integrated lm517 as shown in Fig. 6.

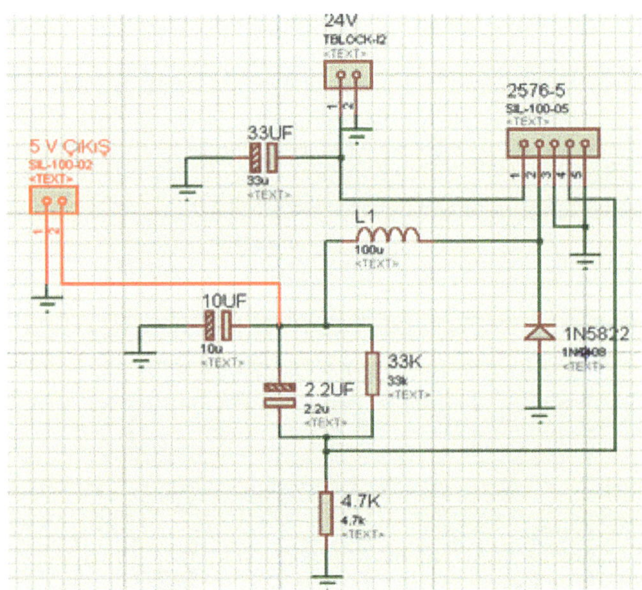

Figure. 6 15–5 V converter circuit

Sensors

Sensors are used to interconnect the electronic circuits in a physical environment and are referred to as sense organs of the control systems. There are many sensors used for measuring different values, such as pressure, temperature, humidity, and sun intensity. The values are generally measured in terms of voltage by changing the resistance on the sensor. Therefore, measurements are made by taking into consideration the relationship between resistance and voltage.

In this study, two photo resistances placed in an east–west direction on the solar tracking systems measuring sun intensity are used. A photo resistance sensor or light-dependent sensor (LDR) is a sensor which changes resistance depending on light and is inversely proportional. The voltage values measured from the sensor output located on the east and west parts of the solar panel are compared. Accordingly, the panel is moved in an east–west direction. Thus, the panel is moved into a position receiving the sun rays perpendicularly providing that the values measured from the sensors are within the tolerance value range.

Photo resistance sensors provide voltage data according to changes in resistance. Values are read from the sensor connected to an analog input pins of an Atmega 328 microcontroller with the aid of correct resistance change between 0 and 1023 due the fact that the input ADC value of the controller is 10 bit. In fact, the equivalent value between 0 and 1023 is 0–5 V. The controller–sensor connection is shown in Fig. 7.

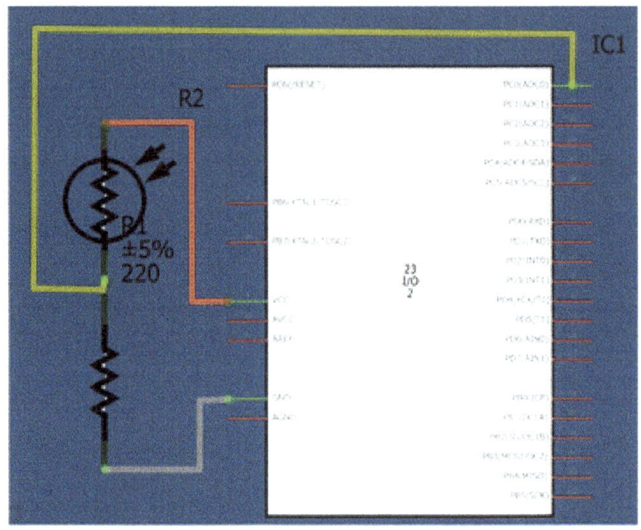

Figure. 7 Atmega 328 LDR connection

PID control system and fuzzy logic controller

PID controllers are commonly used to regulate the time-domain behavior of many different types of dynamic plants (Ahmad et al. 2013). PID is a control technique which aims to reduce errors of a system by undergoing three different mathematical operations and racking the results up producing a control output. PID, an acronym for proportional, integral, and derivative words, constitutes control output from the effect of these three mathematical expressions. The formula for PID is described by (Ozerdem and Zhahin 2014):

$$P + I \times \frac{1}{s} + D \times \frac{N}{1 + N \times \frac{1}{s}}, \tag{1}$$

where: P proportional, I integral, D derivative, N filter coefficient.

Proportional effect has an impact on controller output as errors multiply with a specific gain value. Proportional effect increases the accuracy of static and dynamic response of a system. That is, it has an impact on a system in the way of fast reaction time and a reduction of errors. While the proportional gain factor K increases, the steady-state error of the system decreases. However, despite the reduction, P control can never manage to eliminate the steady-state error of the system. In this study, gain value is found using various automatic tuning methods or by trial and error. In this study, an automatic tuning code is embedded in the microcontroller thanks to feedback obtained from the system. The integral effect has an impact on controller output proportionally to the sum of the errors from the effective operation of the system. While the integral effect increases static accuracy and eliminates error, its dynamic response time reduces. This study focuses on integral effect and the integral coefficient is determined by an automatic tuning system and by the trial and error method.

A derivative effect has an impact on controller output in direct proportion to the change in error. The derivative effect does not contribute to error recovery, while increasing dynamic response.

PID controller has the optimum control dynamics including zero steady-state error, fast response (short rise time), no oscillations, and higher stability. The necessity of using a derivative gain component in addition to the PI controller is to eliminate the overshoot and the oscillations occurring in the output response of the system. One of the main advantages of the PID controller is that it can be used with higher-order processes including more than single energy storage.

In this study, Ziegler–Nicholas method is used for tuning process. Ziegler–Nichols method is one of the most effective methods that increase the usage of PID controllers. The first step in this method is setting the *I* and *D*gains ($K_I$and K_D) to zero, increasing the *P* gain (K_P) until a sustained and stable oscillation (as close as possible) is obtained on the output. Then the critical gain K_C and the oscillation period P_0 is recorded and the *P*, *I*, and *D*values are adjusted accordingly. Critical gain K_C and oscillation period P_0 are used to set the gains as shown:

$$K_P = 0.6K_C, \quad K_I = \frac{2K_P}{P_0}, \quad K_D = \frac{K_P P_0}{8} \tag{2}$$

These steps which written software code embedded into microcontroller.

The transfer function of PID control is given by:

$$G_{\text{PID}}(s) = K_P + \frac{K_I}{s} + K_D\,(s) \tag{3}$$

The term 'fuzzy logic' was introduced with the 1965 proposal of fuzzy set theory by Lotfi A. Zadeh. It is expressed as a mathematical order which is established for the expression of uncertainties and for working with uncertainties (Calvo and Cartwright 1998). The word 'fuzzy' is defined as indefinite, complicated, and imperceptible. As is evident from its name, in this control system, it is intended to obtain approximate values through evolution with reasoned logic in an atmosphere of uncertainty. This control system, which is suitable for many dynamic systems that cannot be modeled in the world, provides an opportunity to view all the states of a system by dealing with, not only 0 and 1, but also all intermediate values contrary to the classical logic of Aristotle.

One-axis solar tracking system generally uses two sensors. LDR sensors measuring light density are placed on solar tracking system by indicating in east and west. The main idea is to read the value from LDR1 and compare it with the value of light density from LDR2. Then, depending on the difference between the two values, the controllers will decide and send commands to the motor and change its angle in order to make the difference equal the tolerance value (LDR1 − LDR2 = tolerance). The tolerance value is determined according to the sensitivity of sensor. Thanks to this it provides that solar rays reflect to solar panels perpendicularly. Control algorithm of the solar tracking system is shown in Fig. 8.

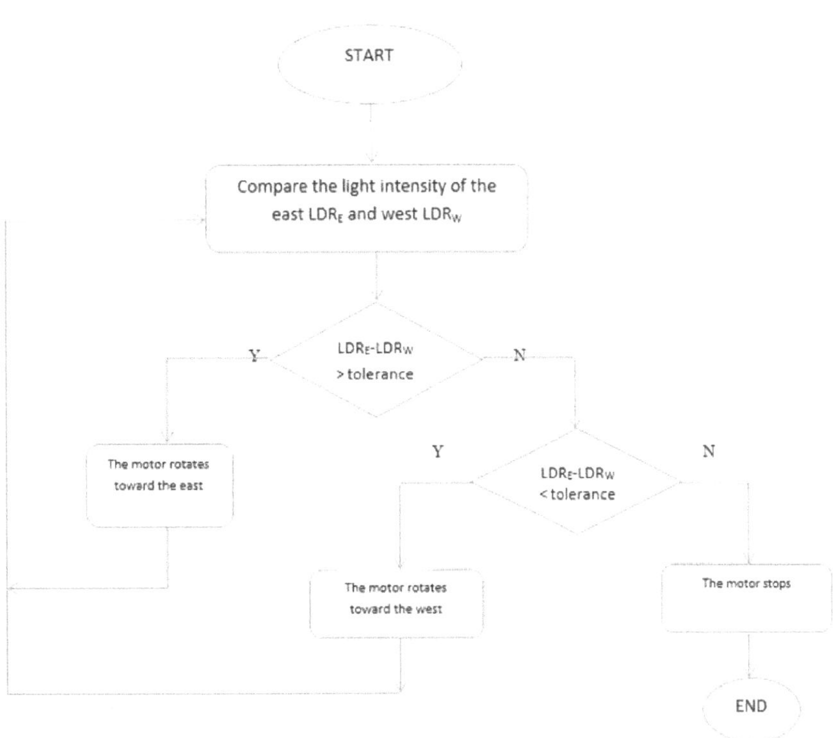

Figure. 8 Control algorithm of solar tracking system

In this study, the maximum revolution of motor used is 4 rpm. Response of the controllers is shown in Fig. 9according to 4 rpm input signal.

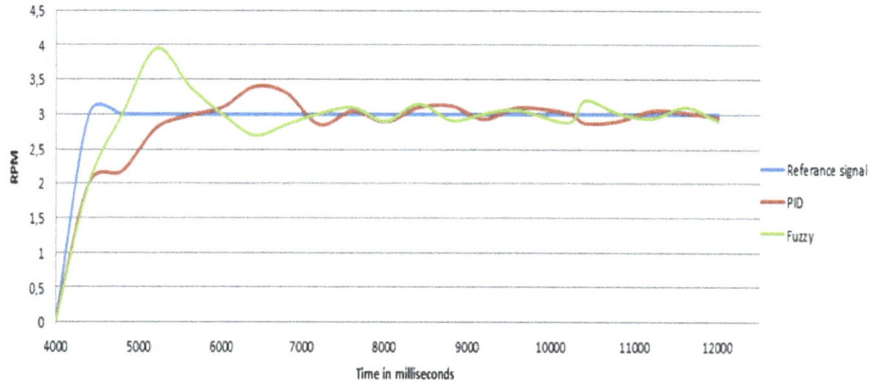

Figure. 9 Response of controller for 4 rpm speed

A simplified block diagram representation of the solar tracking system is shown in Fig. 10.

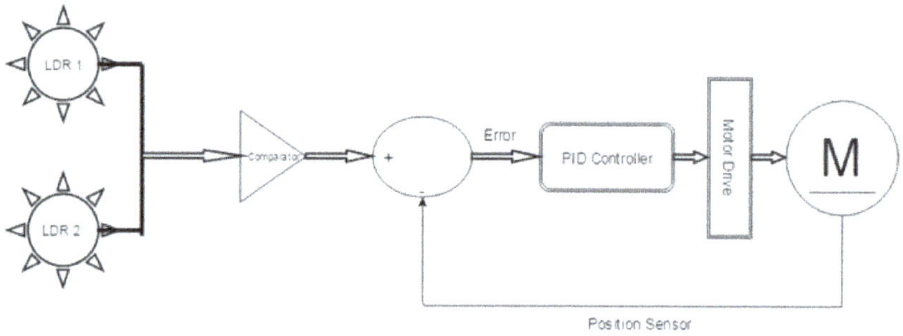

Figure. 10 Simplified block diagram representation of solar tracking system

Here, $u(t)$ is the control signal, $y(t)$ is the output signal, $r(t)$ is the reference signal, and $e(t)$ is the error signal in block diagram. According to block diagram, PID output $u(t)$ is calculated as Eqs. 4 and 5.

$$u(t) = K_P e(t) + K_I \int e(t)\mathrm{d}t + K_D \frac{\mathrm{d}e\,(t)}{\mathrm{d}t} \tag{4}$$

$$e(t) = r(t) - y(t) \tag{5}$$

RESULTS AND DISCUSSION

Voltage and current data of the solar tracking system using a PID controller, are monitored in real time. The difference between systems remaining at a constant angle and systems with changing angle constantly is shown in Fig. 11.

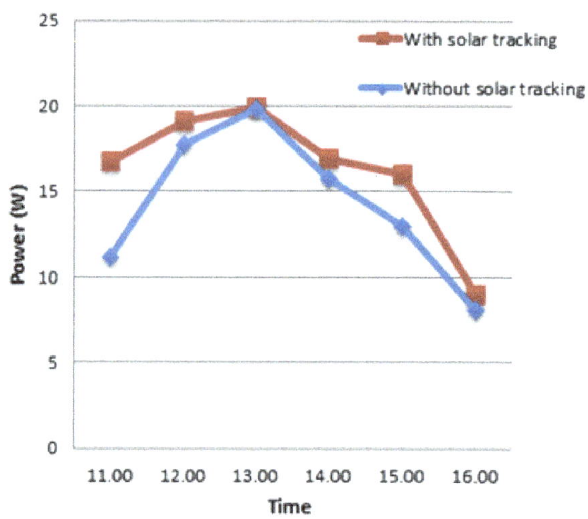

Figure. 11 The graphic of constant and moving angle systems.

The flowchart in Fig. 12 explains the algorithm used for fuzzy-based control of the solar tracking system. Block diagram of the fuzzy controller is shown in Fig. 13.

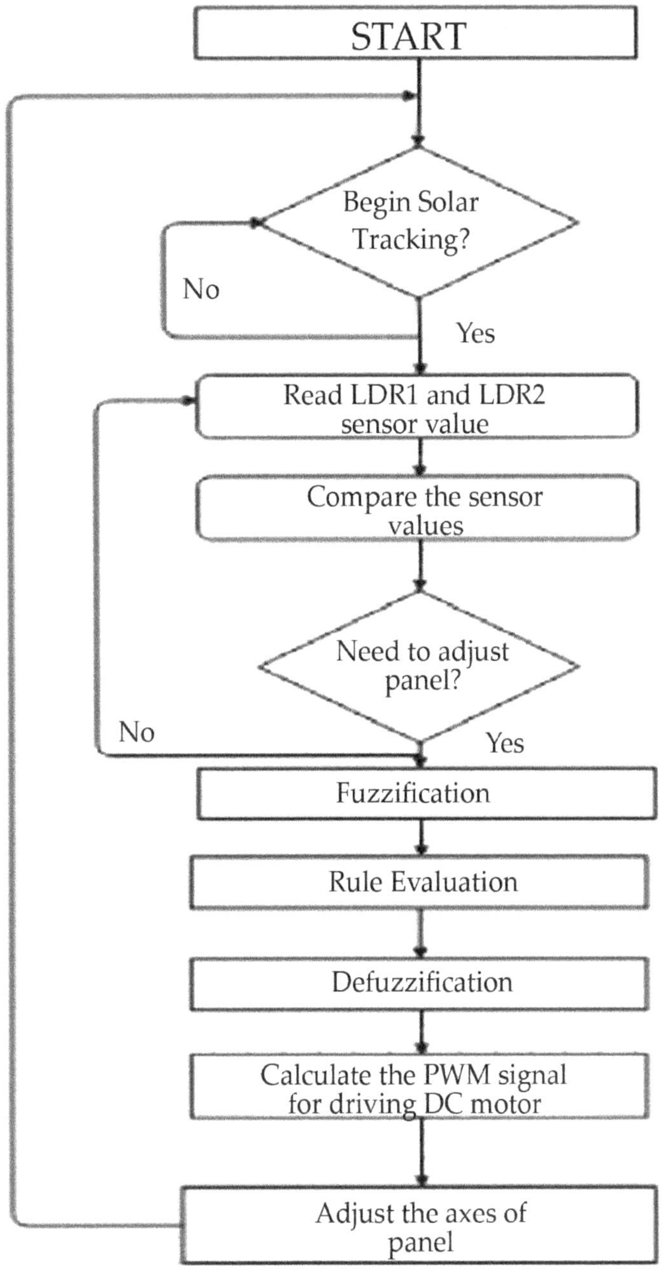

Figure. 12 Flowchart for fuzzy-based control of solar tracking

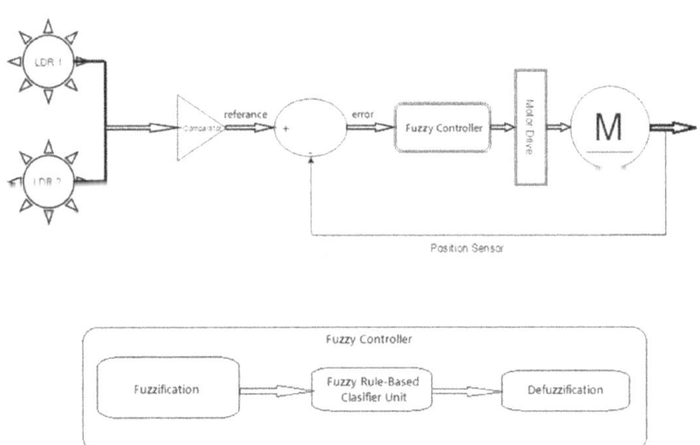

Figure. 13 Block diagram of fuzzy control

In this study, the solar tracking system is controlled using fuzzy logic. Data obtained from two different photo resistance sensors on the system are applied to the control system as input. Applied inputs are fuzzificated by entering in a fuzzification unit that is the first unit of the fuzzy control. In other words, a membership value is assigned to each item of data, and is transformed into a linguistic structure and then sent to the rule processing unit. Data obtained from the rule processing unit are governed by rules which can be modeled in accordance with rule processing information, like 'if……and……then…. else', created in this unit. Input consisting of entries and these rules are sent to the inference unit and the result is obtained. In the next step, the results obtained as a result of the rules are sent to the defuzzification unit and then converted into real numbers according to a scale.

In this study, a triangle type is seen to be appropriate as a membership function. Thus, working load of microcontroller decreases as compared with using of other functions (trapezoidal, gauss etc.). The fact that working load in microcontroller decreases it, takes an important place. Membership functions and fuzzy control system are created via a MATLAB program. The created input and output membership are shown in Fig. 14.

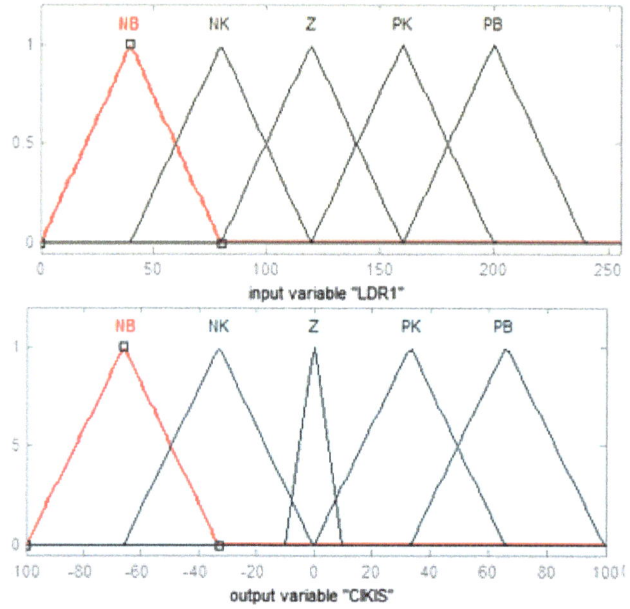

Figure. 14 The first LDR and output membership function

Membership labels defined for the membership function mean that:

- NB = negative big
- NK = negative small
- Z = zero
- PK = positive small
- PB = positive big

Input membership set values are obtained by converting solar intensity data between 0–962 from the photo-resistor sensors to the 0–240 value linearly. The photo-resistor sensors are tested in daylight, and set bound values are obtained from the results of these tests. For example, NB set limits are chosen between 0 and 80 (0–341) because the photo-resistor sensor receives maximum sunlight between these values. The maximum 240 (LDR analog value = 962) is the value read from the photo-resistor sensor at sunset.

Output membership set values are determined as positive and negative in order to adjust the rotational direction of the DC motor. Values represent the PWM duty cycle rate.

Data received from the first and second photo resistance sensor are applied to the input. The duty cycle and direction control of the PWM signal of the DC motor which controls speed and direction as output are adjusted. Values beginning with N allow the motor to rotate counter clockwise, while values beginning with P allow the motor to rotate clockwise. The duty cycle is adjusted by selecting appropriate membership values in order that the motor rotates fast and slowly. There is a rule table of the system in Table 2.

Table 2 Fuzzy control system rule table

LDR2	**LDR1**				
	NB	**NK**	**Z**	**PK**	**PB**
NB	Z	PK	PB	PB	PB
NK	NK	Z	PK	PB	PB
Z	NB	NK	Z	PB	PB
PK	NB	NB	NK	Z	PK
PB	NB	NB	NB	NK	Z

The rule table is created as a result of testing. The system is tested over 15 days in different weather conditions. The surface table showing the output of the rules created in the MATLAB program is shown in Fig. 15.

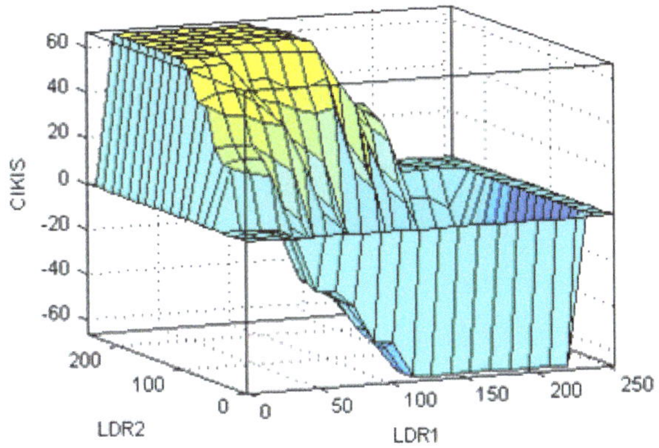

Figure. 15 Matlab rule surface image.

The voltage and current information of the solar energy system are monitored in real time. The difference between the systems remaining at a constant angle and systems with changing angle constantly is shown in Fig. 16.

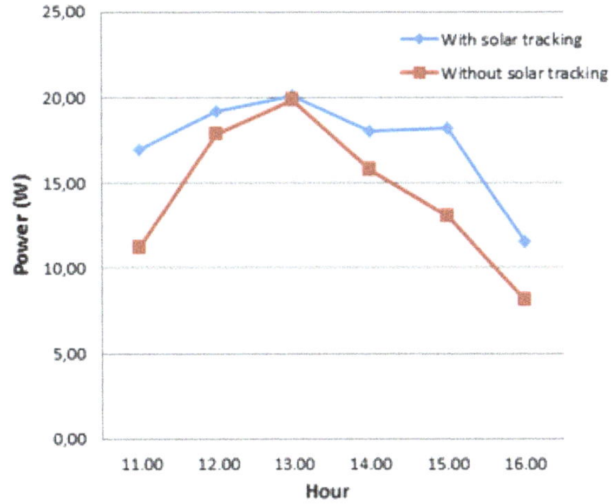

Figure. 16 The graphic of constant and moving angle systems

Voltage, current, and power information measured for the system with a tracking system, and the system without a tracking system, is given in Table 3. As can be seen from these results, the energy derived from the system using the tracking system increases.

Table 3 Voltage, current, and power graphic of constant angle and moving angle

Time	Solar tracking with fuzzy			Without solar tracking		
	V (V)	*I* (A)	*P* (W)	*V* (V)	*I* (A)	*P* (W)
11.00	13	1.3	16.9	11.2	1	11.2
12.00	14.5	1.32	19.14	13.9	1.28	17.79
13.00	15	1.34	20.1	15	1.32	19.8
14.00	14.9	1.21	18.03	14.3	1.1	15.73
15.00	15.1	1.2	18.12	13	1	13
16.00	12.8	0.9	11.52	10	0.81	8.1
Total			103.8			85.6

When taking into consideration the power value obtained for cases where the tracking system is utilized or is not utilized in Table 3, it can be seen that efficiency increases by 21.2 %.

$$\text{Efficiency} = \frac{(103.8 - 85.6) \times 100}{85.6} = 21.2\,\% \tag{6}$$

At the same time, the produced and consumed power information for the solar tracking system, based on fuzzy logic and the PID-based solar tracking system, are calculated. Related graphs are shown in Figs. 17, 18, and 19.

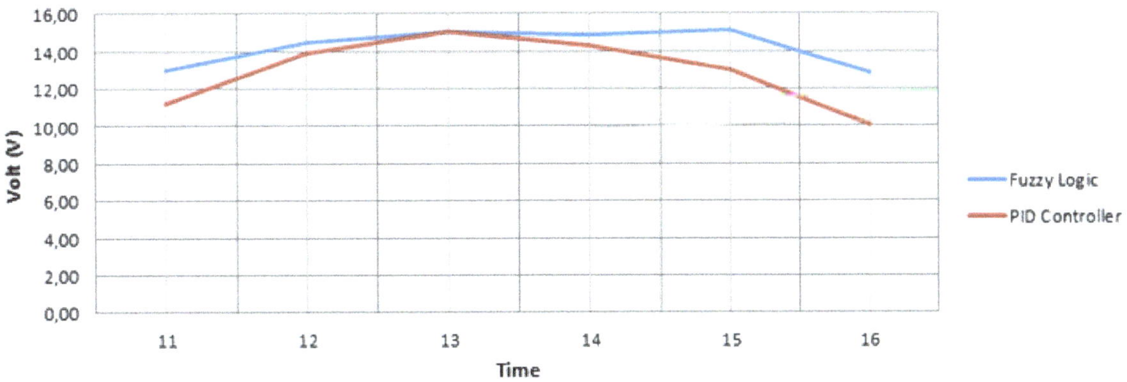

Figure. 17 Voltage–time graph of fuzzy logic-based and PID controller systems

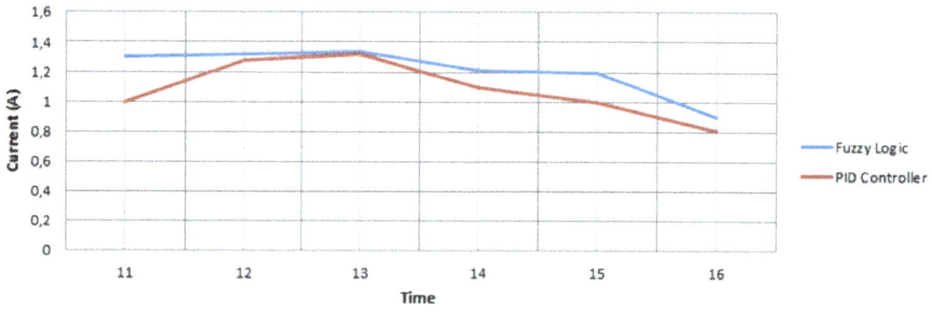

Figure. 18 Current–time graph of fuzzy logic-based and PID controller systems

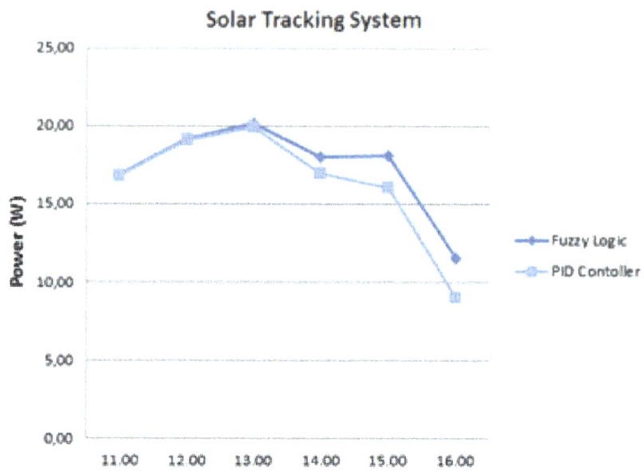

Figure. 19 Power–time graph of fuzzy logic-based and PID controller systems

Voltage data measured at the top 20 s at the beginning for PID and fuzzy controller are shown in Fig. 20. As is seen, at the beginning PID controller steadied the late to settling time compared to fuzzy controller and the energy at full efficiency is not generated.

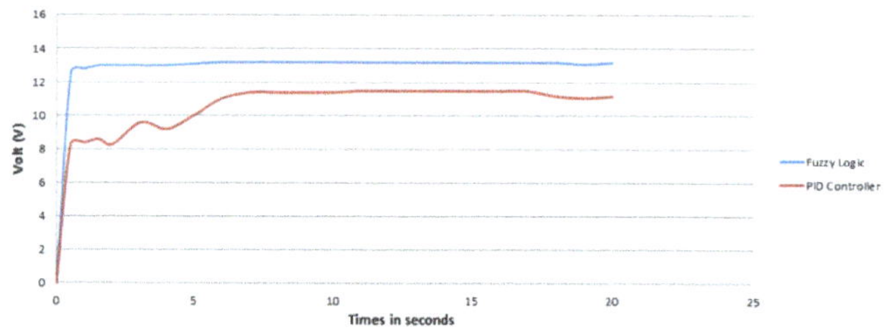

Figure. 20 Voltage–time graph of fuzzy logic-based and PID controller systems

Voltage, current, and power information of the solar tracking systems developed with fuzzy logic-based and PID-based controllers is shown in Table 4. As can be seen from these results, the obtained energy increases in system utilized the fuzzy logic-based solar tracking system.

Table 4 Voltage, current, and power values of the fuzzy logic-based and PID-based solar tracking system

Time	Fuzzy logic controller			PID controller		
	V (V)	*I* (A)	*P* (W)	*V* (V)	*I* (A)	*P* (W)
11.00	13	1.3	16.9	11.2	1	16.8
12.00	14.5	1.32	19.14	13.9	1.28	19.1
13.00	15	1.34	20.1	15	1.32	19.95
14.00	14.9	1.21	18.03	14.3	1.1	18
15.00	15.1	1.2	18.12	13	1	17.01
16.00	12.8	0.9	11.52	10	0.81	11.5
Total consumption			3.6			4.5
Net			100.2			97.86

When the total power and power consumption obtained from the fuzzy logic and PID-based tracking system in Table 4 is considered, the efficiency of the fuzzy logic-based system increased by 2.39 %.

$$\text{Efficiency} = \frac{(100.2 - 97.86) \times 100}{97.86} = 2.39\,\% \qquad (7)$$

CONCLUSION

In this study, a solar tracking system is designed in order to the increase efficiency of solar energy, which is a most important energy source for the future. Fuzzy logic and PID-based controllers are used to control the solar tracking system. The mechanics of the control, motor driver, feed circuit, and a single-axis tracking system are prepared.

As a result, it is observed that the energy obtained from the system using fuzzy logic for a solar tracking system increases by 21.2 %, compared to systems not using fuzzy logic for a solar tracking system. It can be seen that the efficiency can be increased when systems designed for the production of electricity from solar energy are combined with solar tracking systems. Also, according to results of tests conducted on the fuzzy logic-based controller and PID-based controller, the fuzzy logic-based tracking system is found to be 2.39 % more efficient than fixed systems. At the same time, fuzzy controller is steadied to settling time according to PID controller.

It is planned to apply the solar tracking system in two axes. An interface will be designed considering real-time data transmission and the system will be continuously monitored. Additional sensors are added to the available sensor, as an input to the fuzzy logic control system, with sensitivity being optimized.

DECLARATIONS

Authors' contributions

EK participated in the design of the study and performed the PID controller and fuzzy controller for a single-axis solar tracking system. GG conceived of the study, and participated in its design and coordination and helped to draft the manuscript. All authors read and approved the final manuscript.

Competing interests

The author declare that they have no competing interests.

REFERENCES

1. Abdallah, S., & Nijmeh, S. (2004). Two axes sun tracking system with PLC control. *Energy Conversion and Management,45*, 1931–1939.
2. Ahmad, A. M., Ayman, A. A., & Farhan, A. S. (2013). Mechatronics design of a mobile robot system. *International Journal of Intelligent Systems and Applications, 5*(3), 23–36.
3. Akkaya, A. V., Akkaya, E., & Dağdaş, A. (2002). Yenilenebilir Enerji Kaynaklarının Çevresel Açıdan Değerlendirilmesi. IV. Ulusal Temiz Enerji Sempozyumu, Cilt 1, Su Vakfı Yayınları, İstanbul.
4. Al-Mohamad, A. (2004). Efficiency improvements of photo-voltaic panels using a sun-tracking system. *Applied Energy, 79*, 345–354.
5. Arghira, N, & Iliescu, S. S. (2013). Design and implementation of a solar-tracking algorithm. In *Proceedings of the 24th DAAAM international symposium on intelligent manufacturing and automation*.
6. Bawa, D., & Patil, C. Y. (2013). Fuzzy control based solar tracker using Arduino Uno. *International Journal of Engineering and Innovative Technology, 2*(12), 179–187.
7. Calvo, O., & Cartwright, J. (1998). Fuzzy control of chaos. *International Journal of Bifurcation and Chaos, 8*(8), 1743–1747.
8. Colak, İ., Bayındır, R., Sefa, İ., Demirbaş, Ş., & Demirtaş, M. (2005). Güneş Takip Sistemi Tasarım ve Uygulaması. 1. Enerji Verimliliği ve Kalite Sempozyumu, TMMOB Elektrik Mühendisleri Odası Kocaeli Şubesi (pp. 301–305).
9. Figueiredo, J. M. G., & Costa, J. M. G. S., (2008). Intelligent sun-tracking system for efficiency maximization of photovoltaic energy production. In *ICREPQ'08—international conference on renewable energies and power quality symposium, Santander*.
10. Ghassami, A. A., Sadeghzadeh, S. M., & Soleimani, A. (2013). A high performance maximum power point tracker for PV.*Journal of Electrical Engineering, Electrical Power and Energy Systems, 53*, 237–243.
11. Green, M. A., Emery, K., King, D. L., & Igari, S. (2000). Solar cell efficiency tables. *Progress in Photovoltaics: Research and Applications, 8*, 187–196.
12. Hamed, B., & Mohammed, S. E., (2012). Fuzzy controller design using FPGA for photovoltaic maximum power point tracking. *International Journal of Advanced Research in Artificial Intelligence, 1*(3), 14–21.
13. Irina, T., & Cătălin, A. (2010). A study on the tracking mechanisms of the photovoltaic modules. *Fascicle of Management and Technological Engineering, IX*, 59–66.
14. Khaehintung, N., Pramotung, K., Tuvirat, B., & Sirisuk, P. (2007). Application of maximum power point tracker with self-organizing fuzzy logic controller for solar-powered traffic lights. In *Power electronics and drive systems, PEDS'07 Conference* (pp. 642–646).
15. Li, D. H. W., Cheung, G. H. W., & Lam, J. C. (2005). Analysis of the operational performance and efficiency characteristic for photovoltaic system in Hong Kong. *Energy Conversion and Management, 46*, 1107–1118.
16. Nopporn, P., Premrudeepreechacharn, S., & Sriuthaisiriwong, Y. (2005). Maximum power point tracking using adaptive fuzzy logic control for grid-connected photovoltaic system. *Renewable Energy, 30*(11), 1771–1788.

17. Ozerdem, O. C., & Zhahin, A. (2014). A PV solar tracking system controlled by Arduino/Matlab/Sımulink. *International Journal on Technical and Physical Problems of Engineering, 6*(4), 23–36.

18. Peri, V. M., & Simon, D. (2005). Fuzzy logic control for an autonomous robot. In *Fuzzy Information Processing Society, 2005. NAFIPS 2005. Annual Meeting of the North American*, (pp. 337–342). IEEE, 26-28 June 2005.

19. Punithaa, K., Devaraj, D., & Sakthivel, S. (2013). Development and analysis of adaptive fuzzy controllers for photovoltaic system under varying atmospheric and partial shading condition. *Applied Soft Computing, 13*, 4320–4332.

20. Roth, P., Georgiev, A., & Boudinov, H. (2004). Design and construction of a system for sun-tracking. *Renewable Energy,29*, 393–402.

21. Saxena, A. K., & Dutta, V. (1990). A versatile microprocessor based controller for solar tracking. In *21 IEEE photovoltaic specialists conference, Kissimmee.*

22. Shugar, D. S., Hickman, T., & Lepley, T. (1996). Commercialization of a value-engineered photovoltaic tracking system. In*25th IEEE PVSC proceedings* (pp. 1537–1540).

23. Takagi, T., & Sugeno, M. (1985). Fuzzy identification of systems and its applications to modelling and control. *IEEE Transactions on Systems, Man and Cybernetics, 15*(1), 116–332.

24. Visek, E., Mazzrella, L., & Motta, M. (2014). Performance analysis of a solar cooling system using self-tuning fuzz-PID control with TRNSYS. In *2013 ISES solar world congress energy procedia* (Vol. 57, pp. 2609–2618).

CHAPTER 10

VIRTUAL PROTOTYPING AND PERFORMANCE ANALYSIS OF TWO MEMORY ARCHITECTURES

Huda S. Muhammad and Assim Sagahyroon

INTRODUCTION

Due to the rapid advances in circuit integration technology, and to optimize performance while maintaining acceptable levels of energy efficiency and reliability, multicore technology or Chip-Multiprocessor is becoming the technology of choice for microprocessor designers. Multicore processors provide increased total computational capability on a single chip without requiring a complex microarchitecure. As a result, simple multicore processors have better performance per watt and area characteristics than complex single core processors [1].

A multicore architecture has a single processor package that contains two or more processors. All cores can execute instructions independently and simultaneously. The operating system will treat each of the execution cores as a discrete processor. The design and integration of such processors with transistor counts in the millions poses a challenge to designers given the complexity of the task and the time to market constraints. Hence, early virtual system prototyping and performance analysis provides designers with critical information that can be used to evaluate various architectural approaches, functionality, and processing requirements.

In these emerging multicore architecture, the ability to analyze (at an early stage) the performance of the memory subsystem is of extreme importance to designers. The latency resulting by the access of different levels of memory reduces the processing speeds causing more processor stalls while the data/instruction is being fetched from the main memory. Ways in which multiple cores send and receive data to the main memory greatly affect the access time and thus the processing speed. In multicore processors, two approaches to memory subsystem design have emerged in recent years, namely, the AMD DirectConnect architecture and the Intel Shared Bus architecture [2, 3,4, 5]. In the DirectConnect architecture, a processor is directly connected to a pool of memory using an integrated memory controller. A processor can access the other processors' memory pool via a dedicated processor-to-processor interconnect. On the other hand, in Intel's dual-core designs, a single shared pool of memory is at the heart of the memory subsystem. All processors access the pool via an external front-side bus and a memory controller hub.

In this work, virtual system prototyping is used to study the performance of these alternatives. A virtual systems prototype is a software-simulation-based, timing-accurate, electronic systems level (ESL) model, used first at the architectural level and then as an executable golden reference model throughout the design cycle. Virtual systems prototyping enables developers to accurately and efficiently make the painful tradeoffs between that quarrelling family of design siblings functionality, flexibility, performance, power consumption, quality, cost, and so forth.

Virtual prototyping can be used early in the development process to better understand hardware and software partitioning decisions and determine throughput considerations associated with implementations. Early use of functional models to determine microprocessor hardware configurations and architectures, and the architecture of ASIC in development, can aid in capturing requirements, improving functional performance and expectations [6].

Citation: Huda S Muhammad and Assim Sagahyroon, Virtual Prototyping and Performance Analysis of Two Memory Architectures, DOI: 10.1155/2009/984891.

In this work, we explore the performance of the two memory architectures introduced earlier using virtual prototyping models built from parameterized library components which are part of the VisualSim Environment [7]. Essentially, VisualSim is a modeling and simulation CAD tool used to study, analyze, and validate specification and verify implementation at early stages of the design cycle.

This paper is organized as follows: in Section 2 we provide an overview of the two processors and the corresponding memory architectures. Section 3 introduces the VisualSim environment as well as the creation of the platform models for the processors. Simulation Results and the analysis of these results form Section 4 of this paper. Conclusions are summarized in Section 5.

OVERVIEW OF PROCESSORS MEMORY ARCHITECTURE

The AMD Opteron Direct Connect Architecture

The AMD's direct Connect Architecture used in the design of the dual core AMD Opteron consists of three elements:

(i) an integrated memory controller within each processor, which connects the processor cores to dedicated memory,

(ii) a high-bandwidth Hyper Transport Technology link which goes out the computer's I/O devices, such as PCI controllers,

(iii) coherent Hyper Transport Technology links which allow one processor to access another processor's memory controller and Hyper Transport Technology links.

The Opteron uses an innovative routing switch and a direct connect architecture that allows "glueless" multiprocessing between the two processor cores. Figure 1 shows an Opteron processor along with the system request queue (SRQ) and host bridge, Crossbar, memory controller, DRAM controller, and HyperTransport ports [3,8].

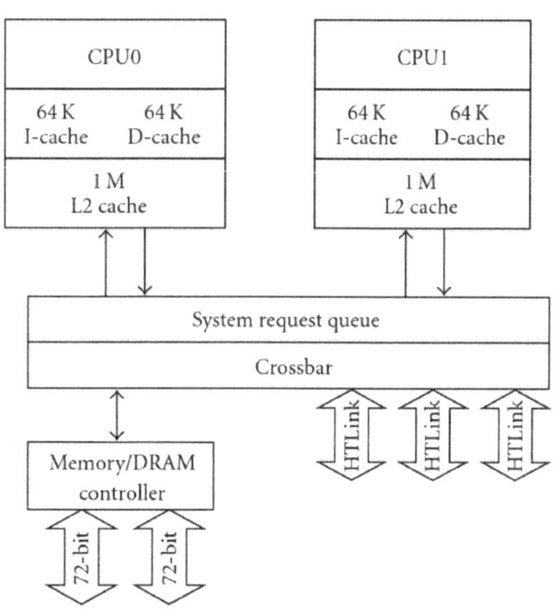

Figure 1 AMD dual core Opteron.

The Crossbar switch and the SRQ are connected to the cores directly and run at the processor core frequency. After an L1 cache miss, the processor core sends a request to the main memory and the L2 cache in parallel. The main memory request is discarded in case of an L2 cache hit. An L2 cache miss results in the request being sent to the main memory via the SRQ and the Crossbar switch. The SRQ maps the request to the nodes that connect the processor to the destination. The Crossbar switch routes the request/data to the destination node or the HyperTransport port in case of an off chip access.

Each Opteron core has a local on-chip L1 and L2 cache and is then connected to the memory controller via the SRQ and the Crossbar switch. Apart from these external components, the core consists of 3 integer and 3 floating point units along with a load/store unit that executes any load or store microinstructions sent to the core [9]. Direct Connect Architecture can improve overall system performance and efficiency by eliminating traditional bottlenecks inherent in legacy architectures. Legacy front-side buses restrict and interrupt the flow of data. Slower data flow means slower system performance. Interrupted data flow means reduced system scalability. With Direct Connect Architecture, there are no front-side buses. Instead, the processors, memory controller, and I/O are directly connected to the CPU and communicate at CPU speed [10].

Intel Xeon Memory Architecture

The Dual-Core Intel Xeon Processor is a 64-bit processor that uses two physical Intel NetBurst microarchitecture cores in one chip [4]. The Intel Xeon dual core processor uses a different memory access technique, by including a Front-Side-Bus (FSB) to the SDRAM and a shared L3 cache instead of having only on-chip caches like the AMD Opteron. The L3 cache and the two cores of the processor are connected to the FSB via the Caching Bus Controller. This controller controls all the accesses to the L3 cache and the SDRAM. Figure 2 below provides an overview of the Intel Dual core Xeon and illustrates the main connections in the processor [5].

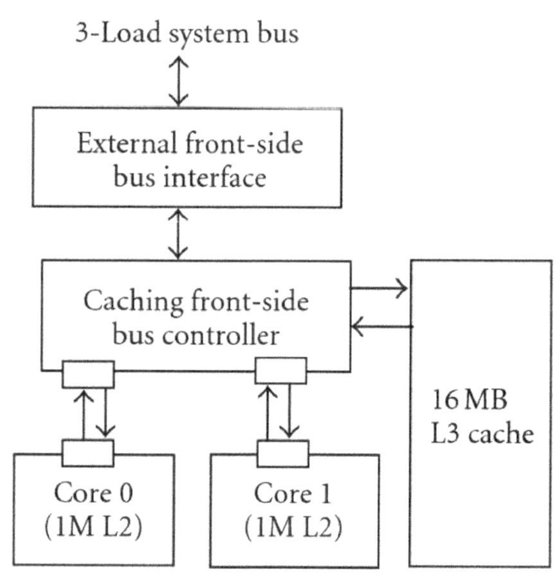

Figure 2 Block diagram of the dual-core Intel Xeon.

Since the L3 cache is shared, each core is able to access almost all of the cache and thus has access to a larger amount of cache memory. The shared L3 cache provides a better efficiency over a split cache since each core can now use more than half of the total cache. It also avoids the coherency traffic between cache in a split approach [11].

VisualSim Simulation Environment

At the heart of the simulation environment is the VisualSim Architect tool. It is a graphical modeling tool that allows the design and analysis of "digital, embedded, software, imaging, protocols, analog, control-systems, and DSP designs". It has features that allow quick debugging with a GUI and a software library that includes various tools to track the inputs/stimuli and enable a graphical and textual view of the results. It is based on a library of parameterized components including processors, memory controllers, DMA, buses, switches, and I/O's. The blocks included in the library reduce the time spent on designing the minute details of a system and instead provide a user friendly interface where these details can be altered by just changing their values and not the connections. Using this library of building blocks, a designer can for example, construct a specification level model of a system containing multiple processors, memories, sensors, and buses [12].

In VisualSim, a platform model consists of behavior, or pure functionality, mapped to architectural elements of the platform model. A block diagram of a platform model is shown in Figure 3.

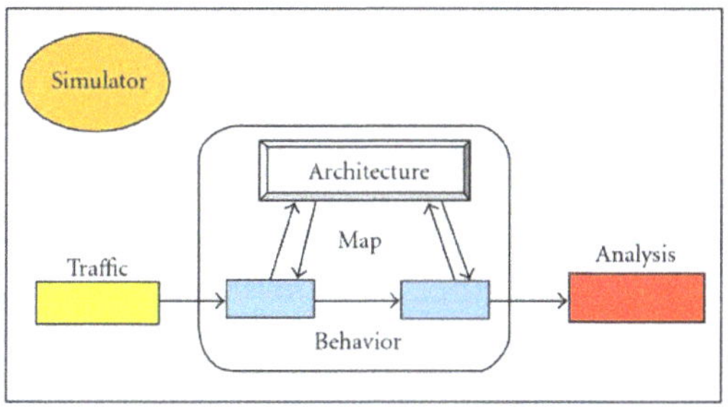

Figure 3 Block diagram of a platform model.

Once a model is constructed, various scenarios can be explored using simulation. Parameters such as inputs, data rates, memory hierarchies, and speed can be varied and by analyzing simulation results engineers can study the various trade-offs until they reach an optimal solution or an optimized design.

The key advantage of the platform model is that the behavior algorithms may be upgraded without affecting the architecture they execute on. In addition, the architecture could be changed to a completely different processor to see the effect on the user's algorithm, simply by changing the mapping of behavior to architecture. The mapping is just a field name (string) in a data structure transiting the model.

Models of computation in VisualSim support block-oriented design. Components called blocks execute and communicate with other blocks in a model. Each block has a well-defined interface. This interface abstracts the internal state and behavior of a block and restricts how a block interacts with its environment. Central to this block-oriented design are the communication channels that pass data from one port to another according to some messaging scheme. The use of channels to mediate communication implies that blocks interact only with the channels they are connected to and not directly with other blocks.

In VisualSim, the simulation flow can be explained as follows: the simulator translates the graphical depiction of the system into a form suitable for simulation execution and executes simulation of the system model, using user specified model parameters for simulation iteration. During simulation, source modules (such as traffic generators) generate data structures. The data structures flow along to various other processing blocks, which may alter the contents of Data Structures and/or modify their path through the block diagram. In VisualSim simulation continues until there are no more data structures in the system or the simulation clock reaches a specified stop time [7]. During a simulation run, VisualSim collects performance data at any point in the model using a variety of prebuilt probes to compute a variety of statistics on performance measures.

This project uses the VisualSim (VS) Architect tool, to carry out all the simulations and run the benchmarks on the modeled architectures. The work presented here utilizes the hardware architecture library of VS that includes the processor cores, which can be configured as per our requirements, as well as bus ports, controllers, and memory blocks.

Models' Construction (Systems' Setup)

The platform models for the two processors are constructed within the VS environment using the parameters specified in Table 1.

Table 1 Simulation models parameters.

	AMD Opteron	Intel Xeon
Core Speed	2 GHz	2 GHz
$Bus_{(Core-to-Cache)}$ Speed	N/A	2 GHz
$Bus_{(Core-to-SRQ)}$ Speed	2 GHz	N/A
Crossbar Speed	2 GHz	N/A
FSB Speed	N/A	1066 MHz (Width = 4 B)
$Bus_{(SRQ-to-RAM)}$ Speed	2 GHz (Width = 4 B)	N/A

L1 Cache Speed	1 GHz	1 GHz
L2 Cache Speed	1 GHz	1 GHz
RAM Speed	1638.4 MHz	1638.4 MHz
I1 Cache Size	64 kB	64 kB
D1 Cache Size	64 kB	64 kB
L2 Cache Size	4 MB (2 MB per core)	4 MB (shared cache)

The Opteron Model

The basic architecture of the simulated AMD dual core Opteron contains two cores with three integer execution units, three floating point units and two loads/stores, and branch units to the data cache [2]. Moreover, the cores contain 2 cache levels with 64 kB of L1 data cache, 64 kB of L1 instruction cache, and 1 MB of L2 cache.

The constructed platform model for the AMD Opteron is shown in Figure 4.

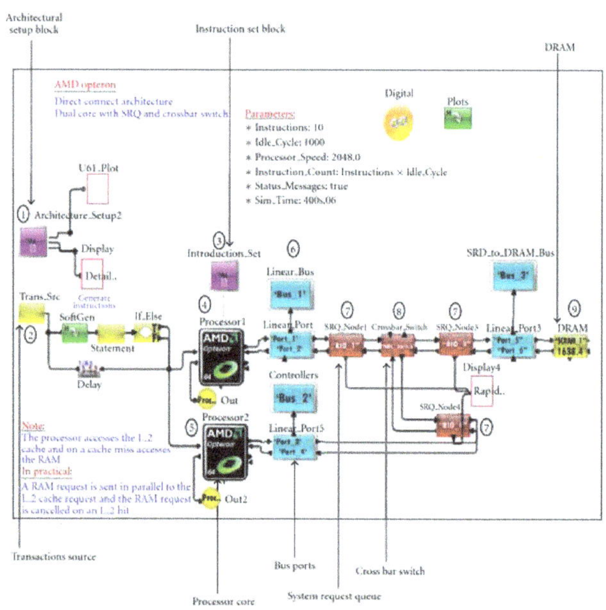

Figure 4 Platform model for the DirectConnect Architecture.

In the above model, the two large blocks numbered 4 and 5, respectively, are the Processor cores connected via bus ports (blocks 6) to the System Request Queue (block 7), and then to the Crossbar switch (block 8). The Crossbar switch connects the cores to the RAM (block 9) and is programmed to route the incoming data structure to the specified destination and then send the reply back to the requesting core.

On the left block 2 components contain the input task to the two cores. These blocks define software tasks (benchmarks represented as a certain mix of floating point, integer and load/store instructions) that are input to both the processors (Opteron and Xeon) in order to test their memory hierarchy performance. The following subsections give a detailed description of each of the blocks, their functionalities, and any simplifying assumptions made to model the memory architecture.

Architecture Setup

The architecture setup block configures the complete set of blocks linked to a single Architecture_Name parameter found in most blocks. The architecture setup block of the model (block 1) contains the details of the connections between the fields mappings of the Data Structure attributes as well as the routing table that contains any of the virtual connections not wired in the model. The architecture setup also keeps track of all the units that are a part of the model and its name has to be entered into each block that is a part of the model.

Core and Cache

Each core of Opteron implemented in the project using VS is configured to a frequency of 2 GHz and has 128 kB of L1 cache (64 kB data and 64 kB instruction cache), 2 MB of L2 cache, and the floating point, integer, and load/store execution units. This 2 MB of L2 cache per core is compatible with the 4 MB of shared cache used in the Intel Xeon memory architecture. The instruction queue length is set to 6 and instructions are included in the instruction set of both the cores, so as to make the memory access comparison void of all other differences in the architectures. These instructions are defined in the instruction block that is further described in a later section.

Certain simplifications have been made to the core of the Opteron in order to focus the analysis entirely on the memory architecture of the processor. These assumptions include the change of the variable length instructions to fixed length micro-ops [9]. Another assumption made is that any L1 cache miss does not result in a simultaneous request being sent to the L2 cache and the RAM. Instead the requests are sent sequentially, where an L1 cache miss results in an L2 cache access and finally an L2 cache miss results in a DRAM access.

Pipeline

The pipeline of the modeled Opteron consists of four stages, mainly the prefetch, decode, execute, and the store. The prefetch of each instruction begins from the L1 cache and ends in a DRAM access in case of L1 and L2 cache misses. The second stage in the pipeline is the decode stage that is implemented by introducing a delay into the entire process. The decode stage does not actually decode the instruction; instead the time required to decode the instruction is added to the time comprising of the delays from the prefetch stage to the end of the execution stage. The third stage, the execution stage, takes place in the five execution units that are present in the cores, and finally after the execution, the write-back stage writes back the specified data to the memory, mainly the L1 cache. The *Pipeline_Stages* (text box in Figure 5) shows the four pipeline stages that have been defined for both cores. It also contains the configuration of one of the cores of the Opteron along with the number of execution units and the instruction queue length. The lower text window depicts the details and actions of the pipeline stages.

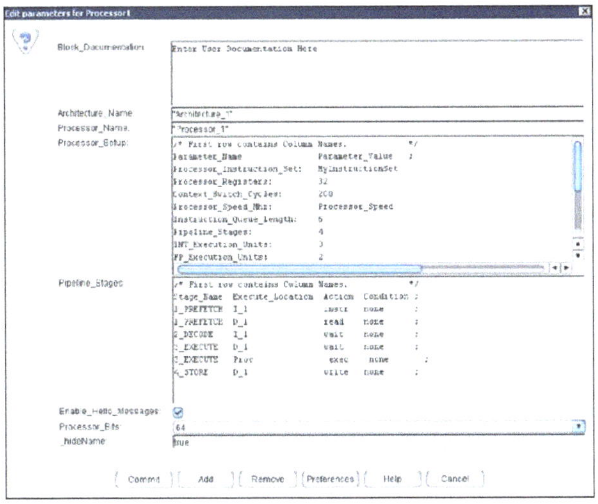

Figure 5 Opteron processor core and pipeline configurations.

Crossbar Switch and the SRQ Blocks

The Crossbar switch configured in the simulation model, is used to route the data packets to the destination specified by the *"A_Destination"* field of the data structure entering the switch. The main memory is connected to the Crossbar switch via the System Request Queue (SRQ) block both of which are implemented using the virtual machine scripting language available in the VisualSim environment. The SRQ accepts only 8 requests in the queue and does not entertain any further requests until there is an empty space in the queue. Each core and the SDRAM are connected to individual SRQ blocks that are in turn connected to the crossbar switch. Figure 6 shows the crossbar switch as the *NBC_Switch* and the SRQ blocks as the *RIO_IO_Nodes* which are linked to the bus ports connected to the SDRAM and the processor cores. Figure 4 provides a general overview of the crossbar switch and the SRQ nodes in context of the entire model.

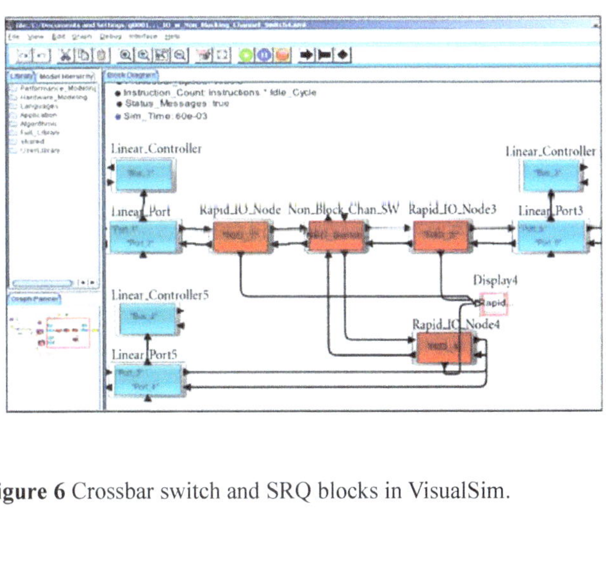

Figure 6 Crossbar switch and SRQ blocks in VisualSim.

Main Memory and the Memory Controller

In the simulation model, the RAM has a capacity of 1 GB and has an in-built memory controller configured to run at a speed of 1638.4 MHz. At this speed and a block width of 4 bytes, the transfer of data from the memory to the cache takes place at a speed of 6.4 GB/s. This rate is commonly used in most of the AMD Opteron processors but can be different depending on the model of the processor. The same rate is also used in the model of the Xeon processor. Each instruction that the RAM executes is translated into delay specified internally by the memory configurations. These configurations are seen in Figure 7 in the *Access_Time* field as the number of clock cycles spent on the corresponding task.

The SDRAM connects to the cores via the SRQ blocks and the Crossbar switch which routes the SDRAM requests from both the cores to the main memory block and then sends a response back to the requesting core, in terms of a text reply. This process requires a certain delay that depends on the type of instruction sent to the SDRAM. In case the SRQ block queue is empty, a single DRAM response time depends on whether the instruction is a memory read, write, read/write, or erase instruction. Each of these instructions takes a fixed number of clock cycles to complete and is determined in the SDRAM configuration as determined by the *Access_Time* field seen in Figure 7. To separate the SDRAM access time from the cache access time, a simplification is made such that the SDRAM access request from the core is not sent in parallel to an L2 cache request as in the actual Opteron; instead, the SDRAM request is issued only after an L2 miss has been encountered.

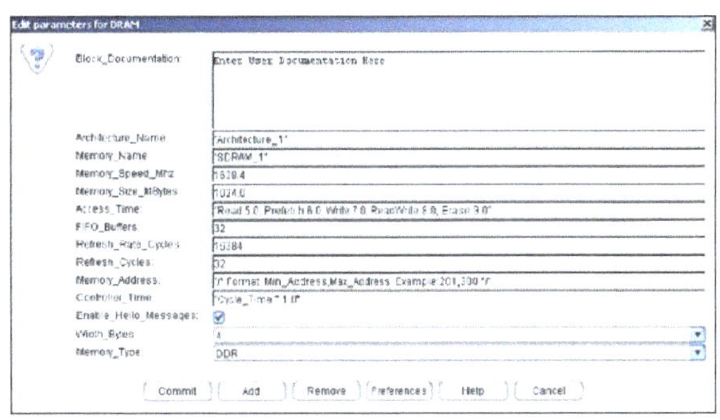

Figure 7 SDRAM and memory controller configurations.

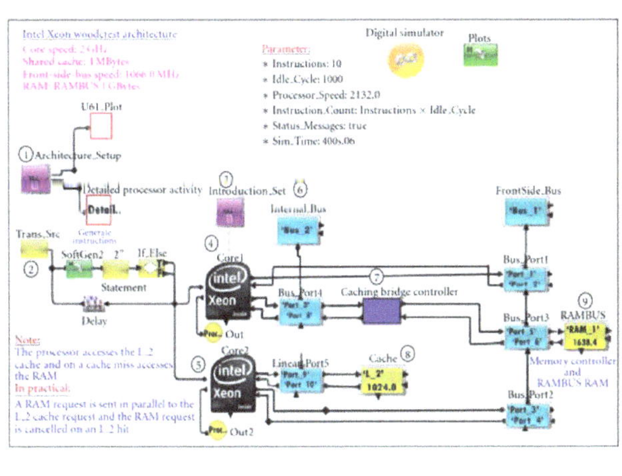

Figure 8 Shared Bus Architecture.

Xeon Model

Basic Architecture

The basic architecture of the Intel Dual Core Xeon is illustrated in Figure 2. The corresponding platform model is depicted in Figure 8. The two cores of the processor are connected to the shared L2 cache and then via the Front-Side-Bus (FSB) interface to the SDRAM. The modeled Intel Xeon processor consists of two cores with three integer execution units, three floating point units, and two loads/stores and branch units to the data cache. The same specifications used to model the Opteron cores in VisualSim are used here as well. Besides, each core is configured with 64 kB of L1 data cache, 64 kB of L1 instruction cache, whereas the L2 cache is a unified cache and is 4 MB in size. The FSB interface, as seen in Figure 8, was constructed using the Virtual Machine block in VS [7] and is connected to the internal bus which links the two cores to the RAM via the FSB. The software generation block (block 2 on the left side of the model) contains the same tasks as the Opteron.

Architecture Setup

The architecture setup block of the model of the Xeon (Figure 8—block 1) is the same as the one implemented in the Opteron and the field mappings of the Data Structure attributes are copied from the Opteron model to ensure that no factors other than the memory architecture affects the results.

Core and Cache

The core implementation of the Xeon is configured using VS to operate at a frequency of 2 GHz and has 128 kB of L1 cache (64 kB data and 64 kB instruction cache), 4 MB of unified and shared L2 cache [5], floating point, integer, and load/store execution units. Here as well, the instruction queue length is set to 6 and instructions are included in the instruction set of both the cores, so as to make the memory access comparison void of all other differences in the architectures. These instructions are defined in the instruction block that is described in a later section.

Certain simplifications have been made to the core of the Xeon in order to focus the analysis entirely to the memory architecture of the processor. The assumption made in accessing the memory is that any L1 cache miss does not result in a simultaneous request being sent to the L2 cache and the RAM. Instead the requests are sent sequentially, where an L1 cache miss results in an L2 cache access and finally an L2 cache miss results in a RAM access.

To simplify the model and the memory access technique, the process of snooping is not implemented in this simulation, and similar to the Opteron, no parallel requests are sent to two memories.

Pipeline

The pipeline of the modeled Xeon consists of four stages (similar to the Opteron model), the prefetch, decode, execute, and the store. The prefetch of each instruction begins from the L1 cache and ends in a RAM access in case of L1 and L2 cache misses. The second stage in the pipeline is the decode stage that is mainly translated into a wait stage. The third stage, the execution stage, takes place in the five execution units that are present in the cores, and finally after the execution, the write-back stage writes back the specified data to the memory, mainly the L1 cache.

Caching Bridge Controller (CBC)

The CBC, block 7 of the model is simply a bridge that connects the L2 shared cache to the FSB [13]. This FSB then continues the link to the RAM (block 9) from which accesses are made and the data/instruction read is sent to the core that requested the data. The CBC model is developed using the VisualSim scripting language and simulates the exact functionality of a typical controller.

Main Memory and the Memory Controller

The dual-core Xeon contains a main memory of type RAMBUS with a speed similar to the memory connected to the Opteron. The size of the RAM is 1 GB and contains a built-in memory controller. This memory controller is configured to run at a speed of 1638.4 MHz. At this speed and a block width of 4 bytes, the transfer of data from the memory to the cache takes place at a speed of 6.4 GB/s. Each instruction that the RAMBUS will carry out implies a certain delay which has been specified internally in the memory configurations. These configurations are seen in Figure 9 in the *Access_Time* field as the number of clock cycles spent executing the corresponding task. The RAM connects to the cores via the CBC and data or instruction requests to the RAM from either core are sent to the main memory block via the FSB. The RAM then sends a response back to the requesting core which can be seen as a text reply on the displays that show the flow of requests and replies. DRAM Access Time is the time taken since a request is made and when the data is made available from the DRAM. It is defined in nanoseconds by the user for each operation like Read, Write, or a Read-Write as an access time parameter in the *Access_Time* field of Figure 9.

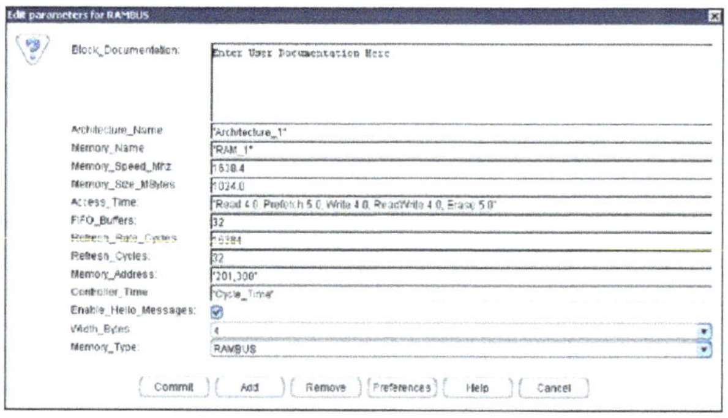

Figure 9 Main memory and controller configurations of the Intel Xeon.

RESULTS AND ANALYSIS

Following a series of experimental tests and numerical measurements using benchmarking software, published literature [14, 15, 16] discusses the performance of the AMD Opteron when compared to the Xeon processor using physical test beds comprised of the two processors. These three references provide the reader with a very informative and detailed comparison of the two processors when subjected to various testing scenarios using representative loads.

In this work, we are trying to make the case for an approach that calls for early performance analysis and architectural exploration (at the system level) before committing to hardware. The memory architectures of the above processors were used as a vehicle. We were tempted to use these architectures by the fact that there were published results that clearly show the benefits of the Opteron memory architecture when compared to the Xeon FSB architecture and this would no doubt provide us with a reference against which we can validate the simulation results obtained using VisualSim.

Additionally, and to the best of our knowledge, we could not identify any published work that discusses the performance of the two memory architectures at the system level using an approach similar to the one facilitated by VisualSim.

Using VisualSim, a model of the system can be constructed in few days. All of the system design aspects can be addressed using validated parametric library components. All of the building blocks, simulation platforms, analysis, and debugging required to construct a system are provided in a single framework.

Synopsys integrated Cossap (dynamic data flow) and SystemC (digital) into System Studio while VisualSim combines SystemC (digital), synchronous data flow (DSP), finite state machine (FSM), and continuous time (analog) domains. Previous system level tools typically supported a single modeling specific domain. Furthermore, relative to prior generations of graphical modeling tools, VisualSim integrates as many as thirty bottom-up components functions into a single system level, easy to use, reusable blocks, or modules.

Finally, it is worth mentioning that results obtained using the VisualSim environment in this work are generally in line with results and conclusions found in the literature [14, 15, 16].

In the work reported here, Simulation runs are performed using a Dell GX260 machine with a P4 processor running at 3.06 GHz, and a 1 Gbyte RAM.

For simulation purposes and to test the performance of both architectures, traffic sequences are used to trigger the constructed models. These sequences are defined data structures in VisualSim; a traffic generator emulates application-specific traffic. The *Transaction_Source* block in Figures 4 and 8 is used to generate tasks that are applied to the processors as input stimuli. These tasks are benchmarks consisting of a varied percentage mix of integer, floating-point, load/store, and branch instructions. The different percentages are inserted into the software generator's *Instruction_Mix* file and supplied to the processor cores. Thirty three tasks (Table 2) were generated and used to assess the performance of the DirectConnect and the Shared Bus architectures.

Table 2 Benchmark tasks [12].

Model task name	**Actual task name**
Task_0	DFT
Task_1	DFT
Task_2	DFT
Task_3	CS_Weighting
Task_4	IR
Task_5	Q_Taylor_Weighting
Task_6	CS_Weighting
Task_7	IR
Task_8	Q_Taylor_Weighting
Task_9	CS_Weighting
Task_10	IR
Task_11	Q_Taylor_Weighting
Task_12	CS_Weighting
Task_13	IR
Task_14	Q_Taylor_Weighting
Task_15	CS_Weighting
Task_16	IR
Task_17	Q_Taylor_Weighting
Task_18	CS_Weighting
Task_19	IR
Task_20	DFT
Task_21	DFT
Task_22	DFT
Task_23	DFT
Task_24	DFT
Task_25	DFT
Task_26	DFT
Task_27	DFT
Task_28	DFT
Task_29	DFT
Task_30	DFT
Task_31	DFT
Task_32	DFT

Table 3 shows the "Task Latencies" and "Cycles/Task" measured for both processors. Since the models were constructed by having similar configurations in almost every aspect except the memory architecture, the values obtained are therefore a reflection of the performance of that architecture. Figure 10 is a graph of the "Latency per Task" results for each configuration. This graph is plotted using the total number of clock cycles that were taken by each task. The instructions contained in each of these tasks are defined in the software_generation block. At the end of the execution, a text file is generated by VisualSim that contains the number of cycles that were taken by each of the task, including the stall time and the delays caused by the components linking the RAM to the cores.

Table 3 Task latencies and Cycles/Task.

Task_Name	$Latency_{(Opteron)}$	$Latency_{(Xeon)}$	Cycles / $Task_{(Opteron)}$	Cycles / $Task_{(Xeon)}$
Task_0	1.21E-7	1.27E-7	247	260
Task_1	7.713E-7	7.99E-7	1579	1636
Task_2	7.235E-7	7.299E-7	1481	1494
Task_3	6.538E-7	6.605E-7	1338	1352
Task_4	8.538E-7	1.027E-6	1748	2103
Task_5	7.358E-7	7.415E-7	1506	1518
Task_6	7.453E-7	7.52E-7	1526	1540
Task_7	8.538E-7	1.0222E-6	1748	2093
Task_8	7.362E-7	7.43E-7	1507	1521
Task_9	6.783E-7	7.051E-7	1389	1444
Task_10	8.538E-7	1.0241E-6	1748	2097
Task_11	6.858E-7	7.09E-7	1404	1452
Task_12	7.1629E-7	7.216E-7	1466	1478
Task_13	8.537E-7	1.0179E-6	1748	2084
Task_14	7.298E-7	7.365E-7	1494	1508
Task_15	7.133E-7	7.196E-7	1460	1473
Task_16	8.537E-7	1.0251E-6	1748	2099
Task_17	7.11299E-7	7.169E-7	1456	1468
Task_18	6.862E-7	6.92E-7	1405	1417
Task_19	1.5795E-6	1.9301E-6	3234	3952
Task_20	7.4229E-7	7.482E-7	1520	1532
Task_21	2.6452E-6	3.3117E-6	5417	6782
Task_22	7.4649E-7	7.769E-7	1528	1591
Task_23	2.6454E-6	3.3002E-6	5417	6758
Task_24	7.4529E-7	7.52E-7	1526	1540
Task_25	2.6451E-6	3.3039E-6	5417	6766
Task_26	7.523E-7	7.58E-7	1540	1552
Task_27	2.645E-6	3.3078E-6	5416	6774
Task_28	7.458E-7	7.524E-7	1527	1540
Task_29	2.6453E-6	3.3313E-6	5417	6822
Task_30	7.9129E-7	7.976E-7	1620	1633
Task_31	2.6453E-6	3.2723E-6	5417	6701
Task_32	7.558E-7	7.6189E-7	1547	1560

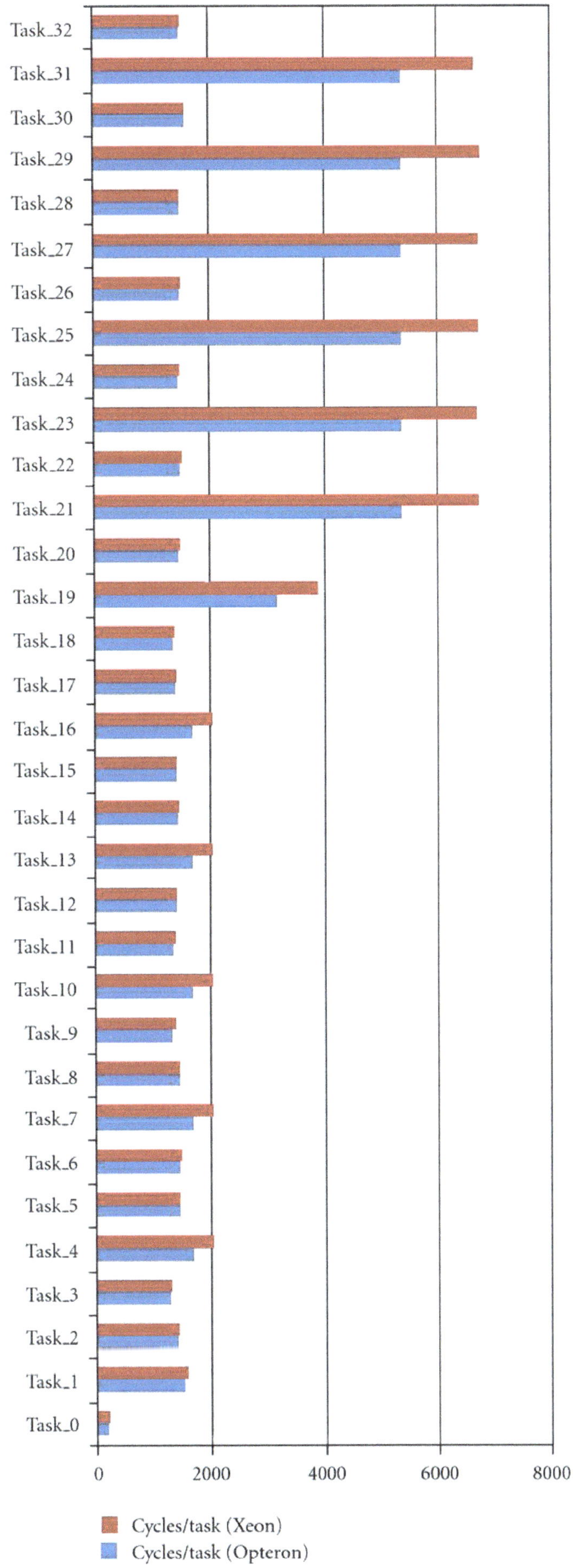

Figure 10 Latency per Task (Cycles per Task).

Cleary, the Xeon has exhibited a higher latency in almost every task. These numbers show that the time taken to execute the same task in the Xeon were more than the Opteron.

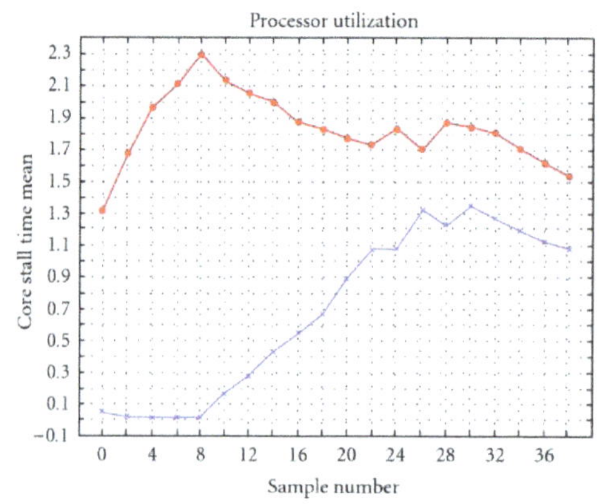

Figure 11 Processor Stall Times (Opteron).

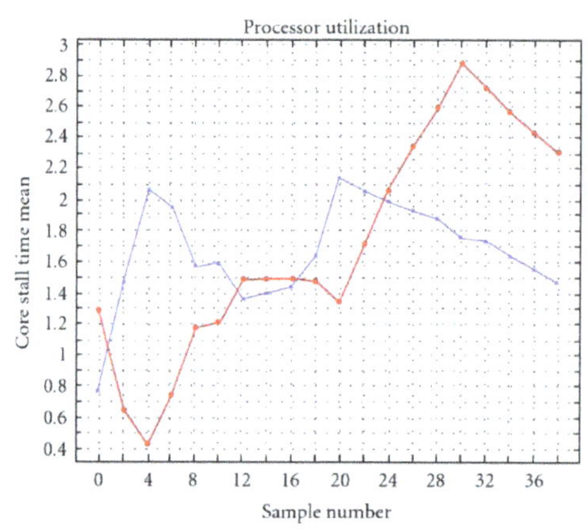

Figure 12 Processor Stall Times (Xeon).

Figures 11 and 12 show a graph of processors' stall times. In both cases, 20 samples are taken during the entire simulation period and the data collected is used in the depicted graphs. During the execution of all the tasks, the maximum time for which the processors stalled was different for each kind of architecture. The maximum stall time for the DirectConnect architecture was 2.3 microseconds whereas for the Shared Bus architecture the maximum stall time was 2.9 microseconds. Due to the shared bus in the Xeon's architecture, delays were greater than the DirectConnect approach of the Opteron, and thus the difference in the stall time.

As the models described earlier suggest, the Opteron follows the split cache approach where each core in the processor has its own L1 and L2 cache; thus no part of the cache is shared between the two cores. On the contrary, the Xeon processor employs the shared cached technique and thus both the cores have access to a larger amount of cache than the ones in the Opteron. Whenever one of the cores in the Xeon is not accessing the shared cache, the other core has complete access to the entire cache which results in a higher hit ratio.

Table 4 shows the hit-ratios of both the architecture models. As the values suggest, the hit-ratios of the Xeon/Shared cache are higher than those of the Opteron/split cache, the reason being discussed above. It is worth mentioning that few simplifying assumptions were made to the memory architectures operation of both processors as discussed in Section 3.

Table 4 Hit ratios.

Hit Ratios (%)	Opteron	Xeon
Processor_1_D_1_Hit_Ratio_Mean	97.23	98.37
Processor_1_I_1_Hit_Ratio_Mean	92.13	95.11
Processor_2_D_1_Hit_Ratio_Mean	98.98	99.92
Processor_2_I_1_Hit_Ratio_Mean	95.21	96.14
L_2_Hit_Ratio_Mean	N/A	96.36

CONCLUSIONS

In this work, we utilized a system modeling methodology above the detailed chip implementation level that allows one to explore different designs without having to write Verilog, VHDL, SystemC, or simply C/C++ code. This approach contributes to a considerable saving in time and allows for the exploration and assessment of different designs prior to implementation.

Since predictability of performance is critical in microprocessors design, simulation models can be used to evaluate architectural alternatives and assist in making informed decisions. Simulation is an acceptable performance modeling technique that can be used to evaluate architectural alternatives and features. In this work, we used Virtual System Prototyping and simulation to investigate the performance of the memory subsystems of both, the Opeteron, and the Xeon dual core processors.

Simulation results indicate that the Opteron has exhibited better latency than the Xeon for the majority of the tasks. In all cases, it either outperformed the Xeon or at least had similar latency. This demonstrates that using an FSB as the only means of communication between cores and memory has resulted in an increase in stalls and latency. On the other hand, in the DirectConnect Architecture, the cores being directly connected to the RAM, via the crossbar switch and the SRQ which were running at processor speed, had minimal delays. Each RAM request from either of the cores was sent individually to the SRQ blocks and they were routed to the RAM that had its memory controller on-chip and the cores did not have to compete for a shared resource.

REFERENCES

1. Balakrishnan S, Rajwar R, Upton M, Lai K: The impact of performance asymmetry in emerging multicore architectures. Proceedings 32nd International Symposium on Computer Architecture (ISCA '05), June 2005, Madison, Wis, USA 506-517.
2. Conway P, Hughes B: The AMD Opteron Northbridge architecture. IEEE Micro 2007,27(2):10-21.
3. Keltcher CN, McGrath KJ, Ahmed A, Conway P: The AMD Opteron processor for multiprocessor servers. IEEE Micro 2003,23(2):66-76. 10.1109/MM.2003.1196116
4. Dual-Core Intel® Xeon® Processor 5000 Series Datasheet May 2006
5. Varada R, Sriram M, Chou K, Guzzo J:Design and integration methods for a multi-threaded dual core 65nm XeonÂ https://static-content.springer.com/image/art%3A10.1155%2F2009%2F984891/MediaObjects/13639_2009_Article_213_IEq10_HTML.gif processor. Proceedings of IEEE/ACM International Conference on Computer-Aided Design, Digest of Technical Papers (ICCAD '06), November 2006, San Jose, Calif, USA 607-610.
6. Prowse-Fosler L: Virtual Systems Prototyping Ensures Reusable Design Platforms. http://www.vastsystems.com/solutions-technical-papers.html
7. Mirabilis Design User's Manual vol. 1, http://www.mirabilisdesign.com/
8. HyperTransport I/O Link Specification http://www.hypertransport.org/
9. http://www.chiparchitect.com/
10. http://www.amd.com/us-en/Processors/
11. Rusu S, Tam S, Muljono H, et al.:A 65-nm dual-core multithreaded Xeon https://static-content.springer.com/image/art%3A10.1155%2F2009%2F984891/MediaObjects/13639_2009_Article_213_IEq11_HTML.gif processor with 16-MB L3 cache. IEEE Journal of Solid-State Circuits 2007,42(1):17-24.
12. Shankar D: Selecting Memory Controllers for DSP Systems. DSP DesignLine, May 2007
13. Tarn S, Rusu S, Chang J, Vora S, Cherkauer B, Ayers D:A 65nm 95W dual-core multi-threaded Xeon https://static-content.springer.com/image/art%3A10.1155%2F2009%2F984891/MediaObjects/13639_2009_Article_213_IEq12_HTML.gif processor with L3 cache. Proceedings of IEEE Asian Solid-State Circuits Conference (ASSCC '06), November 2006, Hangzhou, China 15-18.
14. Peng L, Peir J-K, Prakash TK, Chen Y-K, Koppelman D: Memory performance and scalability of Intel's and AMD's dual-core processors: a case study. Proceedings of the 27th IEEE International Performance, Computing, and Communications Conference (IPCCC '07), April 2007, New Orleans, La, USA 55-64.
15. Guiang CS, Milfeld KF, Purkayastha A, Boisseau JR: Memory performance of dual-processor nodes: comparison of Intel Xeon and AMD Opteron memory subsystem architectures. Proceedings for ClusterWorld Conference & Expo, 2003
16. Advanced Micro Devices: AMD Opteron vs Intel Xeon Processor Performance Comparison http://www.veritest.com/

CHAPTER 11

A QUANTITATIVE EVALUATION OF A NETWORK ON CHIP DESIGN FLOW FOR MULTI-CORE CONSUMER MULTIMEDIA APPLICATIONS

Andreas Hansson and Kees Goossens

Keywords: System on Chip, Network on Chip, Design flow, Case study

INTRODUCTION

Systems on Chip (SoC) grow in complexity with an increasing number of independent applications on a single chip [67]. The applications are realised by hardware and software Intellectual Property (IP), e.g. processors and application code, and many applications have real-time requirements on their end-to-end behaviour [15]. One of the major challenges in SoC design is ensuring that all application real-time requirements are fulfilled under all circumstances [25]. In a firm real-time application, such as a software-defined radio [53], deadline misses are highly undesirable due to standardisation, e.g. upper bounds on the response latency in wireless standards, or steep quality reduction in the case of misses. Soft real-time applications, e.g. an MPEG-2 decoder [55], can tolerate occasional deadline misses with only a modest quality degradation. The growing number of applications and the increasing requirements require a system architecture that is scalable on the physical, architectural and functional level [39]. Networks on Chip (NoC) are proposed to overcome the scalability issues of traditional interconnects like crossbars and buses [18]. NoCs address the physical scalability by splitting the interconnect into multiple segments with relaxed timing requirements [16, 64] and offer architectural scalability through a structured and modular architecture [6]. It is thus possible to implement a network that scales with the number of IPs in the design [43, 72]. NoCs also offer guaranteed services, i.e. latency and throughput bounds, through various resource-reservation schemes [3, 9, 26, 40, 45, 50, 59, 60, 66, 74]. However, physical and architectural scalability coupled with the availability of guaranteed services does not necessarily imply that the interconnect scales with increasing requirements. The difference between architectural scalability and functional scalability lies in the ability to not only provide a hardware capability, but actually translate that capability into useful work, i.e. satisfied requirements. Compare this situation to e.g. a modern general-purpose multi-core computer architecture where advanced compiler technology is needed to exploit the hardware [20]. Similarly for a NoC, we need design tools that can turn requirements into an interconnect instance. While mature tooling exists for the computation and storage elements of the SoC, interconnect design tools is an active research area [25]. Compared to traditional interconnects, a NoC is typically tailored for a given set of applications, and one of the more complex parts of the SoC. As such, it also plays an important role in defining the area and power consumption, and design effort of the overall system [48]. Thus, a operational tool flow is a critical, but often overlooked, part of on-chip interconnect design, and key to functional scalability. As the main contributions of this work, we: (1) describe a complete design flow to instantiate and verify application-specific interconnect instances, (2) quantitatively evaluate the entire architecture and design flow using the communication requirements of two largescale industrial case studies, and (3) investigate the functional scalability of the proposed interconnect by looking at the cost/performance ratio of hundreds of complete interconnect instances as the requirements change. Our two examples are both high-end chips with a high degree of IP integration. The first example resembles a digital TV set-top box ASIC in the line of designs described in [24, 41]. The second example is based on an automotive info-tainment ASIC [51, 61]. In both examples,

---

Citation: Andreas Hansson, Kees Goossens, A quantitative evaluation of a Network on Chip design flow for multi-core consumer multimedia applications, DOI 10.1007/s10617-011-9073-7.

the architecture consists of one or more microcontrollers, supported by several domain- or function-specific hardware cores, needed to achieve a high computational performance at an acceptable power consumption. The remainder of this paper is structured as follows. First we review related work in Sect. 2, also highlighting the important properties of our interconnect architecture. Then, Sect. 3 gives an overview of the proposed design flow, using the digital TV SoC as the running example. We continue with an in-depth evaluation of the TV SoC in Sect. 4, followed by the automotive system in Sect. 5. Finally, we conclude in Sect. 6.

RELATED WORK

We review related work in a bottom-up fashion, based on the levels of scalability. First, we look at how different interconnects address physical scalability, i.e. the ability to scale the interconnect to larger die sizes (transistor count, wire delay) without impairing the performance. Thereafter we raise the level and look at architectural scalability, i.e. the ability to add more IPs without negatively affecting the performance. Lastly, we turn to the functional scalability, i.e. the ability to translate a growing number of (more demanding) application requirements into a hardware and software interconnect instance

Physical scalability

Many networks [26, 45, 50, 60] rely on a (logically) globally synchronous clock. While techniques such as link pipelining have been proposed to overcome link latency [64, 70], the cycle-level synchronicity severely constrains the clock distribution [78] and negatively affects the scalability [12]. A range of asynchronous networks [3, 10, 66] completely remove the need for a clock, and aim to leverage the characteristics of asynchronous circuits to reduce power and electromagnetic emissions. While offering physical scalability, these networks rely on not yet well-established design methodologies and verification techniques. They also complicate the provision of guaranteed services. To overcome the disadvantages of global synchronicity, but still enable a traditional synchronous design style, the networks in [12, 27, 36, 47, 59, 75, 78] use mesochronous links between synchronous network elements. By allowing local phase differences, clock distribution and placement is greatly simplified [78]. Moreover, the global phase difference is allowed to grow with chip size. The constraints on the clock distribution within the interconnect apply also to the (global) control infrastructure. Thus, dedicated interconnects [49, 80] that rely on global synchronicity impair the overall physical scalability and must be avoided. By reusing one and the same interconnect for both user data and control, as proposed in [28, 60], the problem is reduced to providing one scalable interconnect. In addition to relaxing the constraints on the clock distribution within the interconnect and control infrastructure, it is crucial for physical scalability that IPs are able to use independent clocks (from each other as well as the interconnect). While asynchronous networks natively offer a Globally Asynchronous Locally Synchronous (GALS) design approach at the IP level, e.g. through the use of pausable clocks [4], synchronous and mesochronous networks must add Clock Domain Crossings (CDC) at the interfaces to the IPs. Solutions based on bi-synchronous FIFOs are proposed in [11, 26, 29, 59]. Such a clock domain crossing, e.g. [76], is robust with regards to metastability and offers a simple protocol between synchronous modules, while still allowing every module's frequency and voltage to be set independently [42]. In this work we use the logically synchronous Æthereal [26] network with mesochronous links [34]. Figure 1 shows a part of an example instance showing the architectural building blocks, e.g. routers (R), Network Interfaces (NI), shells, and buses. The blocks are presented in detail in [29] and we return to discus their functionality in Sect. 3.1. The mesochronous (and possibly pipelined) links inside the network alleviate the designer from strict requirements on clock skew and link delay, thus enabling an effective distributed placement of the

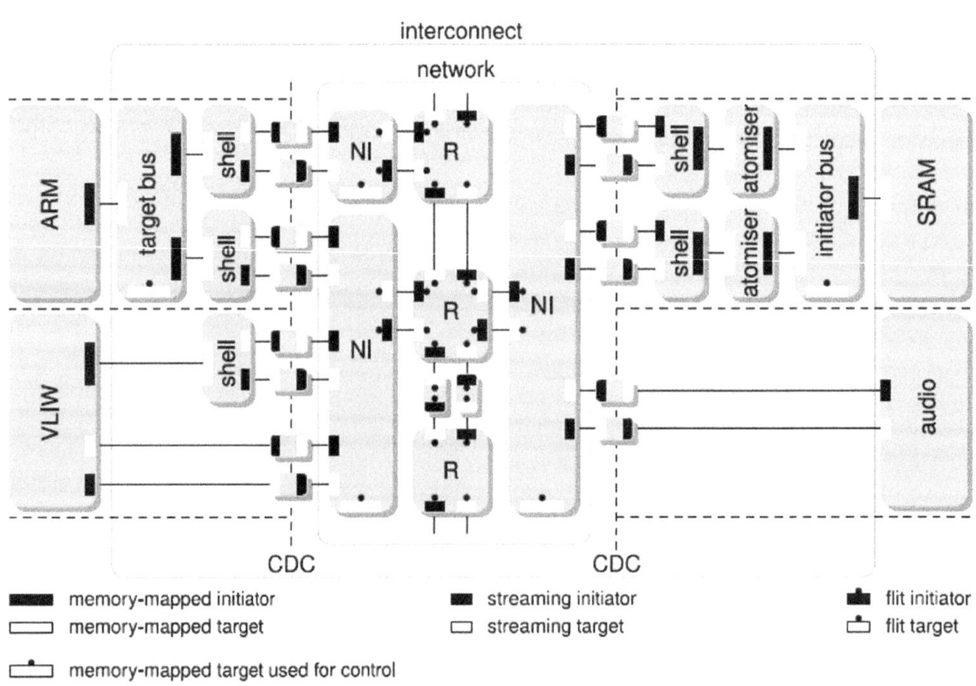

Figure. 1 Interconnect architecture overview.

network components even for larger die sizes. Furthermore, clock domain crossings between the network and the IPs enable a GALS design approach. The same interconnect is used for both data and control, as described in [28].

Architectural scalability

In bus-based interconnects, e.g. the Silicon Backplane [79], the performance decreases as the amount of sharing increases. In crossbar-based interconnects, e.g. Prophid [44], a growing number of IPs lead to a rapid growth in silicon area and wire density. This in turn causes layout issues (e.g. timing and packing) and negatively affects performance [64]. Although the problem can be mitigated by using a partial crossbar [54], the key to architectural scalability is a distributed multi-hop interconnect. In a multi-hop interconnect, e.g. Sonics MX [69] that uses a hybrid shared-bus and crossbar approach, the individual buses and crossbars are not scalable on their own. Rather, the architectural scalability comes from the ability to arbitrarily add more buses and crossbars without negatively affecting the performance of other parts of the interconnect. A segmented bus [16, 62] achieves architectural scalability by partitioning a bus into two or more segments, operating in parallel, and connected by bridges. NoCs extend on the concepts of segmented buses and offer a structured approach to multi-hop interconnects [6, 18]. From the perspective of architectural scalability, there is an important distinction to be made between NoCs where arbitration is performed at every hop (routers), e.g. [3, 10, 59, 64, 66] and those where arbitration only takes place only at the edges (NIs) of the network, as done in Nostrum [50], aSoC [45], Æthereal [26], and TTNoC [60]. The works in [3, 10, 59, 64, 66] use virtual channels to provide connection-based services. As a consequence, the router area grows quadratically with the number of connections passing through the router. Moreover, the maximum frequency of the router is also reduced, and the router is likely to be part of the critical path of the network. In contrast to the NoCs that rely on virtual channels, the NoCs in [26, 45, 50, 60] all have stateless [71] routers. Thus, the router is independent of the number of connections in terms of area (number and depth of buffers) and speed (arbitration). This moves the router out of the critical path and pushes the issue of scalability to the NIs, which can be heavily pipelined without being in the critical path [29]. In this work we use a small and fast router design [34], with no restrictions on the network topology. The architectural scalability stems from the ability to arbitrarily add more links, routers, NIs, shells and buses without negatively affecting the services of the existing components [29]. Memory-mapped initiator ports that use distributed memory communication, e.g. the ARM in Fig. 1, may even use different NIs for the connections to different targets. Similarly, a memory-mapped target port that is shared by multiple initiators may distribute the connections across multiple NIs. Thus, as we shall see in Sects. 4 and 5, it is possible to push the point of contention all the way to the IP ports and offer arbitrarily low latency and high throughput in the interconnect, limited only by the intrinsic throughput and latency of a single link.

Functional scalability

Much work focus on generating, exploring, evaluating, formalising, and comparing NoC architectures and instantiations [2, 5, 8, 14, 22, 25, 38, 39, 45, 48, 52, 63]. However, only a few works address guaranteed services [25, 39, 45], and most works are limited to evaluations of a single small-scale example, or only a few steps of the design flow, as exemplified by the overview in [48]. Moreover, the evaluation of the design flows is predominantly done using high-level simulation or analytical models rather than actual implementations. A few NoC design flows have been demonstrated on FPGA [2, 22, 33, 43, 52], predominantly using simple media-centric applications like M-JPEG [33] and MPEG-2 [43]. In addition to FPGAs, complete NoC-based SoC ASICs are presented in [7, 17, 72] together with a post-layout evaluation. However, neither of the works evaluate the scalability and instead focus on a single design point. Moreover, rather than considering the IPs given, the works in [2, 22, 43, 52, 72] customise the entire system to fit with the NoC, e.g. by modifying the IPs [43] or adding NoC-specific instructions [72]. In an industrial context this is often not possible and the interconnect has to accommodate a wide range of IP interfaces [29]. Most works also do not provide any real-time guarantees. A number of works move from the architecture and connection level and look at performance analysis of actual applications when mapped to the NoC [13, 35, 68]. By modelling the applications, the works also solve the problem of buffer sizing, which is closely linked to the end-to-end performance requirements. Probabilistic analysis, using stochastic selfsimilar traffic models, is used in [13, 68], whereas [35] is based on dataflow models and analysis proposed in [77]. The two approaches are complementary and can be used in parallel in our current work, depending on the application. All works [13, 35, 68] focus on application models and buffer sizing and do not discuss the implications on scalability. The composable and predictable multi-processor SoC presented in [33] demonstrates an interconnect design flow, based on [25], that is able to provide real-time guarantees, also on the application level [35]. The work focuses on the qualitative concepts and studies a few applications in detail. No evaluation is made of the scalability in terms of requirements. In this work, we focus on quantitatively proving functional scalability of NoC design, using large-scale examples that represent state-of-the-art SoCs for consumer multimedia applications. We consider the IPs and the requirements given and demonstrate a complete flow that is able to satisfy the real-time requirements. Using the examples, we evaluate the

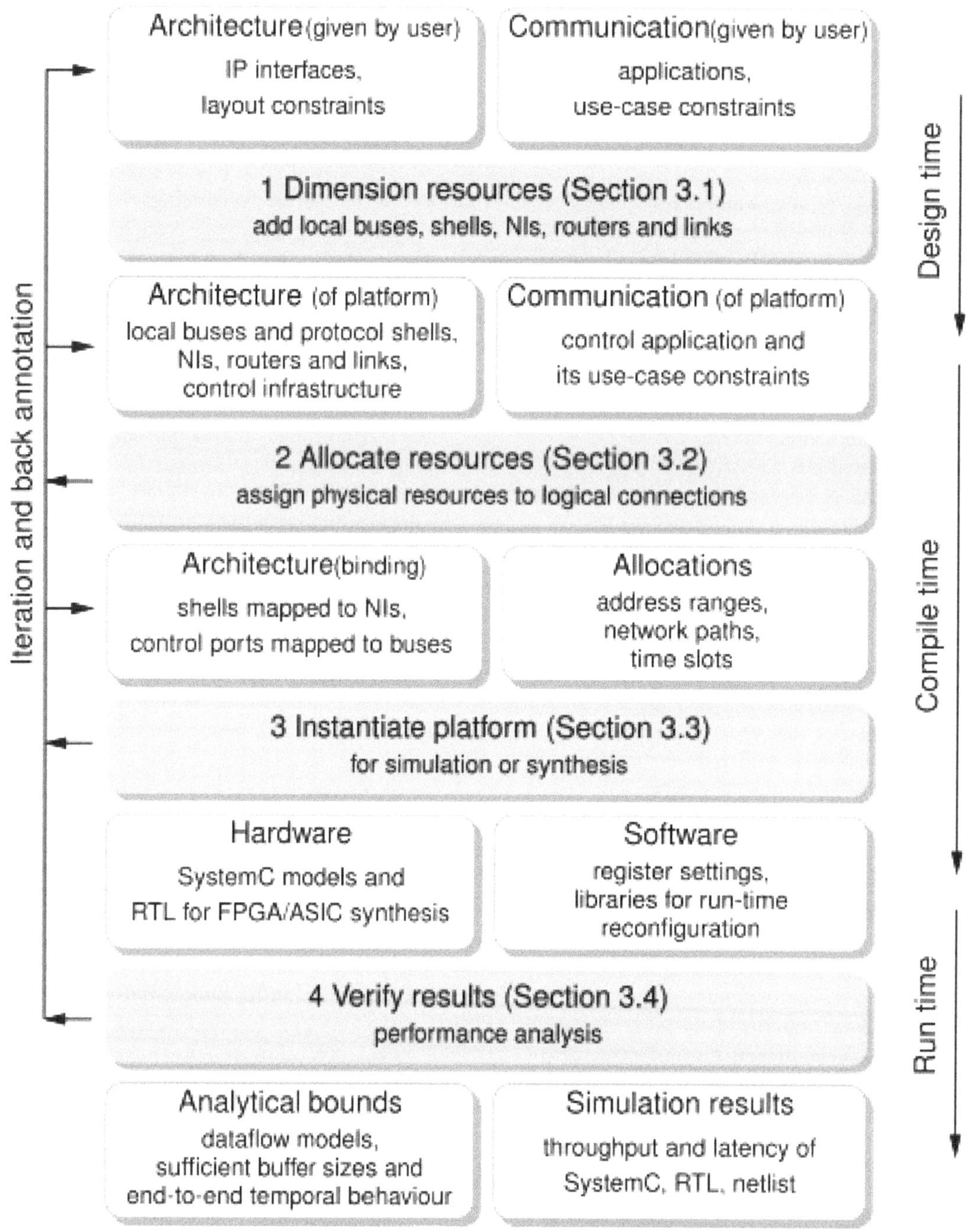

Figure. 2 Interconnect hardware and software design flow.

quantitative aspects of the proposed interconnect and also look in greater detail on the scalability of the hardware architecture as well as the design flow, about which more presently.

Design flow

One of the main contributions of this work is the design flow depicted in Fig. 2, introduced here and discussed in detail in Sects 3.1 through 3.4. Note that many of the steps are previously published in part, but no complete overview is given, and the design flow as a whole is not evaluated. In the figure, white boxes denote specifications (stored in XML files [23]) and the grey boxes denote (collections of) tools that manipulate the specifications. As shown in the figure, the flow takes as its starting point the specification of the physical interfaces that are used by the IPs, i.e. everything outside the interconnect in Fig. 1. An example (based on the digital TV SoC in Sect. 4) is shown in Listing 1. The specification starts by declaring the clocks, giving them symbolic names and setting the period (in ns). This is followed by the various IPs and their ports. Each IP (and potentially port) either specifies a clock name (as done by U2 on line 17) or uses the default global clock (like the host). A port

```
1 <architecture id='85500_inspired'>
2   <!-- Global parameters -->
3   <parameter id='clk' type='string' value='global_533MHz'/>
4
5   <!-- Clocks (period in ns) -->
6   <clk id='global_533MHz' period='1.87' />
7   <clk id='tm_450MHz' period='2.19' />
8   <clk id='vid_200MHz' period='5' />
9   <clk id='mem_200MHz' period='5' />
10  ...
11
12  <!-- IPs (port width in bits) -->
13  <ip id='host' type='Host'>
14    <port id='pi' type='Initiator' protocol='MMIO_DTL' />
15  </ip>
16  <ip id='U2' type='IP'>
17    <parameter id='clk' type='string' value='vid_200MHz' />
18    <port id='s1' type='Initiator' protocol='MMIO_DTL' />
19    <port id='s2' type='Initiator' protocol='MMIO_DTL'>
20      <parameter id='width' type='int' value='8' />
21    </port>
22    ...
23  </ip>
24  <ip id='mem_ddr' type='IP'>
25    <parameter id='clk' type='string' value='mem_200MHz' />
26    <port id='p1' type='Target' protocol='MMIO_DTL'>
27      <parameter id='width' type='int' value='64' />
28    </port>
29    <port id='p2' type='Target' protocol='MMIO_DTL' />
30    ...
31  </ip>
32  ...
33
34  <!-- Placement constraints (as regular expressions) -->
35  <constraint id='group1' port='U2.*' />
36  <constraint id='group2' port='mem_ddr.*' nis='NIx(012)y0n[0-1]'/>
37  ...
38 </architecture>
```

Listing 1 Architecture specification

specifies whether it is an initiator or a target, and what protocol it uses [1, 19, 57]. A data width of 32 bits is assumed for the ports unless the default is overridden by a parameter (line 21 and 27). The last part of the user architecture specification is the placement constraints that enable the user to group IP ports and thereby enforce a common NI (selected from a subset that can also be specified). The constraints use regular expressions to specify both the ports and the NIs (as shown on line 36). The architecture description is complemented by the communication requirements of the applications, specified per application as a set of logical interconnections between IP ports, each with bounds on the maximum latency and minimum throughput. Listing 2 shows a subset of the connections for the digital TV SoC. Each connection is assigned a logical identifier and specifies which initiator port and target port it connects. The temporal behaviour is specified separately for reads and writes (both or only one as shown on line 23). In both cases, the throughput specification is given in Mbyte/s and the latency requirement in ns. Together with the ports' protocol, the (minimum) burst size (given in bytes) is used to bound the overhead of the interconnect serialisation, as we shall see in Sect. 3.2. The last part of the communication specification is the use-case constraints, determining how the applications

```
1 <communication>
2   <!-- Applications -->
3   <application id='storage'>
4     <connection id='0'>
5       <initiator ip='U2' port='s1' />
6       <target ip='mem_ddr' port='p1' />
7       <read bw='150' burstsize='256' latency='1500' />
8       <write bw='100' burstsize='256' latency='900' />
9     </connection>
10    <connection id='1'>
11      <initiator ip='U2' port='s2' />
12      <target ip='mem_ddr' port='p1' />
13      <read bw='100' burstsize='256' latency='1500' />
14      <write bw='100' burstsize='256' latency='900' />
15    </connection>
16    ...
17  </application>
18
19  <application id='vld_dec'>
20    <connection id='0'>
21      <initiator ip='vliw' port='instr' />
22      <target ip='mem2_ddr' port='ptm' />
23      <read bw='50' burstsize='128' latency='50' />
24    </connection>
25    ...
26  </application>
27
28  <!-- Use-case constraints -->
29  <constraint type='allow' appl='storage' with='.*' />
30  <constraint type='allow' appl='vid_dec' with='.*' />
31  ...
32 </communication>
```

Listing 2 Communication specification

may be combined temporally. In this example, all applications are assumed to be running concurrently (e.g. allow to run with all on line 29 and 30), but it is possible to benefit from mutually exclusive applications [31]. The architecture and communication specification are assumed to be given by the application and system designer, referred to as the user in Fig. 2. The interface specifications follow directly from the choice of IPs, and the use-case constraints are part of the early system design. It is less obvious, however, where the temporal requirements on the interconnect, i.e. the latency and throughput requirements of the connections, stem from. Depending on the application, the desired throughput and latency may be analytically computed [35], the outcome of high-level simulation models of the application, or simply guesstimates based on back-of-the-envelope calculations or earlier designs [41]. The design flow makes no distinction of the real-time character, e.g. if the requirement reflects an average case or a worst case. The requirement will be satisfied (assuming the IPs have the right characterisation) and the interconnect is robust with respect to potential over-asking from the IPs. Section 3.1 begins our bottom-up description of the design flow, introducing the designtime dimensioning of the interconnect. This is followed by a discussion on the compile-time allocation of resources in Sect. 3.2. Section 3.3 describes how the resources and allocations come together in an instantiation at run time. Next, Sect. 3.4 outlines how the interconnect hardware and software comes together in an analysis model, and how it is applied in buffer sizing and formal verification of the end-to-end application behaviour. Throughout the design flow, there are many opportunities for successive refinement, as indicated by the iteration in Fig. 2. Iteration is either a result of failure to deliver on requirements or a result of cost assessment, e.g. silicon area or power consumption. It is also possible that the specification of requirements changes as a result of the final evaluation. The outcome of the proposed design flow is a complete SoC interconnect architecture (hardware) and a set of resource allocations (software) that together implement the requirements specified by the user. As we shall see plenty examples of in Sects. 4 and 5, the resulting platform is instantiated (in SystemC or synthesisable HDL), together with the software libraries required to configure and orchestrate use-case switches (in Tcl or ANSI C).

Dimensioning

The dimensioning constitutes Step 1 in Fig. 2 and is responsible for adding and connecting hardware resources, pushing the complexity of allocating and using the resources to the later parts of the design flow. Nevertheless, the dimensioning is split into three steps, corresponding to three tools in the design flow. The first step of the dimensioning (busdim) bridges between the network and the IP ports by dimensioning buses together with atomisers and protocol shells, i.e. everything belonging to the interconnect but not the network in Fig. 1. Initiator and target buses are added to the IPs that make use of shared and distributed memory, respectively. The dimensioning of these buses is driven by the maximum number of concurrent connections for the IPs (based on the communication requirements). In Fig. 1, for example, the ARM has two concurrent connections to distributed memories, and the SRAM is shared by two concurrent connections. Shells are added based on the IP port protocol and the atomisers are placed in front of shared target ports (in combination with initiator buses) to ensure a fixed burst size [29]. Clock domain crossings are inserted where necessary. Listing 3 shows examples of the modules that are added to the architecture, all described in more depth in [29]. Note that the entire specification from Listing 1 is still included and the interconnect components are appended. The initiator bus (line 21) serves to exemplify a parametrisable IP. The ports of the initiator bus also show how identifiers are added to distinguish between the ports. At this point, the architecture also contains a number of connects, representing wires between two (compatible) ports. The following step (netwdim) dimensions the network topology, adding the NIs, the routers and links, as shown in Listing 4. This step is based on a user specification of the router topology, the link pipeline stages and the number of NIs per router. As detailed in [34], the network uses three-word flits, wormhole switching and source routing. A few basic topology types, e.g. meshes, tori and rings, are supported by the dimensioning to simplify the task. It is possible to customise the output or to specify a fully custom topology by editing the XML file. The resource allocation guarantees deadlock freedom both on the network and transaction level [32]. Note that the NIs, at this point, are connected to a router and have a control port, but no ports connecting them to the IPs. The final part of the interconnect dimensioning is the addition of the control infrastructure (ctrldim), as shown in Listing 5. One control bus (line 28) is instantiated for each NI and the control ports of all (user and interconnect) IPs are connected to the control buses. The necessary ports are added to the NIs (line 8 to 15), and shells are added to connect the FIFO ports of the NIs and the memory-mapped ports of the control buses. The appropriate connects are added to enable a host IP access to the control infrastructure and the address map of the control buses is determined according to user specification. This step also adds control connections to the communication requirements [28]. The control connections are automatically added as an independent application (not shown for brevity) and allocated resources like any user connection by the succeeding steps of the design flow.

```
<architecture id='85500_inspired'>
  <!-- Everything from Listing 1 -->
  ...

  <!-- Shells -->
  <ip id='U2_s1_shell' type='TrgtShell'>
    <parameter id='clk' type='string' value='vid_200MHz' />
    <port id='pt' type='Target' protocol='MMIO_DTL' />
    <port id='pst' type='Target' protocol='FIFO_AE' />
    <port id='psi' type='Initiator' protocol='FIFO_AE' />
  </ip>
  <ip id='U2_s2_shell' type='TrgtShell'>
    <parameter id='clk' type='string' value='vid_200MHz' />
    <port id='pt' type='Target' protocol='MMIO_DTL' />
    <port id='pst' type='Target' protocol='FIFO_AE' />
    <port id='psi' type='Initiator' protocol='FIFO_AE' />
  </ip>
  ...

  <!-- Buses -->
  <ip id='mem_ddr_p1_bus' type='InitBus'>
    <parameter id='clk' type='string' value='mem_200MHz' />
    <parameter id='arbiter' type='string' value='CCSP' />
    <port id='pi' type='Initiator' protocol='MMIO_DTL' />
    <port id='pt0' type='Target' protocol='MMIO_DTL'>
      <parameter id='portid' type='int' value='0' />
    </port>
    <port id='pt1' type='Target' protocol='MMIO_DTL'>
      <parameter id='portid' type='int' value='1' />
    </port>
    ...
  </ip>
  ...

  <!-- Clock domain crossings -->
  <ip id='mem_ddr_p1_bus_pt1_shell_pst_cdc' type='CDC'>
    <port id='psi' type='Initiator' protocol='FIFO_AE'>
      <parameter id='clk' type='string' value='mem_200MHz' />
    </port>
    <port id='pst' type='Target' protocol='FIFO_AE' />
  </ip>
  ...

  <!-- Connects -->
  <connect>
     <endpoint ip='mem_ddr_p1_bus' port='pi' />
     <endpoint ip='mem_ddr' port='p1' />
  </connect>
  <connect>
     <endpoint ip='mem_ddr_p1_bus_pt0_shell' port='pi' />
     <endpoint ip='mem_ddr_p1_bus' port='pt0' />
  </connect>
  ...
</architecture>
```

Listing 3 Architecture with shells and buses

At this point, the streaming ports of the IPs and shells are not yet connected to NIs. It is more efficient to do the mapping of IPs to NIs as part of the path finding and slot allocation [30] and we therefore delay this step, about which more next.

```
<architecture id='85500_inspired'>
  <!-- Everything from Listing 3 -->
  ...

  <!-- Routers -->
  <ip id='Rx0y0' type='Router'>
    <port id='p0_in' type='Target' protocol='FLIT_AE'>
      <parameter id='portid' type='int' value='0' />
    </port>
    <port id='p0_out' type='Initiator' protocol='FLIT_AE'>
      <parameter id='portid' type='int' value='0' />
    </port>
    <port id='p1_in' type='Target' protocol='FLIT_AE'>
      <parameter id='portid' type='int' value='1' />
    </port>
    <port id='p1_out' type='Initiator' protocol='FLIT_AE'>
      <parameter id='portid' type='int' value='1' />
    </port>
    ...
  </ip>
  ...

  <!-- Pipelined links -->
  <ip id='Rx0y0_p1_out_0' type='Link'>
    <port id='trgt' type='Target' protocol='FLIT_AE' />
    <port id='init' type='Initiator' protocol='FLIT_AE' />
  </ip>
  ...

  <!-- NIs -->
  <ip id='NIx0y0n0' type='NI'>
    <port id='p0_in' type='Target' protocol='FLIT_AE' />
    <port id='p0_out' type='Initiator' protocol='FLIT_AE' />
    <port id='pgm' type='Target' protocol='MMIO_DTL' />
  </ip>
  ...

  <!-- Connects -->
  <connect>
      <endpoint ip='Rx0y0' port='p0_out' />
      <endpoint ip='Rx0y1' port='p0_in' />
  </connect>
  <connect>
      <endpoint ip='NIx0y0n0' port='p0_out' />
      <endpoint ip='Rx0y0' port='p2_in' />
  </connect>
  <connect>
      <endpoint ip='Rx0y0' port='p1_out' />
      <endpoint ip='Rx0y0_p1_out_0' port='trgt' />
  </connect>
  ...
</architecture>
```

Listing 4 Architecture with network

Allocation

After dimensioning the interconnect, the application requirements area translated into resource allocations, based on the temporal requirements and the use-case constraints. This constitutes Step 2 in Fig. 2 and involves deciding what bus ports, NIs, paths and time

```
<architecture id='85500_inspired'>
  <!-- Everything from Listing 4 -->
  ...

  <!-- NIs -->
  <ip id='NIx0y0n0' type='NI'>
     ...
    <port id='NIx0y0n0_cfgbus_rmt_shell_ps_in' type='Target'
     protocol='FIFO_AE'>
      <parameter id='iqsz' type='int' value='3' />
      <parameter id='portid' type='int' value='0' />
    </port>
    <port id='NIx0y0n0_cfgbus_rmt_shell_ps_out' type='Initiator'
     protocol='FIFO_AE'>
      <parameter id='oqsz' type='int' value='3' />
      <parameter id='portid' type='int' value='0' />
    </port>
  </ip>
  ...

  <!-- Shells -->
  <ip id='NIx0y0n0_cfgbus_rmt_shell' type='TrgtShell'>
    <port id='pt' type='Target' protocol='MMIO_DTL' />
    <port id='pst' type='Target' protocol='FIFO_AE' />
    <port id='psi' type='Initiator' protocol='FIFO_AE' />
  </ip>
  ...

  <!-- Control buses with fixed address decoders -->
  <ip id='NIx0y0n0_cfgbus' type='TrgtBus'>
    <parameter id='default_port' type='int' value='0' />
    <port id='pt' type='Target' protocol='MMIO_DTL' />
    <port id='lcl' type='Initiator' protocol='MMIO_DTL'>
      <parameter id='addr_equal' type='int' value='0x80000000' />
      <parameter id='addr_mask' type='int' value='0xfffffc00' />
      <parameter id='portid' type='int' value='0' />
    </port>
    <port id='rmt' type='Initiator' protocol='MMIO_DTL'>
      <parameter id='addr_equal' type='int' value='0x80004000' />
      <parameter id='addr_mask' type='int' value='0xffffc000' />
      <parameter id='portid' type='int' value='1' />
    </port>
  </ip>
  ...

  <!-- Connects -->
  <connect>
     <endpoint ip='host' port='pi' />
     <endpoint ip='NIx0y0n0_cfgbus' port='pt' />
  </connect>
  <connect>
     <endpoint ip='NIx0y0n0_cfgbus' port='rmt' />
     <endpoint ip='NIx0y0n0_cfgbus_rmt_shell' port='pt' />
  </connect>
  ...
</architecture>
```

Listing 5 Architecture with control infrastructure.

slots to use. Allocations are created and verified at design or compile time. Hence, runtime choices are confined to choosing from the set of fixed allocations. While limiting the run-time choices to a set of predefined use-cases, this is key as it enables us to guarantee, at compile time, that all application constraints are satisfied, and that all the use-cases fit on the given hardware resources. The allocation part of the flow is split into two tools. The first step (busalloc) allocates ports on the initiator and target buses. In Sect. 3.1 we only dimensioned the buses, i.e. we only decided the number of bus ports, and did not yet decide what connection uses what port in what use-case. In case of distributed memory communication, we allocate an initiator port on the target bus. Similarly, we allocate a target port on the initiator bus in case of shared memory communication. The selection of a bus port implicitly selects a shell on each side of the network (as there is a one-to-one mapping between shells and bus ports). Listing 6 shows the output of the bus allocation. The allocation of bus ports is made by a greedy heuristic that iterates over the applications, based on the number of use-cases they span. The intuition behind this ordering is that the ports used by an application cannot be used by any other application that runs concurrently. Hence, the more use-cases an application spans, the more restrictions are imposed on the not yet allocated applications. After the bus allocation, each connection from Listing 2 is split into a request channel and a response channel between two pairs of streaming ports. In case of a memory-mapped protocol, these streaming ports are located on the shells. For a streaming protocol, on the other hand, the ports are those of the IP itself. After performing the bus-port allocation, the succeeding steps of the allocation flow looks at requirements in terms of channels between streaming ports. As seen in Listing 6, each channel has a data rate (in Mbyte/s) that includes the effects of serialisation in the protocol shells. In addition to the user-specified connections the allocation flow also assigns resources to the control connections (line 30 and onwards). Listing 6 also shows the arbiter settings of the initiator bus.

After the bus ports are assigned to connections, the second step (netwalloc) decides the spatial mapping [37, 58] of streaming ports to NIs, i.e. where in the given network topology to connect the IPs, and also determines the channels' paths and time slots. The mapping of IPs to NIs is obviously only possible if the resource allocation is done at design time (since it changes the interconnect hardware). If the allocation is done at compile or run time, the given mapping is left unchanged. The allocation selects paths through the router network and allocates an appropriate set of time slots on the selected paths such that the throughput and latency requirements are met. As discussed in [30], we exploit the close correspondence between the two spatial allocation problems and let the path-selection algorithm take the mapping decisions by incorporating the streaming ports (shells and streaming ports of IPs) in the path-selection problem formulation. Similarly, we include the slot allocation in the path pruning to guarantee that a traversed path is able to satisfy a channel's requirements. The resulting architecture is shown in Listing 7, showing the newly added NI ports and the connects between the shells and NIs. Note that the input and output queues of the NIs at this point have a default size, here 255 words each. The corresponding network resource allocation is shown in Listing 8, with the additions compared to Listing 6 (paths and slots) highlighted. The allocation together with the dimensioning is central to the functional scalability of the interconnect as these steps are responsible for instantiating and then finding appropriate resources to satisfy the application requirements. As we shall see in Sects. 4 and 5 the tools are indeed able to satisfy a growing number of requirements with a constant cost/performance ratio. However, despite the use of heuristics, the core of the path selection is A*Prune which is known to have exponential growth in run time for certain problem instances [46].

```
1 <allocations>
2   <!-- Allocations for user applications -->
3   <allocation application='storage'>
4     <connection id='0'>
5       <channel type='request'>
6         <from ip='U2_s1_shell' port='psi' />
7         <to ip='mem_ddr_p1_bus_pt0_shell' port='pst' />
8         <parameter id='drate' type='double' value='109.375' />
9       </channel>
10      <channel type='response'>
11        <from ip='mem_ddr_p1_bus_pt0_shell' port='psi' />
12        <to ip='U2_s1_shell' port='pst' />
13        <parameter id='drate' type='double' value='150' />
14      </channel>
15      <initbus bus='mem_ddr_p1_bus' port='pt0'>
16        <parameter id='numerator' type='int' value='3' />
17        <parameter id='denominator' type='int' value='51' />
18        <parameter id='initialCredits' type='int' value='51' />
19        <parameter id='priority' type='int' value='0' />
20      </initbus>
21    </connection>
22    ...
23  </allocation>
24  ...
25
```

```
  <!-- Allocations for the control application -->
  <allocation application='control'>
    <connection id='0'>
      <channel type='request'>
        <from ip='NIx0y0n0_cfgbus_rmt_shell' port='psi' />
        <to ip='NIx0y0n1_cfgbus_pt_shell' port='pst' />
        <parameter id='drate' type='double' value='6e-06' />
      </channel>
      <channel type='response'>
        <from ip='NIx0y0n1_cfgbus_pt_shell' port='psi' />
        <to ip='NIx0y0n0_cfgbus_rmt_shell' port='pst' />
        <parameter id='drate' type='double' value='1e-06' />
      </channel>
    </connection>
    ...
  </allocation>
</allocations>
```

Listing 6: Allocation of bus ports.

Note that this does not impair the functional scalability, looking only at the outcome and not the run-time. The complexity is reduced by using a path-selection algorithm based on Dijkstra, as proposed in [30], but this on the other hand compromises the quality of the solution and might cause the allocation to fail. For contemporary systems, with hundreds of IPs and connections, the proposed algorithms are sufficient, but other approaches might be necessary as the problem size grows. After the final allocation step, every connection is allocated an NI on the initiator and target side of the network and the resulting allocation of slots and paths adheres to the channel's throughput and latency requirements. The interconnect hardware and software can thus be instantiated. However, the allocation algorithm takes a network-centric approach and excludes the effects the buses and IPs (and also flow control between these components) have on the end-to-end temporal behaviour. We return to this in Sect. 3.4 where we also

```
<architecture id='85500_inspired'>
  <!-- Everything from Listing 5 -->
  ...

  <!-- NIs -->
  <ip id='NIx0y0n0' type='NI'>
    ...
    <port id='U2_s1_shell_ps_in' type='Target' protocol='FIFO_AE'>
      <parameter id='iqsz' type='int' value='255' />
      <parameter id='portid' type='int' value='1' />
    </port>
    <port id='U2_s1_shell_ps_out' type='Initiator' protocol='FIFO_AE'>
      <parameter id='oqsz' type='int' value='255' />
      <parameter id='portid' type='int' value='1' />
    </port>
    <port id='U2_s2_shell_ps_in' type='Target' protocol='FIFO_AE'>
      <parameter id='iqsz' type='int' value='255' />
      <parameter id='portid' type='int' value='2' />
    </port>
    <port id='U2_s2_shell_ps_out' type='Initiator' protocol='FIFO_AE'>
      <parameter id='oqsz' type='int' value='255' />
      <parameter id='portid' type='int' value='2' />
    </port>
  </ip>
  ...

```

```
  <!-- Connects -->
  <connect>
     <endpoint ip='vliw_instr_shell' port='psi' />
     <endpoint ip='NIx0y0n1' port='vliw_instr_shell_ps_in' />
  </connect>
  <connect>
     <endpoint ip='U2_s1_shell' port='psi' />
     <endpoint ip='NIx1y1n1' port='U2_s1_shell_ps_in' />
  </connect>
  ...
</architecture>
```

Listing 7 Architecture with NI mapping

demonstrate how to incorporate the end points (shells, buses, IPs, etc.) and dimension the NI buffers using dataflow analysis techniques.

INSTANTIATION

The instantiation, corresponding to Step 3 in Fig. 2, takes its starting point in a complete specification of the architecture and resource allocation, and addresses the challenges involved in turning the specification into a hardware and software realisation. Turning highlevel specifications into HDL and C code is a tedious and error prone job. The instantiation flow greatly simplifies the task by fully automating this process. We see examples of the automation in later sections where entire system instances are created in a matter of hours. To instantiate the hardware, the appropriate blocks must be created and interconnected. Many blocks are parametrisable and require the appropriate values to be set. In addition to the instantiation of individual library modules, such as buses and routers, these modules must also be connected according to the architecture description. On the network and interconnect level we thus have to generate instance-specific hardware. All of this is done

```
<allocations>
  <!-- Allocations for user applications -->
  <allocation application='storage'>
    <connection id='0'>
      <channel type='request' slots='0' path='3 2'>
        <from ip='U2_s1_shell' port='psi' />
        <to ip='mem_ddr_p1_bus_pt0_shell' port='pst' />
        <parameter id='drate' type='double' value='109.375' />
      </channel>
      <channel type='response' slots='0 1' path='4 2'>
        <from ip='mem_ddr_p1_bus_pt0_shell' port='psi' />
        <to ip='U2_s1_shell' port='pst' />
        <parameter id='drate' type='double' value='150' />
      </channel>
      <initbus bus='mem_ddr_p1_bus' port='pt0'>
        <parameter id='numerator' type='int' value='3' />
        <parameter id='denominator' type='int' value='51' />
        <parameter id='initialCredits' type='int' value='51' />
        <parameter id='priority' type='int' value='0' />
      </initbus>
    </connection>
    ...
  </allocation>
  ...
```

```
  <!-- Allocations for the control application -->
  <allocation application='control'>
    <connection id='0'>
      <channel type='request' slots='0' path='3'>
        <from ip='NIx0y0n0_cfgbus_rmt_shell' port='psi' />
        <to ip='NIx0y0n1_cfgbus_pt_shell' port='pst' />
        <parameter id='drate' type='double' value='6e-06' />
      </channel>
      <channel type='response' slots='0' path='2'>
        <from ip='NIx0y0n1_cfgbus_pt_shell' port='psi' />
        <to ip='NIx0y0n0_cfgbus_rmt_shell' port='pst' />
        <parameter id='drate' type='double' value='1e-06' />
      </channel>
    </connection>
  </allocation>
</allocations>
```

Listing 8 Allocation of network paths and time slots.

by a tool (vhdlinst) that also generates a set of auxiliary files that are required by the ASIC/FPGA tools for e.g. HDL elaboration, simulation and RTL synthesis. The hardware is of no use without the appropriate (host) software. The resource allocations are thus translated into source code (cinst) and linked with the appropriate run-time libraries that enable a host processor to open and close connections [28]. At run-time, the host then instantiates these allocations as a result of trigger events [56]. The run-time instantiation, in its current form, limits the interconnect scalability as it uses a centralised host. However, the proposed run-time libraries do not inherently limit the scalability, and it is possible to distribute the work across multiple hosts. In addition to the instantiation tools, the design flow also includes an area-estimation tool that quickly shows the total area requirements of the interconnect based on the area models described in [29]. The important observation about the silicon area is that the interconnect is dominated by the NIs, that in turn are dominated by their buffers. Buffer sizing is therefore of utmost importance, as discussed next.

VERIFICATION

The verification constitutes Step 4 in Fig. 2 and serves as a robust design-rule checker based on the given architecture and allocation XML files, which could have been generated, provided or edited by hand. Even in the case where the files are automatically generated by the flow, the allocation of communication requirements only covers the network and excludes the effects of end-to-end flow control. When NI ports are added as part of the dimensioning and allocation, all buffers are given a default size, and until now we have assumed that the buffers are sufficiently large. If this is not the case, the application performance suffers, and conversely, if the buffers are unnecessarily large they waste power and area [11, 21, 35, 65]. In addition to the network, also the clock domain crossings, shells, atomisers and buses affect the performance and must be considered. Furthermore, the communication requirements are merely a result of higher-level application requirements, and it remains to verify that the applications have the expected functional and temporal behaviour when they are executed on the system instance, i.e. taking the interdependencies between the applications, the IPs, and the interconnect into account. Thus, for an existing interconnect instance, we must be able to determine the temporal behaviour of the applications mapped to it, given fixed NI buffer sizes. On the other hand, if we are designing an interconnect specifically for a set of applications, then it is desirable to determine sufficiently large NI buffers. The verification (buffdim) is performed using dataflow models of the interconnect [35] and the IPs. The dataflow models bound the combined behaviour of the interconnect architecture and resource allocation, even for data-dependent applications [77]. Currently our design flow assumes a homogeneous analysis of all applications, using dataflow analysis. However, as discussed in Sect. 2, it is possible to use other buffer-sizing approaches where appropriate, e.g. stochastic traffic models [13, 68]. This diversity enables the designer to capture a wide range of application behaviours and requirements, and is made possible by the composability of our interconnect [33]. Listing 9 shows the resulting architecture specification, highlighting the changes compared to Listing 7. After the buffer dimensioning, the interconnect hardware and software can be instantiated anew. The verification thus completes the interconnect design flow, and we now continue to design interconnects for our two case studies.

DIGITAL TV

We now put the proposed design flow to use with an example inspired by NXP's PNX85500, which is a complete one-chip digital TV, aimed at the cost-sensitive midrange market. The PNX85500 delivers multi-standard audio decoding and multi-standard analogue and digital video decoding. The front-end video processing functions, such as DVB-T/DVB-C channel decoding, MPEG-2/MPEG-4/H.264 decoding, analog video decoding and HDMI reception, are combined with advanced backend video picture improvements. One of the key differentiators is the video enhancement algorithms, e.g. Motion Accurate Picture Processing (MAPP2), Halo reduced Frame Rate Conversion and LCD Motion Blur Reduction. The MAPP2 technology provides state-of-the-art motion artifact reduction with movie judder cancellation, motion sharpness and vivid colour management. High flat panel screen

```
<architecture id='85500_inspired'>
  <!-- Everything from Listing 7 -->
  ...

  <!-- NIs -->
  <ip id='NIx0y0n0' type='NI'>
    ...
    <port id='U2_s1_shell_ps_in' type='Target' protocol='FIFO_AE'>
      <parameter id='iqsz' type='int' value='67' />
      <parameter id='portid' type='int' value='1' />
    </port>
    <port id='U2_s1_shell_ps_out' type='Initiator' protocol='FIFO_AE'>
      <parameter id='oqsz' type='int' value='65' />
      <parameter id='portid' type='int' value='1' />
    </port>
    <port id='U2_s2_shell_ps_in' type='Target' protocol='FIFO_AE'>
      <parameter id='iqsz' type='int' value='67' />
      <parameter id='portid' type='int' value='2' />
    </port>
    <port id='U2_s2_shell_ps_out' type='Initiator' protocol='FIFO_AE'>
      <parameter id='oqsz' type='int' value='66' />
      <parameter id='portid' type='int' value='2' />
    </port>
  </ip>
  ...
</architecture>
```

Listing 9 Architecture with minimal NI buffer sizes.

resolutions and refresh rates are supported, including 1366 × 768 and 1920 × 1080 at 100 Hz/120 Hz. This is coupled with multiplexing of multiple high-definition streams, multiple video outputs and multiple audio outputs. The SoC supports a rich set of I/O standards (e.g. PCI, USB, Ethernet, UART) and opens many possibilities for new TV experiences with IPTV and Video On Demand (VOD). We consider the processor cores and IP blocks given, and a schematic architecture is shown in Fig. 3 [41]. As shown in the figure, the SoC contains multiple programmable processors with caches: a MIPS for control, three TriMedia processors for audio/video, one MIPS DSP, and a 8051 processor for power management and control. The TriMedia VLIWs are used for computation (and communication) intensive image processing, e.g. Halo Reduced Frame Rate Conversion, with on-the-fly compression of video traffic to and from the off-chip SDRAM. In addition to the processors, the SoC contains a large number of functionspecific hardware IP cores. Those accelerator IPs are used for e.g. audio and video decoding, video scaling and enhancement, graphics, and I/O (LVDS, HDMI, Ethernet, USB, Flash, SPI, PCI and I2C). Finally, the SoC has a range of peripherals for clocking, debug, timers, etc. As seen in the block diagram, the SoC is built around a physically centralised memory architecture. Two central off-chip memory controllers connect with two large external SDRAMs, with a 16-bit and 32-bit wide data path respectively. The SoC contains 11 chiplets and five major frequency domains, from 27 MHz (many IPs) to 200 MHz (majority of IPs) to over 450 MHz (processors). The complete XML description of the IP architecture occupies roughly 250 lines, with a subset shown in Listing 1. A major design challenge is the on-chip communication, with 110 memory-mapped ports distributed across the IPs (the majority read-only or write-only), contributing to almost 100 logical connections. Predictability is a design constraint. A majority of the connections have relaxed latency requirements (>1000 ns), as the function-specific hardware IP cores make use of prefetching read accesses and posted write accesses. The deep pipelining makes these

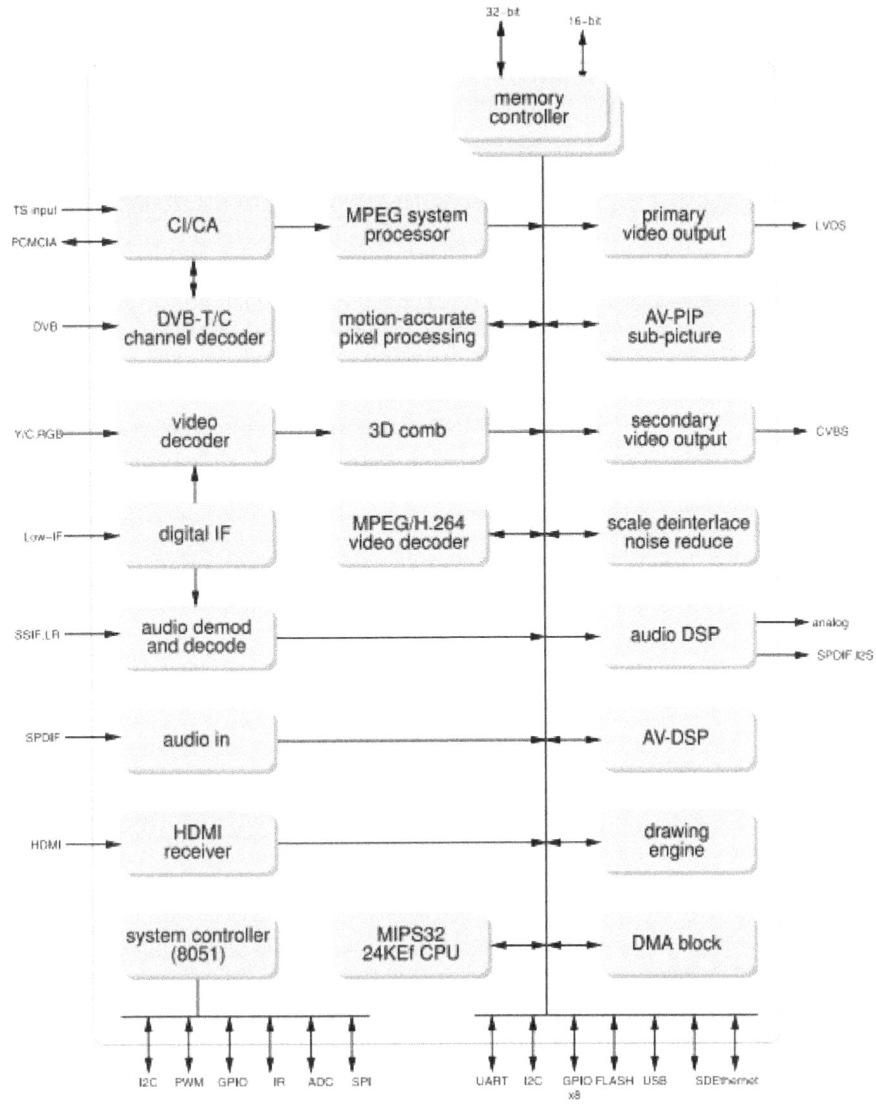

Figure. 3 Digital TV SoC block diagram [41]

connections latency tolerant. Processor instruction fetches (from cache misses), on the other hand, have very low latency requirements (<100 ns) and have relatively small transaction sizes, but require a limited throughput. Some accelerators, e.g. the display controller are also latency constrained. With the most critical low-latency connections implemented through direct connections to the memory controller, and several low-throughput peripheral connections merged, this translates to 53 physical ports, and 45 logical interconnections for the on-chip interconnect. A part of the 260 line specification is shown in Listing 2. Figure 4 shows the distribution of throughput requirements for the 45 connections, with a total of 4402 Mbyte/s (1461 Mbyte/s write, 2941 Mbyte/s read). Note that this is the

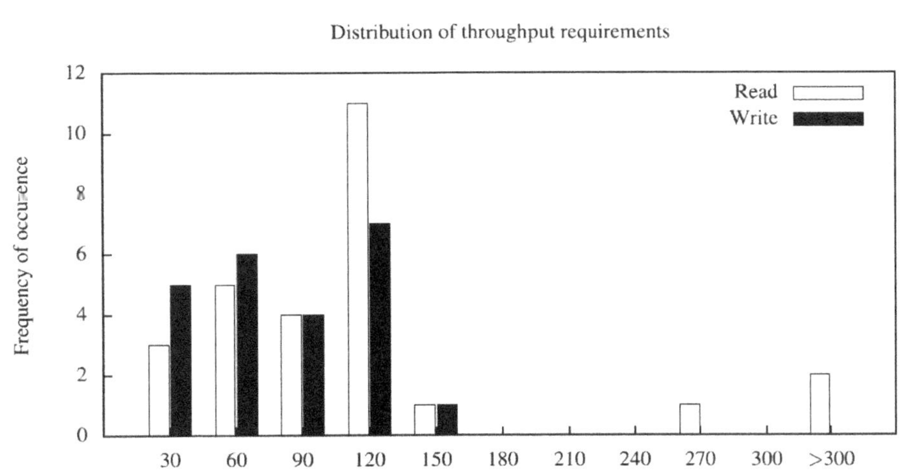

Figure. 4 Write and read throughput distribution (number of connections vs. throughput in Mbyte/s)

data throughput for the memory-mapped communication, not accounting for commands and addresses, byte masks, etc. Burst sizes vary from 64 bytes through 128 and 256 bytes, with the majority of the IPs using the latter (with a mean of 240 bytes). Including the additional address and command signals results in a raw throughput of 4560 Mbyte/s to be delivered by the network. It is worth noting that this particular case study does not make use of the reconfigurability offered by the interconnect. The reason is that the specification is determined without assuming any such functionality (which is unique for NoCs). All connections are thus assumed to be concurrent, and a re-design with reconfigurability in mind could benefit from mutually exclusive applications and connections.

EXPERIMENTAL RESULTS

As discussed in Sect. 3.1, topology selection is not an automated part of the design flow. The tools do, however, aid in translating a high-level topology description into an XML specification. For this example we chose a 2 by 3 mesh network (six routers) with two NIs connected to all routers but one (11 NIs in total). The router arity is thus kept at five for a single router and four for the remaining ones. This topology offers one NI for each chiplet. The network is clocked at 533 MHz, which is the highest IP clock frequency. The latency requirement, as discussed in Sect. 3.2, for low-latency connections is specified to 50 ns (not including the memory scheduler and controller). Given the requirements of the connections and the physical topology, a resource allocation for the single use-case is found with a slot table size of 24. The resulting allocation XML file occupies just over 1000 lines and the corresponding architecture description is almost 4000 lines, all automatically generated by the design flow in a few minutes. As the final part of the dimensioning, the NI buffers are sized using the approach from Sect. 3.4, and the distribution of the NI queue sizes is shown in Fig. 5. The large burst sizes lead to increased buffering in the NIs, and a total of 2249 words (of 37 bits) in output queues and 2232 words in input queues (3060 on the initiator side and 1621 on the target side). In total the NI buffers store roughly 24 kbyte, and the buffers have a big influence on the area requirements which we evaluate next.

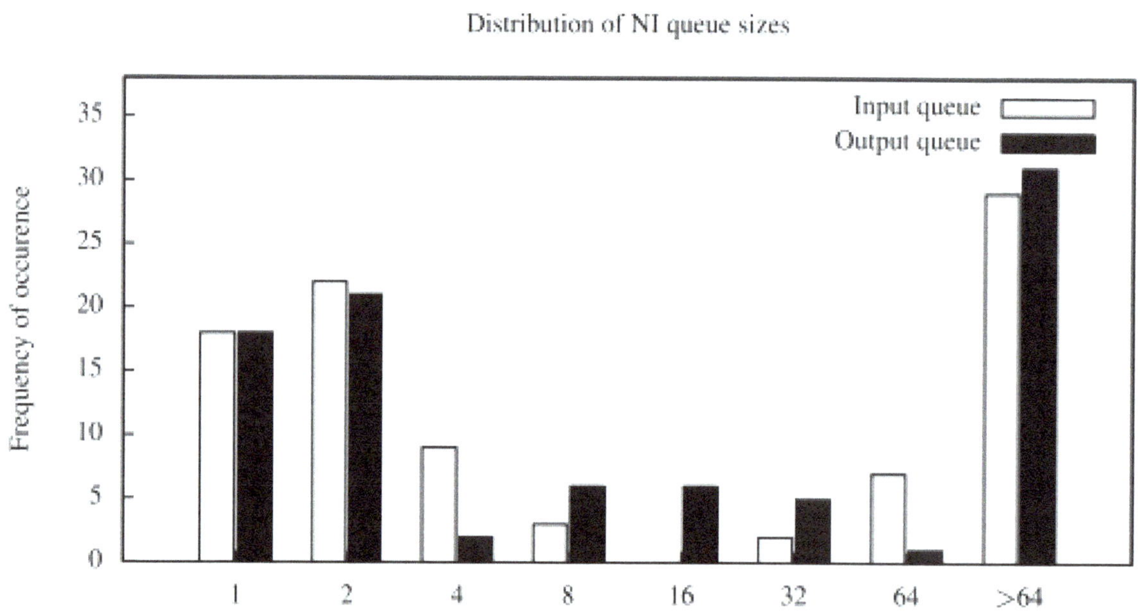

Figure. 5 NI queue size distribution in 37-bit words

Table 1 Area requirements

Module	Instances	Area (mm^2)	Discrepancy
Router	6	0.08	+10%
NI	11	0.41	−10%
Shell	54	0.23	−8%
Bus	6	0.08	+9%

Using the same synthesis setup as described in [29], i.e. Cadence Ambit with Philips 90 nm Low-Power technology libraries, the synthesised interconnect occupies a cell area of 5.5 mm^2 with 4.7 mm2 (85%) in NI buffers. The large fraction in buffers is due to the register-based FIFO implementation, and with FIFOs based on dual-ported SRAMs the area is reduced to roughly 2.6 mm^2 with 1.7 mm2 (65%) in buffers. Further reduction is possible with dedicated FIFOs [76], which results in a total area of 1.6 mm^2 with 0.7 mm2 (44%) in buffers. A break down of the remaining area is shown in Table 1, including a comparison with area estimation models based on the synthesis of the individual blocks in [29]. As seen in the table, the area estimation is within 10% of the result reported by the synthesis tool. Note that the area reported is cell area only, thus before place and route, and clock-tree insertion. To get further insight into the area requirements and the power consumption of the interconnect instance, we synthesise the design using Synopsys Design Compiler Ultra in topographical mode. The synthesis uses TSMC's low-power 65 nm libraries (9 track with a raw gate density of 0.85 Mgate/mm^2) employing mixed Voltage Threshold (VT) synthesis with both high- and low-VT libraries. The synthesis is performed on a flattened design with multi-stage clock gating enabled, using integrated clock gates from the technology library. The power minimisation is driven by activity traces captured in RTL simulation using traffic generators to mimic the IP behaviour. The total area after synthesis is estimated at 2.5 mm2 (9781 cells). Once again the FIFO implementation used is the register-based FIFO, leading to a dominant part of the area spent in buffering. The area of the buffers is, however, reduced by the clock gating as many multiplexers in the FIFOs can be replaced. A total of 7500 clock gating elements are inserted, resulting in 97% of the 251588 registers in the design being gated. At a global (worst-case) operating voltage of 1.08 V the cell power is estimated at 79.5 mW and the net switching power at 43.3 mW. The total dynamic power, under worst-case conditions, is 122.8 mW and the leakage power is estimated at 3.2 mW (or roughly 2.5% of the total power consumption). Roughly 22% of the dynamic power is spent in the clock tree. We can conclude that the lowpower libraries, in combination with mixed-VT synthesis and leads to an implementation where leakage-minimisation techniques like power gating, which is costly in level shifters, are not worthwhile. Next, we evaluate the behaviour of the generated interconnect hardware and software by simulating an instance of the entire system using traffic generators to mimic the IP behaviour. Instrumentation is (automatically) inserted in SystemC and VHDL testbenches respectively to observe the utilisation of the different modules, the latency of individual packets in the network, and the end-to-end latency and throughput between the IPs. The cycleaccurate SystemC simulation executes with a speed of roughly 50 kcycles/s on a modern Linux machine (1 ms of simulated time in less than 20 s). The corresponding VHDL testbench is roughly 200 times slower, with 0.2 kcycles/s. Simulation of the synthesised netlist is again a factor 100 slower, reducing the speed to roughly a cycle per second. First, looking at the critical low-latency connections (specified to 50 ns as previously mentioned) they consistently have an average network latency of 19 ns and a maximum across all the connections of 42 ns. For a specific low-latency connection, the end-to-end latency for a 128 byte read transaction, including the memory controller and memory access, is 280 ns on average and 308 ns for the observed worst case (corresponding to 190 processor cycles at 533 MHz). All the 45 connections have an observed throughput and latency that meets the user-specified requirements, and this behaviour is consistent in both SystemC and VHDL simulation. Comparing SystemC, behavioural VHDL and netlist simulation we see slight differences in the statistical traffic profile. The reason for the discrepancies is the use of different traffic generators for the SystemC and VHDL testbenches. The results, however, looking at the end-to-end latency and throughput, are within ±3%. Next, we look at the utilisation of the interconnect, as observed during simulation. The TDM-based arbitration of the network requires low-latency connections to reserve a larger number of appropriately spaced time slots. As a consequence, these slots are often unused. The utilisation does hence not reflect the reserved slots, but rather the slots that are actually used at run time. We first look at the ingress and egress links of the network, i.e. links that connect an NI to a router. Since the NI is comparatively more expensive, the chosen topology aims to minimise the number of NIs. The ingress and egress links are therefore more likely to have a high utilisation compared to the average over the entire network. For the TV SoC, the average NI link utilisation is 49%, with a minimum of 27% and a maximum of 64%. This is to be compared with an overall network link utilisation of 22%. The link with the lowest utilisation is only used 12% of the time. The relatively low utilisation of the network stresses the importance of using clock gating as part of the logic synthesis. The registerlevel and module-level clock gating ensures that the power consumption is proportional to the usage of the gates rather than the sheer number of gates. As already mentioned, the total interconnect power consumption, under worst-case conditions, is estimated at 125 mW.

Scalability analysis

Having evaluated our proposed interconnect in the context of the contemporary digital TV SoC, we now continue to look at how the interconnect scales with the requirements, and how

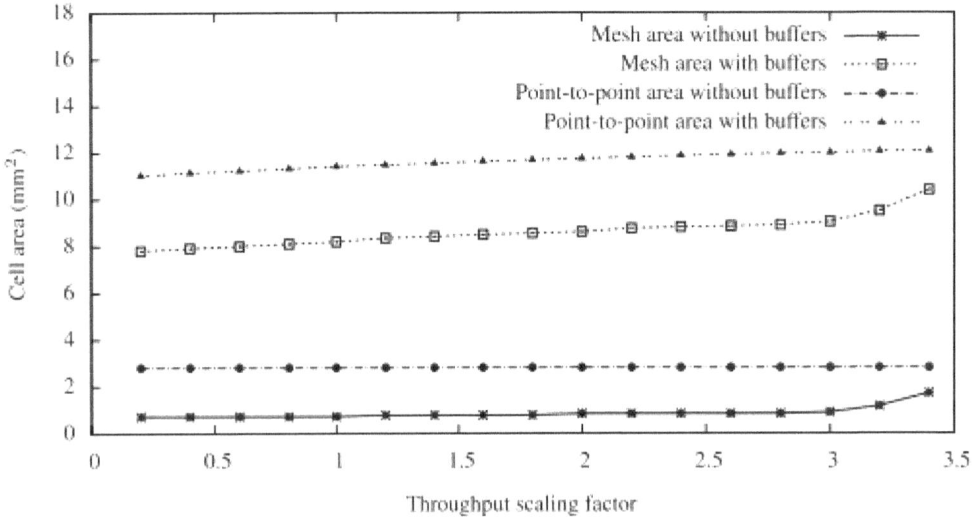

Figure. 6 Scalability of throughput requirements

suitable it is for future generation multimedia applications. As discussed in [41], the number of latency-critical connections has stayed low and fairly constant during the past six generations of TV Socs in the Nexperia platform. The total bandwidth, however, and the number of latency-tolerant function-specific hardware IPs grows over time. These key trends are a result of stable functional subsystems being moved to latency-tolerant implementations as the functionality matures. The latter is a necessary trend to fit more functionality on a single SoC and still be able to serve the latency-critical connections within acceptable bounds. The on- and off-chip throughput is scaling with the generations, but the latency (in absolute time) is fairly constant, leading to growing memory access latencies (measured in processor cycles). With a growing number of latency-critical connections the communication pattern would be inherently non-scalable. Our first scaling experiment fixes the IPs and their 45 connections, and only scales the throughput requirements (leaving the latency untouched). For each design point, we re-run the entire interconnect design flow using the new (scaled) requirements. For each point, we evaluate two different network topologies. First, a minimum mesh, determined by a simple loop that increases the dimensions until a feasible resource allocation (one that satisfies all requirements) is found. For simplicity, the number of NIs is fixed at three per router. Second, a point-to-point topology where each connection is mapped to two unique NIs interconnected by a router (to allow for instantiation of the resource allocation during run-time). A maximum of 32 TDM slots is allowed in both cases. Once a resource allocation is found, we determine sufficiently large buffer sizes (also verifying the performance bounds using dataflow analysis) and estimate the required area using the models from [29]. Thus, for every point we determine two complete hardware and software interconnect instances that are guaranteed to satisfy all the requirements. Figure 6 shows the area, both with and without buffers, for the two topologies as the throughput requirements are scaled from 0.2 to 3.4. The area estimates once again assume a 90 nm process and register-based FIFOs and are thus overly conservative. However, as we are interested in the scaling the absolute numbers are not critical. What is more interesting is the trend as the scaling factor is increased. For scaling factors below three, the total area is growing linearly for both topologies, with the area of the mesh being roughly 25% smaller than the point-to-point topology (75% smaller when excluding the buffers). Although not visible in the figure, the mesh area excluding the buffers is increasing in a stair case as the number of routers and NIs grow. Increased throughput leads to different mapping and routing decisions as the connections are balanced on a larger interconnect. As expected, the point-to-point topology area is constant when the buffers are not considered. With scaling factors above three, the mesh area is increasing faster, approaching the area of the point-to-point topology, and after 3.4 no feasible resource allocation is found for either of the two topologies. This is due to the inherent sharing of the IP ports, i.e. the memory controllers, where the scaled requirement eventually saturate the port capacity. Thus, the point of contention is pushed to the IPs ports and there exist no possible topologies and resource allocations, for any NoC architecture or design flow. We have seen that the interconnect scales linearly with the throughput requirements when the number of connections is kept constant. Now we look at how the interconnect scales with the number of latency-tolerant connections. Rather than extrapolating a next-generation TV SoC, we randomly select a subset of the connections of the original example, and do so for fractions ranging from 0 to 1 in increments of 0.1. For each design point (belonging to a specific fraction) we repeat the random selection 10 times. As a result, for each point we re-run the entire design flow 10 times with (potentially) different subsets of the communication specification. For each run we determine the minimum mesh as before. After finding a feasible resource allocation the buffers are sized and the interconnect area estimated for each one of the 100 instances. Figure 7 shows the area estimates with and without buffers, with each point showing the average of the 10 runs together with the standard deviation. Once again, the absolute numbers are not particularly important as they depend heavily on the chosen technology node and buffer implementation. What is important is that both Figs. 7(a) and 7(b), show a clear linear trend. This suggests that the interconnect architecture and design flow is providing functional scalability and thus able to accommodate a growing number of connections with a roughly constant cost/performance ratio.

Automotive radio

Our second example is a software-defined radio SoC, aimed at the automotive industry. Although mobility, connectivity, security and safety drive the automotive industry, there is an accelerated adoption of consumer features in cars, with home entertainment also in your car. The main applications include high quality, real-time, interactive audio and video, with standards like DAB, DVB-H, DVB-T, T-DMB, HD Radio etc. While delivering a true multimedia experience, the SoC must also meet automotive quality demands. Due to the nature of the SoC, there is an essential need for a powerful and flexible multi-standard, multi-channel software-defined radio, with efficient channel decoder processing. Similar to the digital TV SoC, we consider the processor cores and IP blocks given. The example is inspired by a SoC that contains three programmable processors with caches: an ARM Cortex for control, and two Embedded Vector Processors [73] (EVP). The EVPs constitute the core of the software-defined radio. In addition to the processors, the SoC contains a large accelerator subsystem. The accelerators are involved in signal-processing and I/O (several DACs and ADCs, DMA, and USB). In contrast to the digital TV SoC, this system is not making use of a physically centralised memory architecture, as it has many onchip (more than 10) SRAMs in addition to the external SDRAM. As we shall see, the result is less inherent contention in the communication patterns. The SoC contains 7 chiplets and

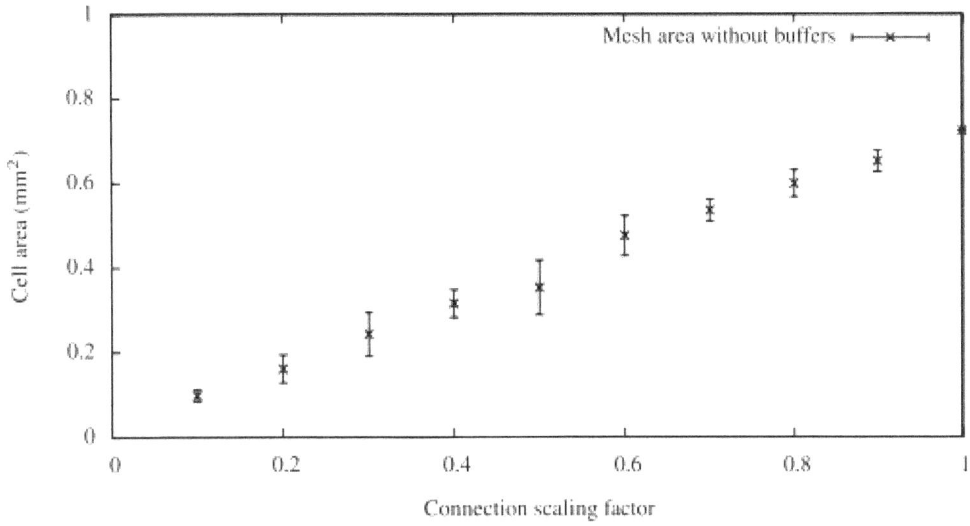

(a) Area without buffers

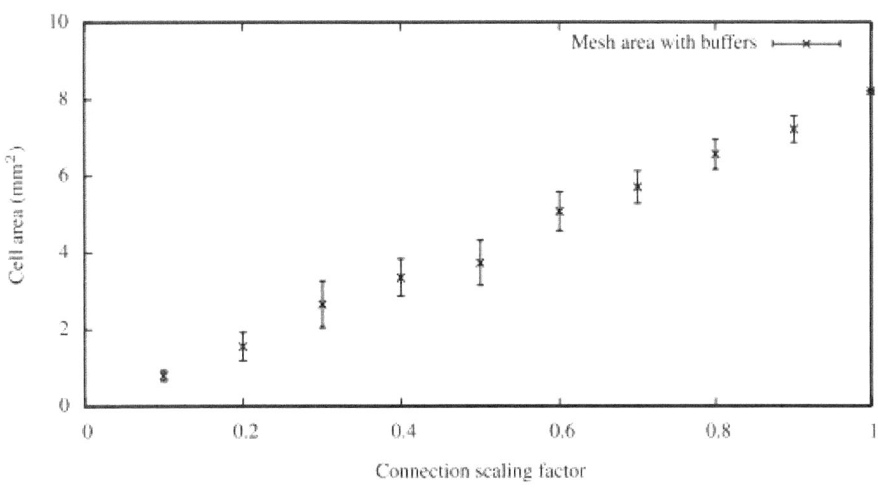

(b) Area with buffers

Figure. 7 Scaling the number of connections.

9 clock domains, ranging from 100 MHz to 350 MHz with the majority of the IPs running at either 100 MHz or 200 MHz. Communication is central to the SoC with 68 IP ports using both memory-mapped and streaming communications with a total of 52 logical connections. Many of the IPs do not use posted/pipelined transactions and their communication is therefore latency-critical. In contrast to the digital TV it is thus not only cache traffic that is latency critical. Due to the real-time nature of the application, predictability is crucial for all on-chip communication. After merging a few low-throughput connections to peripherals, the on-chip interconnect must accommodate 43 ports (distributed across 12 IPs), with 32 logical interconnections. Roughly half of these connections have latency requirements below 100 ns.

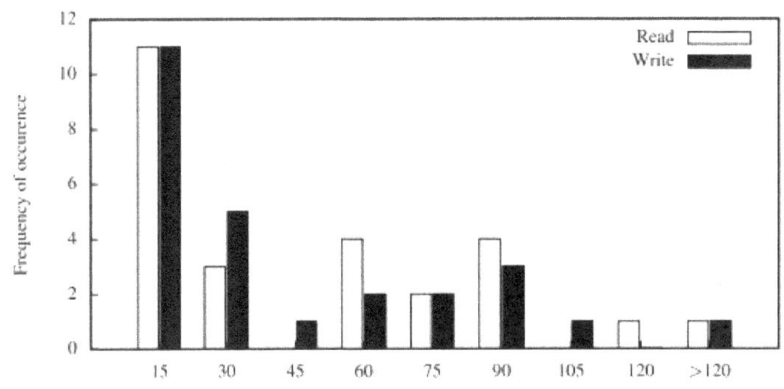

Figure. 8 Write and read throughput distribution (number of connections vs. throughput in MByte/s)

Figure 8 shows the distribution of throughput requirements, with a total of 2151 Mbyte/s (997 Mbyte/s write, 1154 Mbyte/s read). Including command and address this amounts to 4183 Mbyte/s to be delivered by the network. Note that this is almost the double due to the small burst sizes, varying between 4 and 128 bytes (with a mean of 31 bytes). Compare this to the digital TV where we almost exclusively have large bursts, making the raw throughput requirement of the two case studies very similar. Although the throughput requirements are not much different, we shall see that the large difference in burst size has a tremendous impact of the NI buffers, and consequently the interconnect area and power. Similar to the TV SoC, this case study does not exploit the support for multiple use-cases.

Experimental results

For the automotive SoC, we use a 3 by 3 mesh (nine routers) with two NIs connected to the corner routers, one NI for the remaining peripheral routers, and no NI for the central router (12 NIs in total). The router arity is thereby kept low and the topology is still large enough to enable the IPs to be placed around the interconnect, grouped according to chiplets and clock domains. For this case study, the network is clocked at 400 MHz which is double the frequency of most IPs. Given the requirements, an interconnect instance is generated with a slot table size of 21 and the buffers are sized using the proposed design flow. The distribution of NI queue sizes is shown in Fig. 9, with a total of 442 words in input queues, and 587 words in output queues (453 on the target side and 576 on the initiator side). The total capacity is roughly 5 kbyte. Compared to the TV SoC this is roughly four times less buffering per connection, largely due to the big difference in average burst size. The difference in buffer size also has a big impact on the interconnect area. Using the same 90 nm libraries and the same setup as for the TV SoC, the total cell area for the automotive SoC interconnect is 2.13 mm2 with 1.41 mm2 (66%) in register-based NI buffers. With dual-ported SRAMS this is reduced to 1.22 mm2 with 0.5 mm2 (40%) in buffers, or as little as 0.9 mm2 with 0.22 mm2 (23%) in buffers when implemented using dedicated FIFOs [76]. An area break down is shown in Table 2, including both the results from synthesis and the high-level area-estimation models. Just as for the TV SoC, the area estimation is

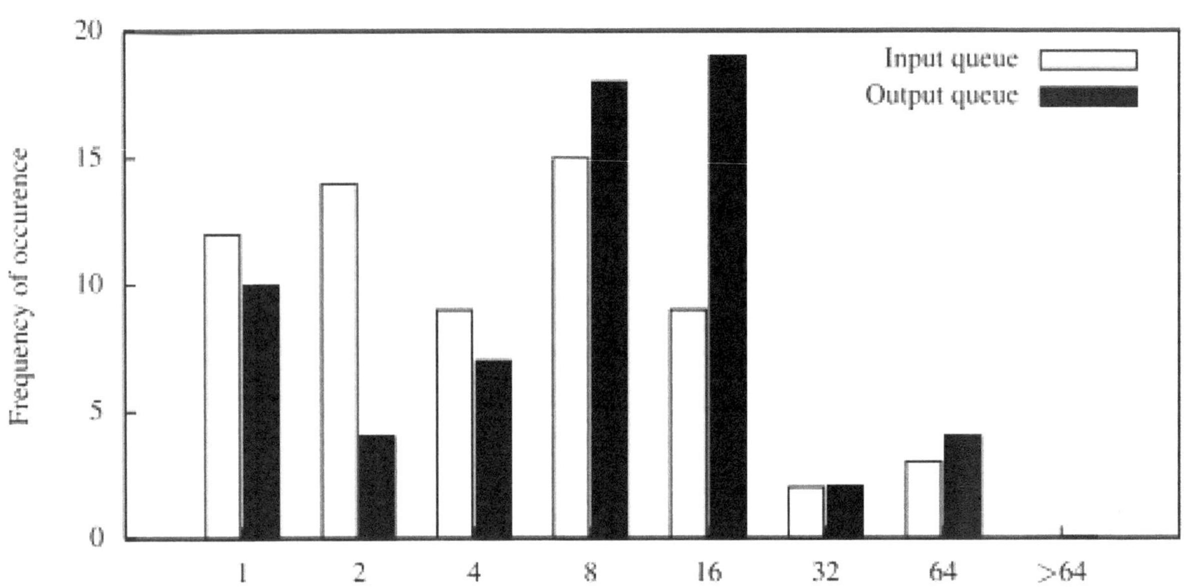

Figure. 9 NI queue size distribution in 37-bit words

Table 2 Area requirements

Module	Instances	Area (mm^2)	Discrepancy
Router	9	0.13	+10%
NI	12	0.32	−8%
Shell	43	0.18	0%
Bus	8	0.07	−4%

within 10%, suggesting the models can be used reliably for early design-space exploration and evaluation. We synthesise also this interconnect using Synopsys Design Compiler Ultra, aiming to minimise the power consumption. The details are the same as described in Sect. 4. The total area after synthesis is estimated at 0.78 mm2 (4784 cells) and 90% of the 76474 registers are clock gated. At worst-case conditions, the cell power is estimated at 41.2 mW and the net switching power at 15.4 mW, resulting in a total dynamic power consumption of only 57 mW. The leakage is reported to be less than 1 mW, once again demonstrating the abilities of the low-power libraries and mixed-VT synthesis. Roughly 27% of the total power is spent in the clock tree. Similar to the TV SoC, we perform SystemC and HDL simulations on the instantiated interconnect. Once again, all throughput and latency requirements are satisfied, and this behaviour is consistent for all the different instantiations. Looking at the utilisation during simulation, the automotive SoC has an average NI link utilisation of 30% with a minimum of 7% and a maximum of 75%. The low utilisation of the NI links is due to the very strict latency requirements which limits the amount of sharing. The overall network link utilisation is also low, only 20%. As previously discussed, the relatively low utilisation highlights the importance of clock gating, as used in our experiments.

Scalability analysis

We continue our evaluation with two different scaling experiments, similar to what is described for the TV SoC in Sect. 4. First, scaling the throughput requirements between 0.2

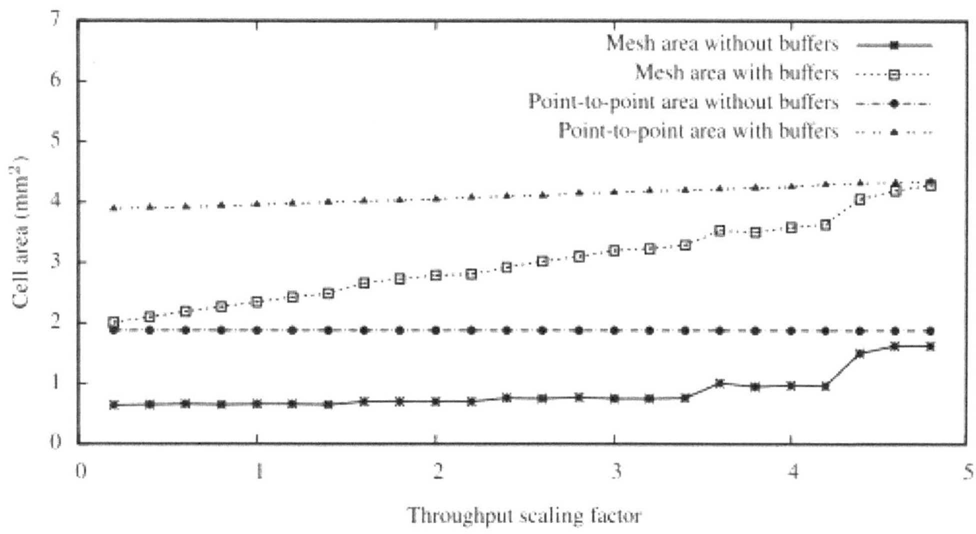

Figure. 10 Scalability of throughput requirements.

and 4.8, we compare the area of a minimum mesh and a point-to-point topology. Figure 10 shows the result, both with and without buffers. The heuristics used in the mesh sizing results in an area that is growing in a stair case, which is now visible as the total area is less dependent on the buffers. The total area is growing linearly for both topologies until they saturate at a scaling factor of 4.8 where the requirements can no longer be allocated. At this point the mesh area is almost equivalent to the point-to-point topology, using a 5 by 4 mesh with 3 NIs per router. Comparing the results with the TV SoC we clearly see the consequences of having a more distributed communication pattern (more memories) as the throughput can be scaled further (4.8 compared to 3.4). Once again this factor is an inherent limit caused by the sharing of the IP ports and not influenced by the choice of a NoC architecture or design flow. Next we evaluate the scalability with respect to the number of connections. In contrast to the TV SoC we now include all connections (latency critical and latency tolerant) in the selection. Again we look at 10 different design points and run the entire design flow 10 times per point. Figure 11 shows the area with and without buffers. Although there is a considerable variation in the results, the linear trend is clear. We have thus shown that also for this example, with a larger number of latency-critical connections, but a more distributed communication pattern, the design flow offers functional scalability. Comparing the results of the two examples we can conclude that the area and power is very much influenced by the temporal requirements and the burst size used by the IPs. However, despite the differences in requirements and behaviours, the proposed design flow is able to scale the two examples, both in terms of requirements and number of connections, for several hundreds of complete system instances.

Conclusions

Systems on Chip (SoC) integrate growing number of applications, often with real-time requirements. This calls for an on-chip interconnect that is able to provide throughput and latency bounds to an increasing number of logical connections. Doing so requires physical,

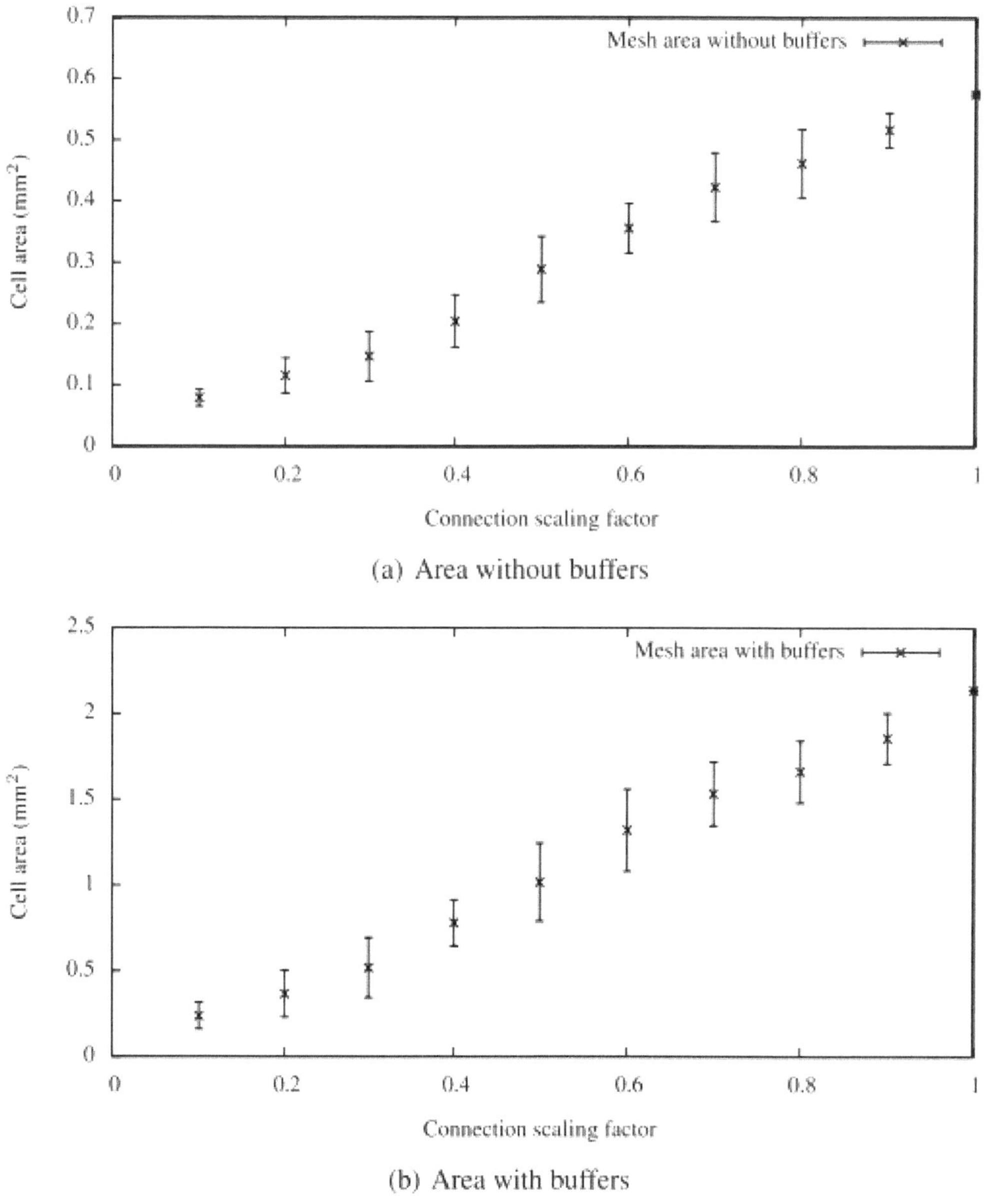

Figure. 11 Scaling the number of connections.

architectural as well as functional scalability. While many Networks on Chip (NoC) address the physical and architectural scalability, the functional scalability depends heavily on the interconnect design flow which is often overlooked. In this work we describe a fully operational design flow used to generate application specific NoC instances for multiple real-time applications. We show how to go from requirement specification to a complete hardware and software instance, using two large-scale industrial case studies. We shown the large impact burst sizes have on the interconnect area and power consumption, and quantify the influence latency-critical connections have on the network utilisation. The relatively low utilisation (around 25%) stresses the importance of clock gating in the logic synthesis to minimise dynamic power. We show that low-power mixed-voltage-threshold libraries lead to leakage power being less than 3% in a 65 nm technology. We evaluate the scalability of the interconnect using the two case studies as our starting point, and show that the proposed interconnect is able to accommodate a growing number of requirements with a constant cost/performance ratio.

REFERENCES

1. AXI (2003) AMBA AXI protocol specification. ARM Limited
2. Bartic T, Desmet D, Mignolet JY, Marescaux T, Verkest D, Vernalde S, Lauwereins R, Miller J, Robert F (2004) Network-on-chip for reconfigurable systems: from high-level design down to implementation. In: Proc FPL
3. Beigne E, Clermidy F, Vivet P, Clouard A, Renaudin M (2005) An asynchronous NOC architecture providing low latency service and its multi-level design framework. In: Proc ASYNC
4. Beigné E, Clermidy F, Miermont S, Vivet P (2008) Dynamic voltage and frequency scaling architecture for units integration within a GALS NoC. In: Proc NOCS
5. Benini L (2006) Application specific NoC design. In: Proc DATE
6. Benini L, de Micheli G (2002) Networks on chips: a new SoC paradigm. IEEE Comput 35(1):70–80

7. Beraha R, Isask'har WI, Kolodny A (2010) Leveraging application-level requirements in the design of a NoC for a 4G SoC. In: Proc DATE

8. Bertozzi D, Jalabert A, Murali S, Tamhankar R, Stergiou S, Benini L, Micheli GD (2005) NoC synthesis flow for customized domain specific multiprocessor systems-on-chip. IEEE Trans Parallel Distrib Syst 16(2):113–129

9. Bjerregaard T, Sparsø J (2005) A router architecture for connection-oriented service guarantees in the MANGO clockless network-on-chip. In: Proc DATE

10. Bjerregaard T, Sparsø J (2005) A scheduling discipline for latency and bandwidth guarantees in asynchronous network-on-chip. In: Proc ASYNC

11. Bjerregaard T, Mahadevan S, Grøndahl Olsen R, Sparsø J (2005) An OCP compliant network adapter for GALS-based SoC design using the MANGO network-on-chip. In: Proc SOC

12. Bjerregaard T, Stensgaard M, Sparsø J (2007) A scalable, timing-safe, network-on-chip architecture with an integrated clock distribution method. In: Proc DATE

13. Bogdan P, Kas M, Marculescu R, Mutlu O (2010) QuaLe: a quantum-leap inspired model for nonstationary analysis of noc traffic in chip multi-processors. In: Proc NOCS

14. Bolotin E, Cidon I, Ginosar R, Kolodny A (2003) QNoC: QoS architecture and design process for network on chip. J Syst Archit 50(2–3):105–128

15. Buttazo GC (1977) Hard real-time computing systems: predictable scheduling algorithms and applications. Kluwer, Dordrecht

16. Chen J, Jone W, Wang J, Lu HI, Chen T (1999) Segmented bus design for low-power systems. IEEE Trans Very Large Scale Integr 7(1):25–29

17. Clermidy F, Bernard C, Lemaire R, Martin J, Miro-Panades I, Thonnart Y, Vivet P, Wehn N (2010) A 477 mw NoC-based digital baseband for MIMO 4G SDR. In: Proc ISSCC

18. Dally WJ, Towles B (2001) Route packets, not wires: on-chip interconnection networks. In: Proc DAC

19. DTL (2002) Device Transaction Level (DTL) Protocol Specification. Version 2.2. Philips Semiconductors

20. Eichenberger A, OBrien J, OBrien K, Wu P, Chen T, Oden P, Prener D, Shepherd J, So B, Sura Z et al (2006) Using advanced compiler technology to exploit the performance of the cell broadband engine architecture. IBM Syst J 45(1):59–84

21. Gangwal O, Radulescu A, Goossens K, Pestana S, Rijpkema E (2005) Building predictable systems on ˇ chip: an analysis of guaranteed communication in the Æthereal network on chip. In: Dynamic and robust streaming in and between connected consumer-electronics devices. Kluwer, Dordrecht

22. Genko N, Atienza D, Micheli GD, Mendias J, Hermida R, Catthoor F (2005) A complete network-onchip emulation framework. In: Proc DATE

23. González Pestana S et al (2004) Cost-performance trade-offs in networks on chip: a simulation-based approach. In: Proc DATE

24. Goossens K, Gangwal OP, Röver J, Niranjan AP (2004) Interconnect and memory organization in SOCs for advanced set-top boxes and TV—evolution, analysis, and trends. In: Nurmi J, Tenhunen H, Isoaho J, Jantsch A (eds) Interconnect-centric design for advanced SoC and NoC. Kluwer, Dordrecht, pp 399–423, Chap 15

25. Goossens K, Dielissen J, Gangwal OP, González Pestana S, Radulescu A, Rijpkema E (2005) A design ˇ flow for application-specific networks on chip with guaranteed performance to accelerate SOC design and verification. In: Proc DATE

26. Goossens K, Dielissen J, Radulescu A (2005) The Æthereal network on chip: concepts, architectures, ˇ and implementations. IEEE Des Test Comput 22(5):21–31

27. Halfhill TR (2006) Ambric's new parallel processor. Microprocessor Report

28. Hansson A, Goossens K (2007) Trade-offs in the configuration of a network on chip for multiple usecases. In: Proc NOCS

29. Hansson A, Goossens K (2009) An on-chip interconnect and protocol stack for multiple communication paradigms and programming models. In: Proc CODES+ISSS

30. Hansson A, Goossens K, Radulescu A (2005) A unified approach to constrained mapping and routing ˇ on network-on-chip architectures. In: Proc CODES+ISSS

31. Hansson A, Coenen M, Goossens K (2007) Undisrupted quality-of-service during reconfiguration of multiple applications in networks on chip. In: Proc DATE

32. Hansson A, Goossens K, Radulescu A (2007) Avoiding message-dependent deadlock in network-based ˇ systems on chip. VLSI Des 2007:1–10

33. Hansson A, Goossens K, Bekooij M, Huisken J (2009) Compsoc: a template for composable and predictable multi-processor System on Chips. ACM Trans Des Autom Electron Syst 14(1):1–24

34. Hansson A, Subburaman M, Goossens K (2009) Aelite: a flit-synchronous network on chip with composable and predictable services. In: Proc DATE

35. Hansson A, Wiggers M, Moonen A, Goossens K, Bekooij M (2009) Enabling application-level performance guarantees in network-based Systems on Chip by applying dataflow analysis. IET Computers and Design Techniques

36. Hoskote Y, Vangal S, Singh A, Borkar N, Borkar S (2007) A 5-GHz mesh interconnect for a teraflops processor. IEEE MICRO 27(5):51–61

37. Hu J, Marculescu R (2003) Exploiting the routing flexibility for energy/performance aware mapping of regular NoC architectures. In: Proc DATE

38. Jalbert A et al (2004) ×pipesCompiler: a tool for instantiating application specific networks on chip. In: Proc DATE

39. Jantsch A (2006) Models of computation for networks on chip. In: Proc ACSD

40. Kavaldjiev N (2006) A run-time reconfigurable network-on-chip for streaming dsp applications. PhD

41. thesis, University of Twente

42. Kollig P, Osborne C, Henriksson T (2009) Heterogeneous multi-core platform for consumer multimedia applications. In: Proc DATE
43. Krstic M, Grass E, Gürkaynak F, Vivet P (2007) Globally asynchronous, locally synchronous circuits: ´ overview and outlook. IEEE Des Test Comput 24(5):430–441
44. Lee H, Chang N, Ogras U, Marculescu R (2007) On-chip communication architecture exploration: a quantitative evaluation of point-to-point, bus, and network-on-chip approaches. ACM Trans Des Autom Electron Syst 12(3):1–20
45. Leijten J, van Meerbergen J, Timmer A, Jess J (2000) Prophid: a platform-based design method. J Des Autom Embed Syst 6(1):5–37
46. Liang J, Swaminathan S, Tessier R (2000) aSOC: a scalable, single-chip communications architecture. In: Proc PACT
47. Liu G, Ramakrishnan KG (2001) A*Prune: an algorithm for finding K shortest paths subject to multiple constraints. In: Proc INFOCOM
48. Mangano D, Locatelli R, Scandurra A, Pistritto C, Coppola M, Fanucci L, Vitullo F, Zandri D (2006) Skew insensitive physical links for network on chip. In: Proc NANONET
49. Marculescu R, Ogras U, Peh LS, Jerger N, Hoskote Y (2009) Outstanding research problems in NoC design: system, microarchitecture, and circuit perspectives. IEEE Trans Comput-Aided Des Integr Circuits Syst 28(1):3–21
50. Marescaux T, Mignolet J, Bartic A, Moffat W, Verkest D, Vernalde S, Lauwereins R (2003) Networks on chip as hardware components of an OS for reconfigurable systems. In: Proc FPL
51. Millberg M, Nilsson E, Thid R, Jantsch A (2004) Guaranteed bandwidth using looped containers in temporally disjoint networks within the Nostrum network on chip. In: Proc DATE
52. Moerman K (2007) Embedded vector processor is one way to tune software-defined radios. In: EE Times
53. Moraes F, Calazans N, Mello A, Möller L, Ost L (2004) HERMES: an infrastructure for low area overhead packet-switching networks on chip. Integr VLSI J 38(1):69–93
54. Moreira O, Valente F, Bekooij M (2007) Scheduling multiple independent hard-real-time jobs on a heterogeneous multiprocessor. In: Proc EMSOFT
55. Murali S, De Micheli G (2005) An application-specific design methodology for STbus crossbar generation. In: Proc DATE
56. Nandi A, Marculescu R (2001) System-level power/performance analysis for embedded systems design. In: Proc DAC
57. Nollet V, Marescaux T, Avasare P, Mignolet JY (2005) Centralized run-time resource management in a network-on-chip containing reconfigurable hardware tiles. In: Proc DATE
58. OCP (2007) OCP Specification 2.2. OCP International Partnership
59. Ogras UY, Hu J, Marculescu R (2005) Key research problems in NoC design: a holistic perspective. In: Proc CODES+ISSS
60. Panades I, Greiner A, Sheibanyrad A (2006) A low cost network-on-chip with guaranteed service well suited to the GALS approach. In: Proc NANONET
61. Paukovits C, Kopetz H (2008) Concepts of switching in the time-triggered network-on-chip. In: Proc RTCSA
62. Penning de Vries R (2008) Ic innovations in automotive. In: Plenary session international conference on solid-state and integrated-circuit technology
63. PIBus (1994) PI-Bus Standard OMI 324. Siemens AG, ver. 0.3d edn
64. Pinto A et al (2003) Efficient synthesis of networks on chip. In: Proc int'l conference on computer design (ICCD)
65. Pullini A, Angiolini F, Murali S, Atienza D, De Micheli G, Benini L (2007) Bringing NoCs to 65 nm. IEEE MICRO 27(5):75–85
66. Radulescu A, Dielissen J, Goossens K, Rijpkema E, Wielage P (2005) An efficient on-chip network ˘ interface offering guaranteed services, shared-memory abstraction, and flexible network programming. IEEE Trans CAD Integrated Circuits Syst 4–17
67. Rostislav D, Vishnyakov V, Friedman E, Ginosar R (2005) An asynchronous router for multiple service levels networks on chip. In: Proc ASYNC
68. Rutten M, Pol EJ, van Eijndhoven J, Walters K, Essink G (2005) Dynamic reconfiguration of streaming graphs on a heterogeneous multiprocessor architecture. In: IS&T/SPIE electron imag, vol 5683
69. Scherrer A, Fraboulet A, Risset T (2006) Automatic phase detection for stochastic on-chip traffic generation. In: Proc CODES+ISSS
70. SonicsMX (2005) SonicsMX Datasheet. Sonics, Inc. Available on www.sonicsinc.com
71. Stergiou S, Angiolini F, Carta S, Raffo L, Bertozzi D, de Micheli G (2005) ×pipes Lite: a synthesis oriented design library for networks on chips. In: Proc DATE
72. Stoica I, Zhang H (1999) Providing guaranteed services without per flow management. In: Proc SIGCOMM
73. Tota S, Casu M, Roch M, Macchiarulo L, Zamboni M (2009) A case study for NoC-based homogeneous MPSoC architectures. IEEE Trans Very Large Scale Integr 17(3):384–388
74. van Berkel K, Heinle F, Meuwissen P, Moerman K, Weiss M (2005) Vector processing as an enabler for software-defined radio in handheld devices. EURASIP J Appl Signal Process 2005:2613–2625
75. Weber WD, Chou J, Swarbrick I, Wingard D (2005) A quality-of-service mechanism for interconnection networks in system-on-chips. In: Proc DATE
76. Wentzlaff D, Griffin P, Hoffmann H, Bao L, Edwards B, Ramey C, Mattina M, Miao CC, Brown JF, Agarwal A (2007) On-chip interconnection architecture of the tile processor. IEEE MICRO 27(5):15–31

77. Wielage P, Marinissen E, Altheimer M, Wouters C (2007) Design and DfT of a high-speed area-efficient embedded asynchronous FIFO. In: Proc DATE

78. Wiggers MH, Bekooij MJ, Smit GJ (2008) Buffer capacity computation for throughput constrained streaming applications with data-dependent inter-task communication. In: Proc RTAS

79. Wiklund D, Liu D (2003) SoCBUS: switched network on chip for hard real time embedded systems. In: Proc IPDPS

80. Wingard D, Kurosawa A (1998) Integration architecture for system-on-a-chip design. In: Proc CICC

81. Wolkotte P, Smit G, Rauwerda G, Smit L (2005) An energy-efficient reconfigurable circuit-switched network-on-chip. In: Proc IPDPS

CHAPTER 12 AN IEEE 1149.X EMBEDDED TEST COPROCESSOR

Ukbagiorgis Iyasu Gebremeskel and José Manuel Martins Ferreira

Keywords: Built-In Test, Boundary-Scan, Embedded Coprocessors, MicroBlaze, IEEE 1149.1, IEEE 1149.7

INTRODUCTION

After nearly 25 years of industry acceptance, starting immediately after its approval as a standard in 1990 [1], the IEEE 1149.1 Boundary-Scan (BS) Architecture and Test Access Port is now being used in a range of application areas that largely exceeds its original scope, which is restricted to structural testing of digital printed circuit boards. Newer application domains where BS is being used, such as in-circuit debugging, require more efficient and powerful data transfer mechanisms, and lead many designers to implement workarounds that compromised full 1149.1-compliance. In recognition of the need for a more powerful test standard, compatible with all the investment made in 1149.1, the IEEE 1149.7 Standard for Reduced-Pin and Enhanced-Functionality Test Access Port and Boundary-Scan Architecture was approved in 2009 [2] , and offered a far more powerful solution to test and applications debug. The development of solutions that demonstrate its ability to coexist with 1149.1 applications, while at the same time supporting the more powerful scan transfer formats required for debugging multi-core architectures, represents a fundamental step to unlock the power of the new standard and to promote its wider acceptance [3] -[5] .

This paper presents a microprogrammed architecture for an embedded coprocessor dedicated to IEEE 1149.1/ 1149.7 test operations. Its current instruction set supports all IEEE 1149.1 test operations (state transition, shift operations, shift and compare), and an additional subset of IEEE1149.7 test operations designed as proof of concept for the proposed architecture (optimized scan format Oscan1, zero-bit DR scan and escape sequences). The following section briefly describes the evolution from IEEE 1149.1 to 1149.7. Section 3 describes the micro-programmed architecture of the proposed embedded test coprocessor, and Section 4 explains the test command design process. Section 5 presents implementation cost and performance data. The main conclusions of this work are presented in the last section.

IEEE 1149.X STANDARDS AND TEST COMMAND SET

Several IEEE 1149.x standards and proposed standards were developed by the test community since the mid- 1980s, when the Joint Test Action Group (JTAG) initiated the development of the test technology that became known as the IEEE 1149.1 Standard Test Access Port and Boundary-Scan Architecture. The approval of IEEE 1149.1 in 1990 marked the beginning of a series of IEEE 1149.x test standards that include successful and unsuccessful attempts to provide industry-accepted production test technologies [6] . The IEEE 1149.5 Standard for Module Test and Maintenance Bus (MTM-Bus) Protocol [7] , and the IEEE 1149.4 Standard for a Mixed-Signal Test Bus [8] , are two examples of proposals that met little or no industry acceptance, in spite of having become recognized IEEE standards. The IEEE 1149.6 Standard for Boundary-Scan Testing of Advanced Digital Networks [9] deserved better attention, mostly because it overcomes a specific limitation of boundary-scan—the testing of AC-coupled differential interconnections. Far more successful that all its 1149.x successors, the IEEE 1149.1 boundary-scan test technology is now being used to address application areas that go far beyond its original development

Citation: Gebremeskel, U. and Ferreira, J. (2014) An IEEE 1149.x Embedded Test Coprocessor. Circuits and Systems, 5, 170-180. doi: 10.4236/cs.2014.57019

scope. The use of the standard 4-pin test access port (TAP) to control in-circuit debugging is particularly worth mentioning, both because it gained wide acceptance since the early years of boundary-scan, and also because it has led to the creation of TAPs whose operation includes behavior that deviates from the behavior specified by IEEE Std 1149.1-2001 [2] .

The IEEE 1149.7 standard, approved in 2009, offers a framework that ensures interoperability between 1149.1 and 1149.7 components, while enabling enhanced test and debug features, and a reduced 2-pin TAP (where mode select, data in and data out information are multiplexed in a single TMSC pin). Besides supporting several types of advanced scan formats (MScan, OScan, SScan), which enable a variety of tradeoffs between capability and performance, the IEEE 1149.7 test architecture also supports a power-down mode to reduce consumption when the test and debug logic is not in use. The original BS architecture, illustrated in Figure 1, remains at the core of the new IEEE 1149.7 standard, and helps us to understand why most IEEE 1149.x test operations can be implemented with a reduced set of basic test commands, that are able to control state transition and to carry out shift, and shift and compare operations.

Table 1 presents test command set that is supported by our embedded test coprocessor. Besides covering all the basic 1149.1 requirements, it includes four additional commands to illustrate 1149.7 test operations. This set offers close equivalents to the main Serial Vector Format (SVF)[10] commands, which are accepted by most test equipment manufacturers. The following section discusses the alternatives available to implement its control path, and offers a detailed description of the micro-programmed architecture that was chosen for this purpose.

A MICRO-PROGRAMMED ARCHITECTURE FOR AN IEEE 1449.7 COPROCESSOR

The first step for implementing a controller architecture supporting the set of commands referred in the previous section consists of building a formal representation of their functionality, bringing into evidence the corresponding control and data flow operations. The data flow operations will determine the blocks required in the coprocessor data path, which will be implemented with regular sequential circuits. There is a higher degree of freedom in what concerns the implementation of the control path, where a hardwired or a microprogrammed architecture can be used to implement the corresponding control flow operations. Notice also that the formal representation of each test command will be influenced by the decision of implementing it in Moore or Mealy form. A Moore machine will only update its outputs (the control signals to the data path) upon the rising edge of the system clock, while Mealy machines can update their outputs at any moment. To illustrate the influence of this choice, Figure 2(a) shows an excerpt of an ASMD chart that illustrates a Moore specification for the SHF1 test command.

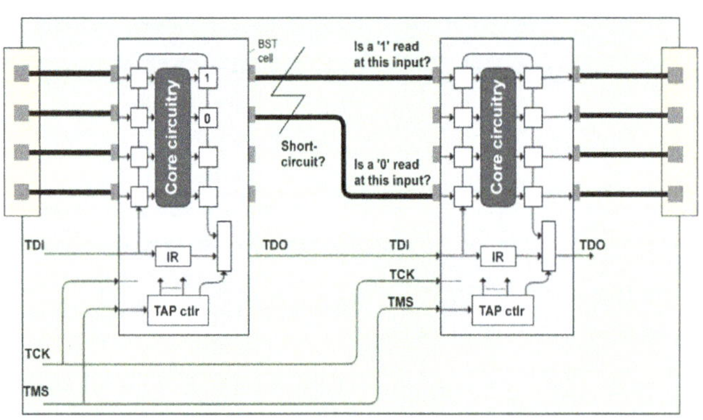

Figure 1. Original IEEE 1149.1 boundary-scan test infrastructure for structural fault detection.

Table 1. IEEE 1149.1/1149.7 test command set supported by the proposed embedded test coprocessor.

Test command	Description
RESET	Takes the boundary-scan logic to Test-Logic-Reset (equivalent to SVF "STATE RESET")
TMS0, TMS1	Set TMS to 0/1 and generate one TCK clock pulse (enabling the implementation of any SVF "STATE" command)
MTCK N	Sets TMS to 0 and generates N TCK pulses (enables the multiple TCK tests carried out by SVF "RUNTEST")
SHF1 N X	Shifts an N-bit bitstream (X) into the [instruction | selected data]register(s) (data bits shifted out are ignored)

SHFCP1 N X, Y, Z	Shifts an N-bit bitstream (X) into the [instruction | selected data]register(s), and compares the output bitstream with its expected response (Y), in all positions indicated by the mask bitstream (Z) (enables the scan & compare "SDR" and "SIR" SVF commands)
ZBS N	Issues N zero bit scan sequences by counting N TMS edge counts (used to generate IEEE 1149.7 commands and to set the control level)
ESC N	Issues N escape sequences (used to select, deselect, reset, and set up other IEEE 1149.7 specific operations)
SHF7O1 N X	Equivalent to SHF1 above, but using the 1149.7 advanced scan format OScan1
SHFCP7O1 N X, Y, Z	Equivalent to SHFCP1 above, but using the 1149.7 advanced scan format OScan1

States 2 and 3 in Figure 2(a) include conditional output boxes used to define Mealy outputs that depend on the condition indicated in the preceding decision box. States 2, 4 and 5 comprise non-empty state boxes used to define Moore outputs, which will be asserted while the controller remains in the corresponding state. The implementation of the same functional behavior in the form of a Moore machine will increase the number of states, since each conditional output box will have to be converted into an equivalent state box, generating a corresponding extra state. The formal representation for the same ASMD chart excerpt, now in the form of a Moore machine, is illustrated in Figure 2(b).

The conversion into a Moore machine increased the number of states by a factor of 2.25 (from 4 to 9). However, this higher number of states does not imply a proportional degradation of cost or performance, be it in the number of logic gates or microprogram memory positions (spatial resources), or in the number of clock cycles (speed). A simple benchmarking of the execution speed can take place by considering the time required to move from state 2 to state 4, assuming that: 1) conditions B and C hold true, and 2) condition D holds false for five times.

The corresponding state transition for the Mealy representation of Figure 2(a) will be 2-3-5-3-5-3-5-3-5-3-5- 3-4, requiring a total of 12 clock cycles. Considering the same initial and final states, and the same assumptions, the equivalent transition for the Moore representation will in this case go through states 2-3-8-5-9-5-9-5-9-5-9-

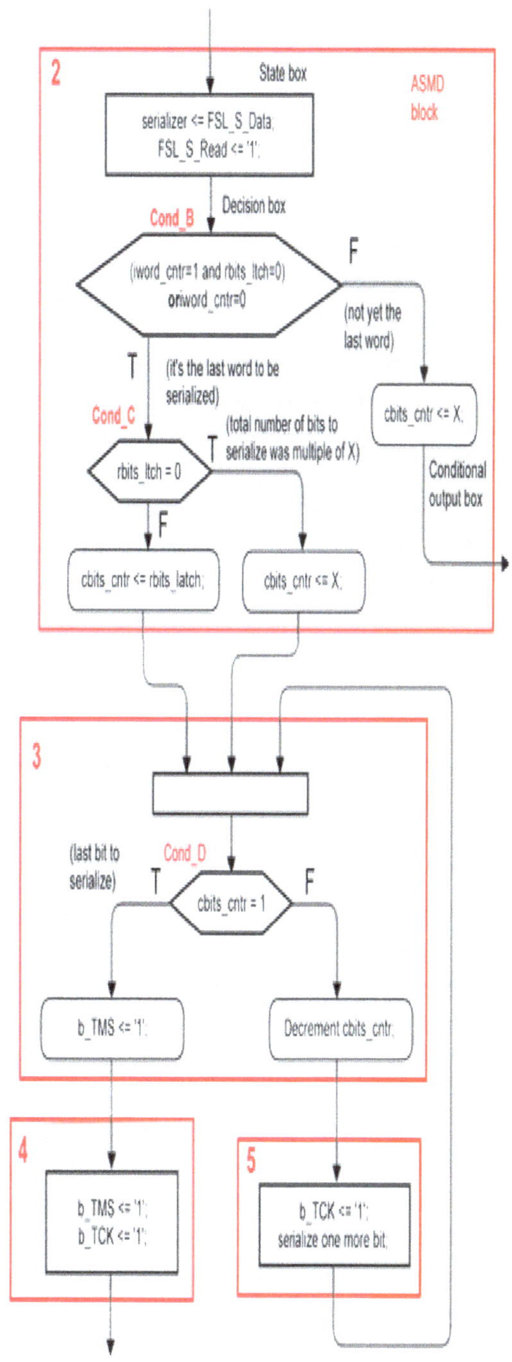

(a)

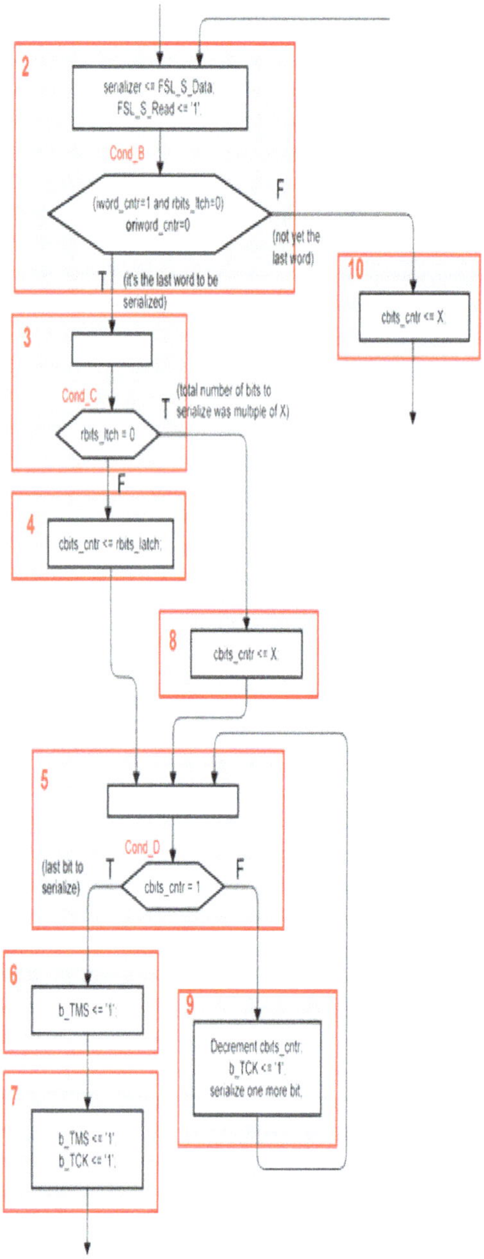

(b)

Figure 2. Partial representation of the ASMD chart for SHF1. (a) Mealy representation. (b) Moore representation.

5-9-5-6-7, and requires a total of 15 clock cycles (that is to say, 1.25 times more than the corresponding Mealy representation).

In order to compare the corresponding implementations in terms of logic resources/FPGA floorspace, a choice will have to be made between a hardwired or a microprogrammed control path architecture. The following reasons explain our choice of a microprogrammed coprocessor architecture:

- A hardwired controller needs to be completely redesigned when a new test command is required, and the designer will have to code each new ASMD chart in the chosen hardware description language (e.g. VHDL, Verilog).
- Even if major differences are not to be expected, a hardwired architecture will correspond to a new sum-ofproducts (or a similar canonical form) structure for each set of test commands, meaning that there will be variations in the critical path and maximum propagation delay (on the contrary, there will be no variation in the control elements of a microprogrammed architecture, since any additional commands correspond solely to further positions added to the microprogram memory).
- Assuming that the "micro-operation set" is able to cover the primitive constructs required to implement new test commands, the designer does not have to write VHDL/ Verilog code whenever a new command is added, but simply to translate the respective ASMD chart into the corresponding microprogram memory bank positions.

The basic control path architecture for a microprogrammed implementation is shown in Figure 3(adapted from [11]). In this case, the most significant address bits of the microprogram memory are determined by the test command opcode (which is loaded into the Bank_reg latch), and the ASMD state encoding defines the least significant address bits.

The architecture illustrated in Figure 3 is best in terms of simplicity, and enables a very straightforward implementation of any test command—each state in the ASMD chart will correspond to one word in the microprogram memory, and the operation flow can simply be specified as a sequence of CONTINUE, JUMP or

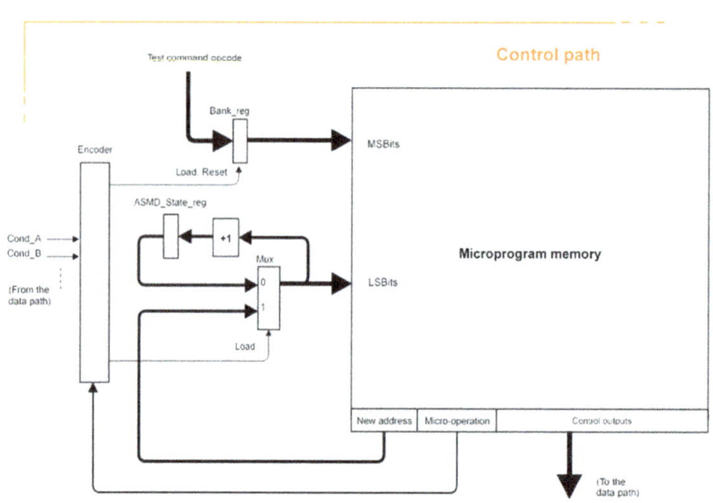

Figure 3. Basic microprogrammed architecture for a Moore control path.

BRANCH IF microinstructions, directing the state transition from beginning to end. However, it can only implement Moore machines, since each ASMD state selects a single microprogram memory position. Since the ASMD blocks frequently contain conditional output boxes, or more than one decision box, this means that the simplicity of the architecture shown in Figure 3 comes at the price of preprocessing the ASMD chart, in order to ensure a pure Moore behavior. The total number of states increases, and so does the number of microprogram memory positions, as well as the number of system clock cycles needed to complete the execution of the corresponding test command (one clock cycle per ASMD chart state).

The general rule for preprocessing the ASMD charts consists of eliminating all conditional output boxes (which will become unconditional outputs specified in a state of their own), and splitting the state when more than one decision box is present. In addition, and since the most significant bits come directly from the test command opcode, the number of microprogram memory positions used to implement each command is fixed. This represents a waste of FPGA floorspace, since the most complex command, with the longest ASMD chart representation, will dictate the number of positions that will be used for all other commands. The control path architecture illustrated inFigure 3 is able to implement any Moore ASMD chart, and would take 9 memory positions to implement the excerpt represented in Figure 2(b). If this was the highest number of states in any command, we would need 4 least significant bits, meaning that the total storage requirements would be given by $2^4*O = 16*O$, where O is the number of opcodes comprised in the test command set.

If we want to enable a Mealy implementation, the least significant address bits will have to be driven from data path conditions. This solution enables any number of decision boxes per state, provided that the corresponding conditions are used to drive the least significant address bits of the microprogram memory. The main drawback is that the number of microprogram memory positions will be equal to $S*2^D$, where S is the number of states, and D is the maximum number of decision boxes existing in a single state (16 memory positions for the example presented earlier). For a test command set with O opcodes, we would have a total microprogram memory storage requirement given by $S*2^D*O$. While this modification means that we will now have two (or a power of two) microprogram memory positions for each ASMD chart state, the end result is not necessarily an explosion of the microprogram memory space, since state decomposition is restricted to the need of preventing multiple decision boxes per ASMD state (a situation that is rather seldom).

The non-limited simple Mealy architecture enables the fastest implementation. On the other hand, it is the most expensive in terms of microprogram memory storage. The intermediate simple Mealy solution limited to one decision box per state is likewise an intermediate solution in terms of speed vs. microprogram memory storage, while the Moore representation is cheapest in terms of microprogram memory storage. It is also the slowest, although the number of clock cycles is not proportional to the number of states (instead it is dictated by the path through the ASMD chart).Table 2 summarizes the pros and cons of three alternatives: Moore (no conditional outputs allowed, maximum state decomposition), Mealy 1 (one-level Mealy, enabling one decision box and its corresponding conditional output boxes per state), and Mealy 2 (two-level Mealy, enabling up to two decision boxes and four conditional output boxes). Since preprocessing takes place only once, and the difference in speed is small (cf. the simple benchmarking of execution speed that was presented in Figure 2(a) and Figure 2(b)), we adopted the Moore architecture that is represented in Figure 4. Its main advantage over the basic Mealy and basic Moore (Figure 3) architectures consists of eliminating the waste of memory positions that takes place in those two solutions, since the microcode storage requirements are in this case limited to the number of states in the ASMD chart (instead of the being fixed, and dictated by the ASMD chart with the highest number of states).

The specific nature of scan test infrastructures dictates that the number of required primitive constructs (micro-operations) is very small. The data path architecture will therefore have a small number of elements, consisting only of counters, latches and serializers. The conditions associated with the operation of these data path elements are easy to typify, and are limited to detecting if the latches and counters reached one or zero. The proposed test coprocessor architecture may be represented as shown in Figure 5.

Table 2. Pros and cons of Moore, Mealy-1, and Mealy-2 microprogrammed control path implementations.

Topology	Pros	Cons	Comments
Moore	Maximum simplicity	Lowest speed	Preprocessing required
Mealy 1	Good speed upgrade	Higher number of memory positions	Minimum preprocessing
Mealy 2	Marginal speed upgrade over Mealy 1	Highest number of memory positions, most of which will not be used	No preprocessing required (if no. of decision boxes per state is ≤ 2)

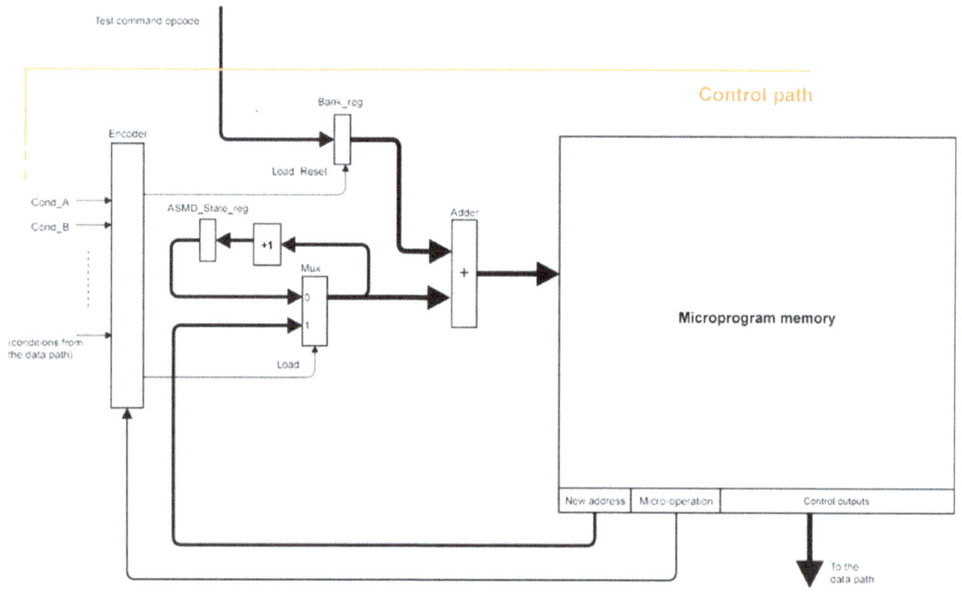

Figure 4. Microprogrammed architecture adopted for the proposed embedded test coprocessor.

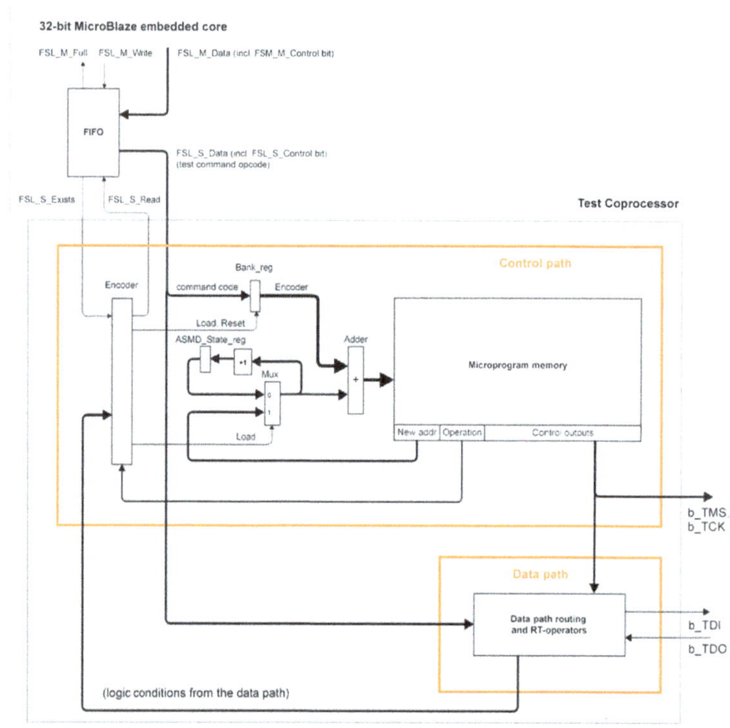

Figure 5. The proposed test coprocessor architecture and interface method.

TEST COMMAND DESIGN

Table 3. Content of the microprogram memory for the MTCK test command.

ASMD state	µprogram ROM		µprogram ROM: control outputs									
	New_Addr	µoperation	iL	iD	rL	cL	cD	cM	sL	sS	bS	bK
0	X	CONT	0	0	0	1	0	0	0	0	0	0
1	Offset to pos. 3	BIF/A	0	0	0	0	0	0	0	0	0	0
2	Offset to END pos.	JUMP	0	0	0	0	0	0	0	0	0	1
3	Offset to pos. 1	JUMP	0	0	0	0	1	0	0	0	0	1

This section presents the micro-operation set that is supported by the microprogrammed control path, and explains the sequence of steps that will enable any designer to add further commands, in order to expand the application domain of the proposed test coprocessor (e.g. further IEEE 1149.7 operations). The microprogrammed control path architecture illustrated in Figure 4 is able to implement the control flow associated to any ASMD chart that is specified in the form of a Moore machine. Each state in the ASMD chart corresponds to one position in the microprogram memory, which comprises the three fields shown in Figure 4 and Figure 5:

- The leftmost bits represent the new address, to be used whenever the next state encoding is different from the current state encoding plus 1 (in which case we'll jump into a new address, instead of incrementing the current address). The number of bits in this field is dictated by the maximum number of states in the ASMD chart of a single test command.
- The middle field contains the micro-operation that represents the required control flow. The implementation of any ASMD chart can be carried out using three basic types of micro-operations: 1) CONTINUE (i.e. increment the current microprogram memory address); 2) JUMP TO ADDRESS (i.e. load the "new address" that is represented in the leftmost bits of the current microprogram memory position); and 3) BRANCH IF CONDITION TO ADDRESS (jumps to the indicated address if the condition is true, continues to the next address otherwise). This third type actually generates a variety of different micro-operations, i.e. BRANCH IF CONDITION_A TO ADDRESS is formally different from BRANCH IF/CONDITION_A TO ADDRESS (branch if not_condition). Each new test command to be supported will most likely require additional BRANCH IF micro-operations, to cope with the specific conditions associated with its execution. The number of bits in this field is determined by 2 (CONTINUE, JUMP TO ADDRESS) plus the total number of BRANCH IF micro-operations required.
- The rightmost field comprises all the control bits that determine the data flow operations indicated in the ASMD chart. In a horizontal microprogrammed architecture, such as the one that was adopted in this work, the number of bits in this field is equal to the total number of control bits required by the data path elements, plus all additional bits that are directly connected to test access port pins (e.g. board TMS and board TCK).

In order to expand the test command set supported by the coprocessor, the designer shall proceed as follows:

1) Draw the ASMD chart specifying the operation of the required test command.
2) If the ASMD chart contains conditional output boxes (Mealy machine), split those states so as to convert all conditional output boxes into state boxes (convert from Mealy to Moore).
3) Once the Moore ASMD chart is ready, fill in the microprogram memory template with the micro-operations and control bit patterns corresponding to each ASMD state.
4) Update the content of the microprogram memory.

As an illustrative example, Figure 6(a) shows an initial ASMD chart that describes the operation of a MTCK N test command. This command is needed to carry out various types of built-in self-test functions (e.g. all those that rely on pseudo-random pattern generation and parallel signature analysis modes), and generates N test clock (b_TCK) pulses, while keeping b_TMS at "0". MTCK starts by loading the number of required b_TCK pulses (present in the FSL_S_Data bus) into one of the data path counters (cbits_cntr in this case), which will be decremented for each b_TCK pulse. As represented in Figure 6(a), state 1 requires two microprogram memory positions, since the decrement control signal for cbits_cntr may, or may not, be active in this state, according to the condition shown. This example also shows that splitting a state, in order to ensure that a conditional output box is converted into a corresponding state box, does not necessarily increase the number of states in the ASMD chart. Figure 6(b) represents the equivalent Moore ASMD chart, which contains the same number of states. Each state in the chart represented in Figure 6(b)corresponds to a single microprogram memory position. We are now ready to move into the third test command design step, where the microprogram memory template is filled to represent all control and data flow operations associated to the execution of MTCK. Table 3 shows the content of the four microprogram memory positions that are needed to specify the execution of this test command set.

As this example shows, expanding the test command set simply consists of updating the content of the microprogram memory, releasing the designer from the need to understand and modify the VHDL code that describes the control path architecture.

IMPLEMENTATION COST AND PERFORMANCE

To evaluate the overall performance of the proposed embedded test coprocessor, Table 4 and Table 5 show the data that were collected for each test command separately:

- The columns showing data for each test command (the four rightmost columns) correspond to the implementation of a single test command, without the FSL interface.
- Since the FSL interface is predesigned and independent of the proposed microprogrammed architecture, all the tables include two columns showing the implementation data for all test commands when no FSL interface is present, and when a single 1-word 32-bit FIFO interface is added from the MicroBlaze to the test coprocessor.

Table 4 shows the comparison of the logic resources needed by various IEEE 1149.1 test commands with a 4-pin TAP test infrastructure. TMS1 and TMS0 belong to the same group of results, and the same happens with MTCK and RESET, so each second command was omitted in all the tables to improve readability. It is important to notice that the usage of logic resources imposed by the most complex test command (SHFCP1) dictates the cost of the full implementation—the sole implementation of SHFCP1 requires practically the same resources as the full implementation of all the IEEE 1149.1 test commands that are presented in Table1 The number of

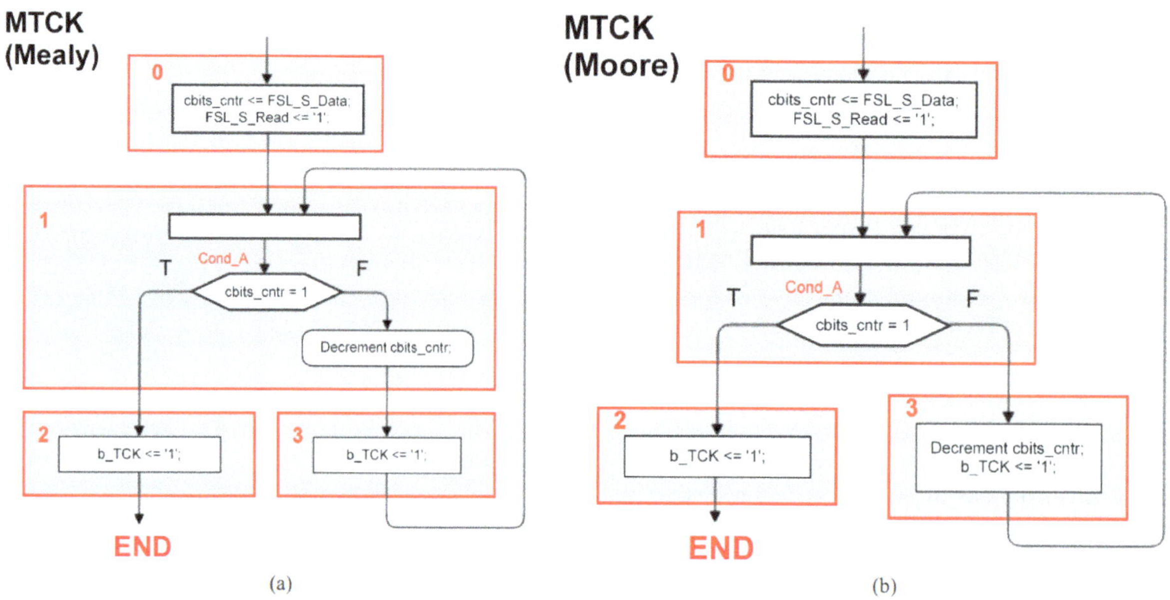

Figure 6. ASMD chart for MTCK. (a) Mealy representation. (b) Moore representation.

(iL: iwordL; iD: iwordD; rL: bitsL; cL: cbitsL; cD: cbitsD; cM: cbitsM; sL: serL; sS: serS; bS: b_TMS; bK: b_TCK).

Table 4. Logic resources usage by the IEEE 1149.1 test commands.

Parameter	All (no FSL)	All (one FSL)	Only TMS1	Only MTCK	Only SHF1	Only SHF-CP1
No. of slice registers	388 (1%)	374 (2%)	88 -	120 -	274 -	338 -
No. of slice LUTs	383 (4%)	424 (4%)	125 -	207 -	307 -	380 -
No. of memory positions	52	52	4	7	16	23

Table 5. Logic resources usage by the IEEE 1149.7 test commands.

Parameter	All (no FSL)	All (one FSL)	Only ESC	Only ZBS	Only SHF7O1	Only SHFCP7O1
No. of slice registers	160 (1%)	377 (2%)	75 -	75 -	144 -	160 -
No. of slice LUTs	232 (4%)	461 (5%)	179 -	146 -	178 -	224 -
No. of memory positions	98	98	9	23	32	40

microprogram memory positions when the full set of commands is implemented, can be calculated by adding the equivalent value for each command individually, but taking into account that the first two positions are common to all the test commands.

The same comparison has been done for the IEEE 1149.7 test commands, and the summary of the collected data is shown in Table5 This table shows that the usage of logic resources imposed by the most complex test command (SHFCP7O1) again dictates the cost of the full implementation—the sole implementation of SHFCP7O1 requires almost the same resources as the full implementation of all the IEEE 1149.7 test commands that are presented in Table1 The number of microprogram memory positions required when implementing the full set of test commands can be calculated as indicated previously for the 1149.1 commands.

With respect to timing performance, we again notice that the most complex command is the main contributor to determine the minimum period/maximum frequency of the embedded test coprocessor operation. Table 6 shows the timing performance of the implemented IEEE 1149.1 test commands with a 4-pin TAP. The full implementation, in the case of "All (no FSL)", is circa 30% slower than what would correspond to SHFCP1 alone. The minimum period is very close to the maximum combinational path delay, in the case of the sole implementation of the test commands TMS1, TMS0, MTCK, or RESET, but it is less so as we move into the more complex test commands (SHF1 and SHFCP1).

Table 7 shows the timing performance of the implemented IEEE 1149.7 test commands with a 2-pin TAP. One can notice that, as we move to the right, the minimum period/maximum frequency is increasing. This shows that the most complex test command (SHFTCP7O1) is the major contributor to determine the minimum period/ maximum frequency of the operation of the proposed embedded test coprocessor. The data presented in Table 7 shows that the proposed embedded test coprocessor is able to run at TCK frequencies slightly above 70% of the FPGA system clock (which in the case of the Nexys 3™ board runs at 100 MHz).

CONCLUSIONS

Several IEEE 1149.1 test controller solutions have been developed over the years [12] -[17] , but little attention has been given to controller architectures proposed as embedded coprocessors. The work described in this paper offers two main contributions with this respect, by proposing:

1) An embedded test coprocessor that has the capability to test any IEEE 1149.x-compatible system. This coprocessor supports a wide range of testing scenarios for embedded systems based on single or multi-core 32-bit MicroBlaze CPUs, from built-in self-test to online fault detection and diagnosis.
2) A microprogrammed control path architecture for the test coprocessor, enabling a straightforward expansion of the test command set to cope with additional application domains, e.g. those made possible by IEEE 1149.7. The simplicity of the selected architecture enables a relatively fast execution of the test commands, reaching above 70% of the Nexys 3™ board system clock (100 MHz).

Table 6. Timing performance for the IEEE 1149.1 test commands.

Parameter	All (no FSL)	All (one FSL)	Only TMS1	Only MTCK	Only SHF1	Only SHFCP1
Min. period Max. freq.	12.276 ns 81.457 MHz	12.315 ns 81.199 MHz	6.053 ns 165.211 MHz	6.249 ns 160.037 MHz	7.722 ns 128.662 MHz	11.246 ns 88.917 MHz
Max. comb. path delay	5.157 ns	8.102 ns	5.157 ns	5.385 ns	5.385 ns	5.519 ns

Table 7. Timing performance for the IEEE 1149.7 test commands.

Parameter	All (no FSL)	All (one FSL)	Only ESC	Only ZBS	Only SHF7O1	Only SHF-CP7O1
Min. period Max. freq.	12.657 ns 79.005 MHz	13.655 ns 73.235 MHz	7.756 ns 128.929 MHz	8.302 ns 120.449 MHz	11.469 ns 87.192 MHz	12.249 ns 81.639 MHz
Max. comb. path delay	-	5.94 ns	-	-	-	-

An FSL interface is used to interact with the MicroBlaze for exchanging the command opcode, the arguments of the various functions, and the test result. The FSL is a very popular coprocessor interface method that uses a simple hardware protocol. It is supported by a flexible, yet dedicated instruction set to write to the output and read from the input port. The 32-bit wide FSL bus interface used offers a dedicated point-to-point data streaming interface. Its point-to-point nature and its minimum hardware requirement are the greatest advantages of this interface method. Because the FSL channels are dedicated, no arbitration or bus mastering is required, ensuring an extremely fast interface with very low data latency.

The proposed architecture has the capability to cope with emerging requirements of test and debug standards based on scan test infrastructures. The microprogrammed control path architecture made the embedded test coprocessor scalable, and the FSL interface enabled an optimized solution for IEEE 1149.x test infrastructures, where the most time-consuming functions were executed in hardware, and the remaining functions were implemented by software at a higher abstraction level.

REFERENCES

1. (2001) IEEE Standard Test Access Port and Boundary Scan Architecture. IEEE Std 1149.1-2001, 212 p.http://dx.doi.org/10.1109/IEEESTD.2001.92950
2. (2009) IEEE Standard for Reduced-Pin and Enhanced-Functionality Test Access Port and Boundary-Scan Architecture. IEEE Std 1149.7-2009, 985.http://dx.doi.org/10.1109/IEEESTD.2010.5412866 [Paper reference:2]
3. Ley, A.W. (2009) Doing More with Less—An IEEE 1149.7 Embedded Tutorial: Standard for Reduced-Pin and Enhanced-Functionality Test Access Port and Boundary-Scan Architecture. International Test Conference, Austin, 1-6 November 2009, 1-10.http://dx.doi.org/10.1109/TEST.2009.5355572
4. Williams, M. (2009) Low Pin-Count Debug Interfaces for Multi-Device Systems. ARM White Paper. http://goo.gl/KyyVvP
5. Fernandes, F.R., Machado, R.J.S., Ferreira, J.M.M. and Gericota, M.G.O. (2012) IEEE Std 1149.7: What, Why, Where? Proceedings of the 27th Conference on Design of Circuits and Integrated Systems, Avignon, 28-30 November 2012, 118-123.
6. Ungar, L.Y., Bleeker, H., McDermid, J.E. and Hulvershorn, H. (2001) IEEE-1149.x Standards: Achievements vs. Expectations. IEEE Systems Readiness Technology Conference, Valley Forge, 20-23 August 2001, 188-205.http://dx.doi.org/10.1109/AUTEST.2001.948964
7. IEEE Standard for Module Test and Maintenance Bus (MTM-Bus) Protocol (1996) IEEE Std 1149.5-1995. http://dx.doi.org/10.1109/IEEESTD.1996.80835
8. IEEE Standard for a Mixed-Signal Test Bus (1999) IEEE Std 1149.4-2010 (Revision of IEEE Std 1149.4-1999). http://dx.doi.org/10.1109/IEEESTD.2011.5738198
9. IEEE Standard for Boundary-Scan Testing of Advanced Digital Networks (2003) IEEE Std 1149.6-2003. http://dx.doi.org/10.1109/IEEESTD.2003.94249
10. Serial Vector Format Specification (1994) Revision E. http://goo.gl/bQQkGI
11. Lynch, M.A. (1993) Microprogrammed State Machine Design. CRC Press, Boca Raton.
12. Yau, C.W. and Jarwala, N. (1990) The Boundary-Scan Master: Target Applications and Functional Requirements. Proceedings of the IEEE International Test Conference, Washington DC, 10-14 September 1990, 311-315.
13. Ferreira, J.M., Matos, J.S. and Pinto, F.S. (1992) Automatic Generation of a Single-Chip Solution for Board-Level BIST of Boundary Scan Boards. Proceedings of the European Design Automation Conference, Brussels, 16-19 March 1992, 154-158. http://dx.doi.org/10.1109/EDAC.1992.205913
14. SN54ACT8990, SN74ACT8990, IEEE STD 1149.1 (JTAG) TAP Masters with 16-Bit Generic Host Interfaces (1997) Texas Instruments Datasheet.
15. Chen, S.J., Zhou, Y., Zhu, D.X. and Guo, S.L. (2011) Design of High-Speed Boundary-Scan Master Controller Base on SOPC. Proceedings of the 2nd International Conference on Mechanic Automation and Control Engineering, Hohhot, 15-17 July 2011, 1195-1198.http://dx.doi.org/10.1109/MACE.2011.5987152.
16. Kaur, M. and Singh, B. (2012) VHDL Implementation of Test Access Port Controller. International Journal of Engineering Science and Technology (IJEST), 4.
17. Cabral, C.J. (2012) Design and Implementation of an IEEE 1149.7-Compliant cJTAG Controller for Debug and Trace Probe. Master of Science in Engineering Report, University of Texas at Austin, Austin, 55 p.

CHAPTER 13

MODELLING FIELD BUS COMMUNICATIONS IN MIXED-SIGNAL EMBEDDED SYSTEMS

Mohamad Alassir, Julien Denoulet, Olivier Romain, Abraham Suissa and Patrick Garda

INTRODUCTION

Nowadays, the design of embedded systems is a key issue for the electronics industry [1]. Many ongoing research projects address this issue by increasing the design abstraction level from the circuit to the system level [2]. However, most embedded systems gather analog and digital electronics with embedded processor cores running large amounts of embedded software. A number of issues are unfortunately discovered after the conception of embedded systems prototypes. These issues include for example coupling between analog and digital functions or unexpected hazard resulting from unexpected low level effects. Of course, discovering such issues at the prototyping step and revising the design and its implementation to get rid of them is time and money consuming. Addressing these issues at the design time requires a mixedsignal model gathering digital and analog functions [3]. A number of approaches exist to provide these mixed-signal models. They are considering different levels of abstraction and different models of computations. These levels and models are simulated with different tools and cosimulation is used to cover the relations between them. To be more specific, we observe that neither VHDL-AMS nor SystemC are perfectly well suited to model the architecture of these embedded systems. On the one hand, VHDLAMS is perfect to model analog and digital electronics, but the simulation speed of processor cores is too slow in VHDL for the simulation of embedded software [4]. On the other hand, SystemC is perfect to model digital electronics and processor cores executing embedded software, but it cannot model analog electronics [5]. These observations led to the proposal of SystemC-AMS, for which beta releases of the language and the simulation engine are now available [6]. SystemC-AMS provides, in addition to the SystemC discrete-event simulation kernel, continuous-time models of computations for conservative and nonconservative system modelling [7–9]. Moreover, a specific AMS working group was created by OSCI [10] in 2006 to specify SystemC-AMS [11]. We want to investigate the potential of SystemC-AMS to provide a unified framework for the design of embedded systems. It could give the advantages of multiple levels and multiple computational models in a single language based on an open standard. Of course, as SystemC-AMS is still under specification, it is unclear as of today if this is the most valid approach, or if it provides significant gains over other approaches, but the objective of our work presented in this paper is to contribute to this specification and to provide the

AMSWG with some applications of mixed-signal embedded systems. Considering the long-term challenge we are addressing and the need for specific performances data to provide a significant insight, we chose a pragmatic step-by-step research strategy, and we begun with a rather simple case of mixed-signal embedded system. In many different applications, (automotive, industrial control, etc.), embedded systems are typically formed by a set of nodes connected through field busses such as I2C or CAN. The nodes are typically mixed analog and digital SOCs, including one (or several in the near future) processor core(s) and an array of peripheral interfaces. In its most usual form, the node is simply a microcontroller, connected to various sensors or actuators. In this paper, we show how to model the field bus communications

Citation: Mohamad Alassir, Julien Denoulet, Olivier Romain, Abraham Suissa, and Patrick Garda, Modelling Field Bus Communications in Mixed-Signal Embedded Systems, doi:10.1155/2008/134798.

between the nodes of an embedded system. Our methodology is based on three main parts: firstly, a generic architecture of an I/O controller realizing the interface between the processor cores and an I2C field bus; secondly, a generic architecture for the analog part of this interface, and a software tool to get the parameters of this interface from industrial IBIS files; thirdly, a generic model of the bus lines. Hence, our method is generic and it can be applied to any field bus. This paper is based on our previous work on the design and modelling of an I2C bus controller in VHDL [12] and SystemC-AMS [13, 14]. It is an extended revision of the (unpublished) paper presented in [15]. For didactical purposes, the paper is organised around a platform presented in Figure 1. It represents a common situation, where a microcontroller is connected to a Systemon-Chip through a field bus. Our model includes discreteevent models for the digital cores, implemented in SystemC, and continuous-time models for the analog interfaces and for the bus lines, implemented in SystemC-AMS. Whereas this methodology is generic, we will present it in the specific case of the I2C and CAN busses. The paper is divided into three main sections. Section 2 describes the architecture of the digital I/O controller. Section 3 presents the architecture of the analog I/O, the method to include data from an IBIS model into an equivalent SystemC-AMS description of a circuit's I/O buffers, and the electrical model of the bus lines. Section 4 gives several simulation results of field bus communication on the platform and estimates the overhead of the analog parts of the models compared to a totally digital simulation.

DIGITAL MODELS

The modelling platform discussed below is presented in Figure 2. It is centered on a field bus that connects four nodes: among them two can be masters on the bus, while the remaining two can only be slaves. The two masters are a 8051 microcontroller in association with a bus controller and a MIPS-based system on Chip. The two slaves are memory devices which will be accessed in the course of simulation by the two masters. Each node is modeled with a mixed SystemC/SystemC-AMS description, where a device's digital core is modeled in SystemC and its analog interface

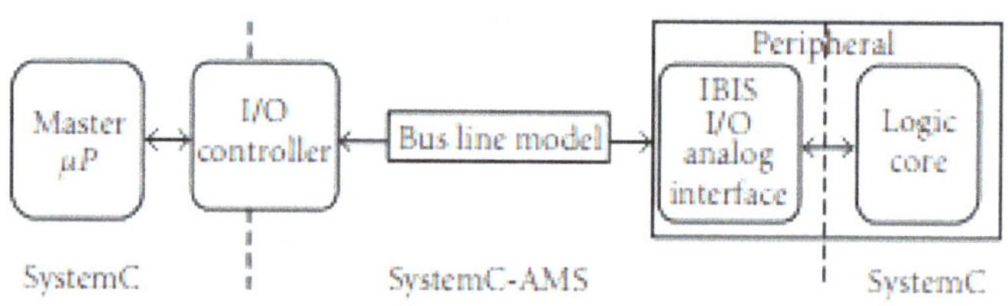

Figure 1: General structure of the modelling platform.

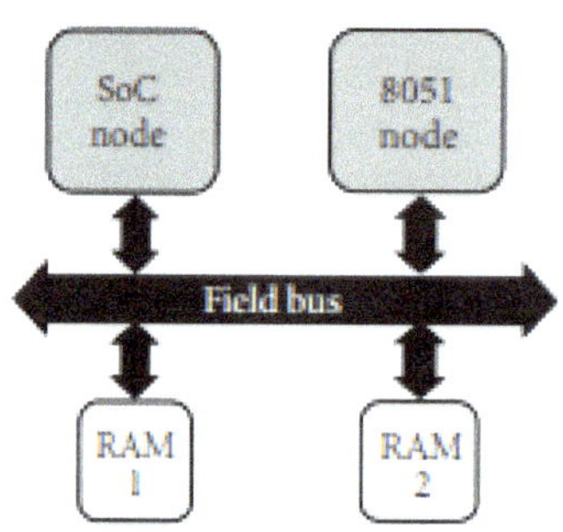

Figure 2: Field-bus based modelling platform.

is modeled in SystemC-AMS. This section introduces the digital architecture of the platform components, especially focusing on the bus controller.

Bus controller

We first decided to model our platform around an I2C bus, mainly because this standard provided a simple protocol and an easy-to-implement mixed-signal interface. However, it seemed appropriate to work on a generic structure that could be applied to different protocols with only a minimum of changes. Thus, this subsection presents our controller architecture applied to the case of an I2C bus

I²C protocol

Historically, the first I2C (Inter-Integrated Circuit) controllers and the I2C protocol were designed by Philips at the beginning of the eighties for television applications [16]. The I2C bus was invented to provide communication on a two-wire bidirectional bus—a serial data (SDA) and clock (SCL)—between a small number of devices (sensors, LCD, microcontroller, etc.). Transfer frequency is up to 100 kbits/s for standard mode, and up to 3.4 Mbits/s in high-speed mode. Data frame for the standard mode is made of a start bit, a 7 bits address, a Read/Write bit, an acknowledge bit (ACK), and a sequence of data bytes. Each data byte is followed by an acknowledge bit issued by the target device. A stop bit finalizes the transmission (Figure 3). Each bit is transmitted on SDA in conjunction with the SCL clock.

1. Transfers are initiated by a START condition. It happens when a falling transition occurs on the SDA line while SCL is high.
2. Transfers end with a STOP condition. It happens when a rising transition occurs on the SDA line while SCL is high (Figure 4).

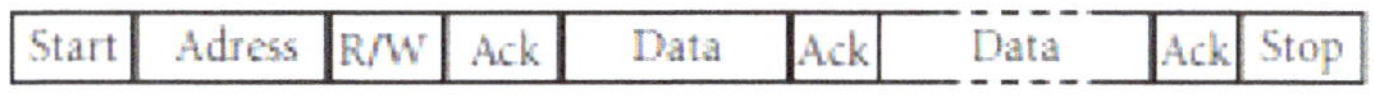

FIGURE 3: I²C data frame.

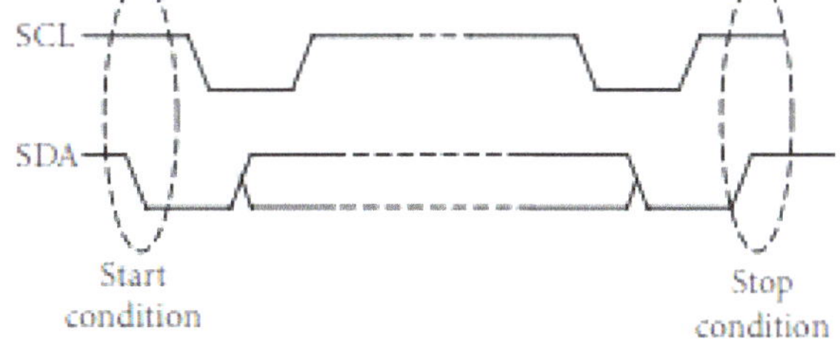

- Data is considered valid during the high state of SCL. Therefore, SDA signal must remain stable during this half period (Figure 5).
- I2C allows multimaster communications and features an arbitration management protocol for such transmissions. Arbitration takes place on the SDA line, while the SCL line is at the high level. Bus control is given to the master which transmits a low level while the others transmit a high level. A master which loses arbitration switches to high impedance its data output stage (Figure 6). As a specification for our I2C controller model, we chose the PCF8584 designed by Philips/NXP [17].

I²C controller digital block architecture

The architecture of the digital block is divided into three blocks (Figure 7).

(i) A microprocessor interface handles communication with the microprocessor core. It is built around a FIFO that stores the successive requests coming from the microprocessor bus (8051 bus, VCI, AMBA, etc.). When the controller has finished a transaction on the bus, and if a request is stored in the FIFO, the FIFO is read and the interface extracts the information needed by the sequencer (type of operation, address, data) to perform the new communication. The interface also includes an interrupt line connected to the microprocessor core so that it can read a data received from the I2C bus.

(ii) The core of the controller is a sequencer which translates the request from the master into a detailed sequence respecting the I2C protocol (frame generation, byte transmission, or reception, etc.).

(iii) Finally, a signal generator module manages or drives the SCL and SDA bus lines (Figure 8). This block manages the acknowledge generation/detection depending on the operating mode (transmitter or receiver). A shift register either serializes data to be sent to the bus line in transmitting mode or collects information from the bus in reception mode. It also sets the transmission frequency by dividing the system clock with a user-defined constant. The SDA and SCL signals from this block will be interfaced with the analog model presented in Section 3.

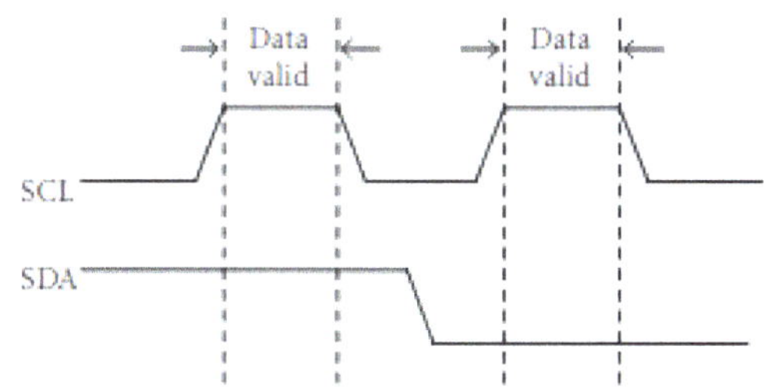

FIGURE 5: Data validity period.

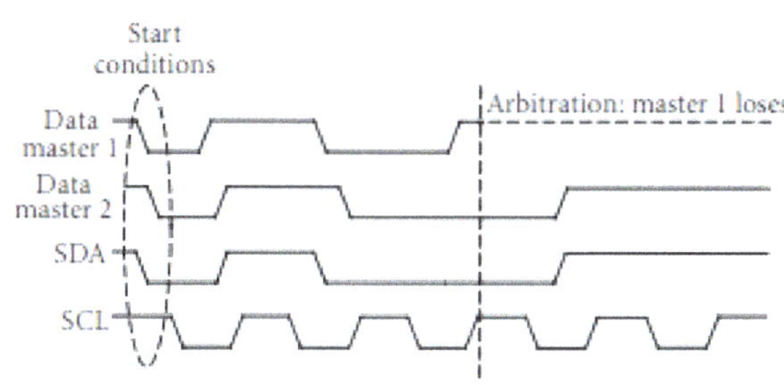

Figure 6: Arbitration procedure with two masters.

Other devices

To develop the SoC node of the platform, we used a virtual prototyping platform for system-on-chips called SoCLib [18]. SoCLib provides a number of IP SystemC models (including processor cores, VCI interconnect, memory devices, I/O controllers, timers, etc.), to easily simulate an application on a user-assembled SoC or MPSoC architecture built around a virtual component interface (VCI) bus [19]. We added our controller model to the SoCLib library and used it to connect an MIPS processor core to an external field bus. To do so, a SystemC wrapper has been developed to interface our controller to the VCI bus provided by SoCLib. The platform also includes a RAM component used by the MIPS to store its code and data and a TTY terminal for debug purposes. A SoCLib-generated platform synopsis can be found in Figure 9. We will not detail the remaining models, the I2C RAM, and the 8051 microcontroller [20]. The first of this model is a basic description of a memory component while the second one is a standard description of this well-known circuit.

ANALOG MODELLING

We introduce in this section the analog modules of our platform. All these devices were modeled in SystemC-AMS v.0.15. First, we show the analog interface of the bus controller, designed according to the requirements of the I2C protocol. Then, for the slave devices, we present an interface retrieved from an IBIS model of a standard component and show how such IBIS models can be automatically translated into a SystemC-AMS description. Finally, we focus on the transmission lines, introducing a model that takes into account the wires' physical imperfections as well as the mutual influence of adjacent transmission lines.

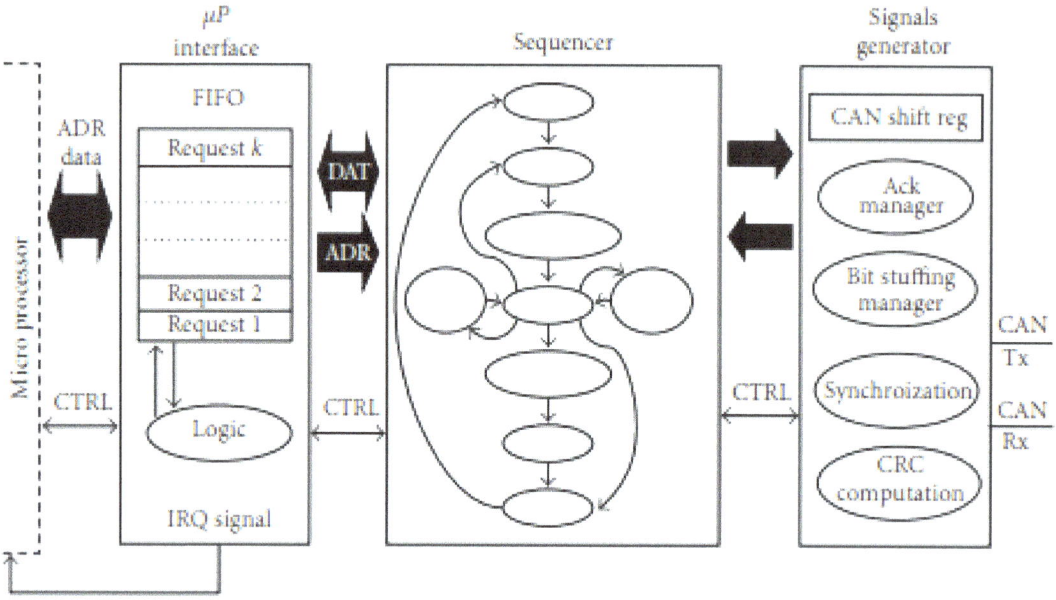

Figure 7: Controller digital block architecture.

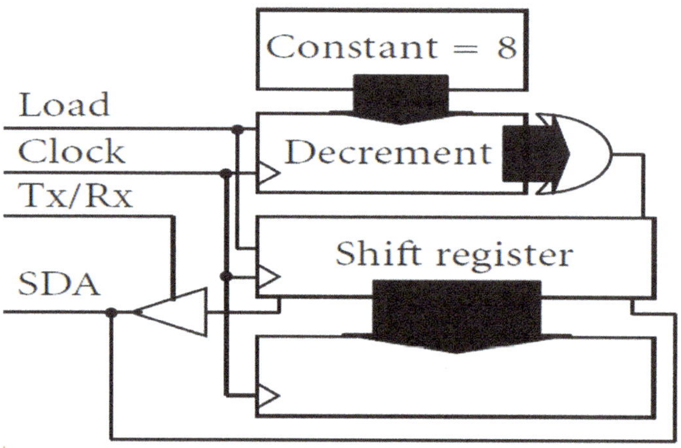

Figure 8: SDA line manager.

Bus controller

Specifications of the I2C protocol indicate that the devices must have open-drain or open-collector outputs (depending on technology) in order to perform a wired-AND function to manage multimaster mode. Both lines also feature a pullup resistor to VCC (Figure 10(a)). We modeled this analog interface with the SystemCAMS 0.15 library, by instantiating linear elements (resistors, capacitors, voltages sources, etc.). Figure 10(b) represents the interface model for the SDA line. It translates the logic levels sent by the digital block to voltages across the bus lines. The SCL model is similar. In this model, the transistor is represented with an interrupter and a resistor as no transistor model is available in SystemC-AMS 0.15. A 20 pF capacitor is used to manage the rising and falling times of the SDA signal. Finally, a pullup resistor sets the line at a high state when no command is applied on the bus. To read a value coming from the bus, a threshold detection is applied to the SDA line and it provides a logical value to the digital block. Algorithm 1 gives a SystemC-AMS sample code of the analog interface for either line of the bus. The electrical parameters of the devices (including the output transistor) were chosen to be in accordance with specifications found in NXP's PCF8584 controller datasheet [17].

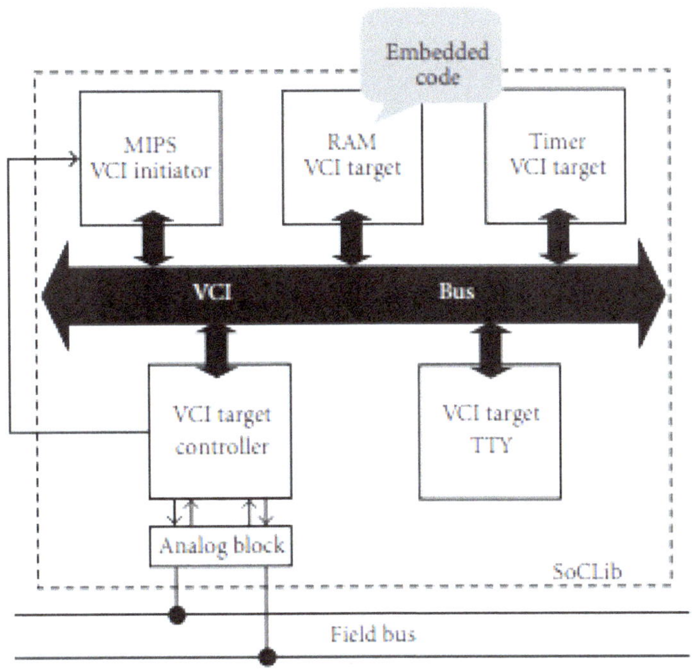

Figure 9: SoCLib simulation platform.

IBIS to SystemC-AMS conversion

In order to accurately model the analog behaviour of the components connected on the bus, and integrate real parameters of industrial components to our SystemC-AMS models, we developed an IBIS-to-SystemC-AMS conversion tool. It parses a standard IBIS source file to produce a SystemC-AMS module we can directly instantiate in our platform. This section briefly presents the IBIS standard and then shows how these models can be used in our simulation platform.

IBIS specifications

The input/output buffer information specification (IBIS) [21] standard was originally introduced to give a behavioural description of a circuit's inputs/outputs. This behaviour is based on voltage/current curves obtained from measurements or full simulation. It especially takes into account the influence of the chip's package on the I/O signal form. An IBIS model is presented as a text file which includes for each input or output pad an R, L, C equivalent circuit modelling the influence of the package as well as another circuit representing the response of the pad to a rising or falling edge transition. Figure 11 presents a model of an output buffer (input buffers are modeled in a similar way). The voltage/current curves included in the IBIS file give the response of the CMOS inverter to a logic transition. However, the IBIS standard specifies that this response is equivalent to another R, L, C circuit. As a result, a SystemC-AMS module of the electrical scheme presented in Figure 12 can accurately model the behaviour of the output buffer. The figure also shows how the module can be interfaced with a SystemC description of the logic core of the component.

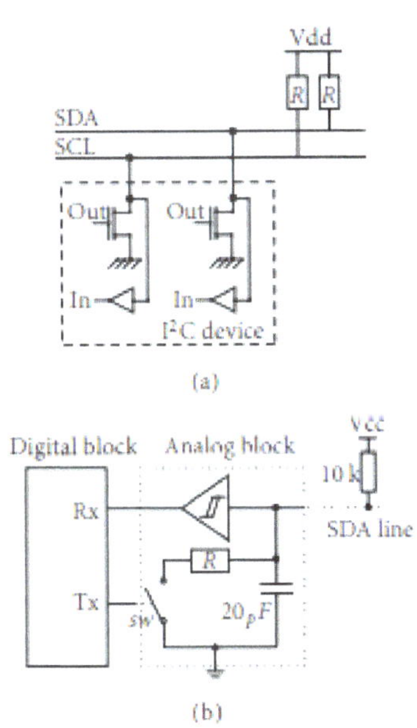

Figure 10: (a) I2C electrical scheme (b) SystemC-AMS model.

IBIS-to-SystemC-AMS conversion

To add IBIS models in our SystemC-AMS simulation platform, we developed an IBIS file analyzer program which parses the IBIS file to extract the I/O circuits useful information and then automatically creates a SystemC-AMS module using these parameters. An IBIS file is made of several sections that first present general information about the chip, display the R, L, C values of the circuit's package, enumerate the different I/O pins, and specify their type (input, output, GND, etc.), then they describe the behaviour of each I/O type found in the chip. A sample of an IBIS file is presented in Algorithm 2. Each section is specifically introduced by a keyword (such as [Package] or [Pin] in the file sample). The aim of our tool

```
SCA_SDF_MODULE(ADC){    // Threshold Detector
  sca_sdf_in<double> in;
  sc_out<bool> out;

  void init(){}
  void sig_proc(){ if(in.read() > 2.5) out = 1;
                   else out = 0;}

  SCA_CTOR(ADC){}
};

SCA_SDF_MODULE(data_conv){
  sca_sdf_in<double> portin;
  sc_out<double> portout;

  double portout_;
  void init(){ portout = 1.0e3;}
  void sig_proc(){ portout_ = portin.read();
                   portout = portout_ * 3.0e2;}

  SCA_CTOR(data_conv){}
};

SC_MODULE(SDA_Mgr){
  sc_in<bool> sda_in;    // From digital block
  sc_out<bool> sda_out;  // To digital block
  sca_sdf_signal<double> sda_sdf; // To I2C Bus

  sc_signal<double> r_value;
  sc_signal<bool> sig_;
  sca_elec_port out;   // Output port
  sca_elec_port gnd;   // Ground port
  sca_elec_node y;     // Internal node

  sca_sc2r *r1;            sca_c *c1;
  sca_rswitch *sw1;        sca_vd2sdf *conv1;
  data_conv *data_conv1;   can *can1;

  SC_CTOR(SDA_Mgr){

    ADC1 = new ADC("ADC1");   // Threshold
         ADC1->in(sda_sdf); ADC1->out(sda_out);

      sw1 = new sca_rswitch("sw1");   // Switch
      sw1->off_val = true; sw1->p(y);
      sw1->n(gnd); sw1->ctrl(sda_in);

    r1 = new sca_sc2r("r1");   // Resistor
        r1->ctrl(r_value); r1->p(out); r1->n(y);

    data_conv1 = new data_conv("data_conv1");
      data_conv1->portin(sda_sdf);
      data_conv1->portout(r_value);

    c1 = new sca_c("c1");   // Capacitor
      c1->value = 20.0e - 12;
      c1->p(out); c1->n(gnd);

    conv1 = new sca_vd2sdf("conv1"); // Output voltage
      conv1->p(out); conv1->n(gnd);
      conv1->sdf_voltage(sda_sdf);
  }
};
```

Algorithm 1: I2C controller analog block code sample.

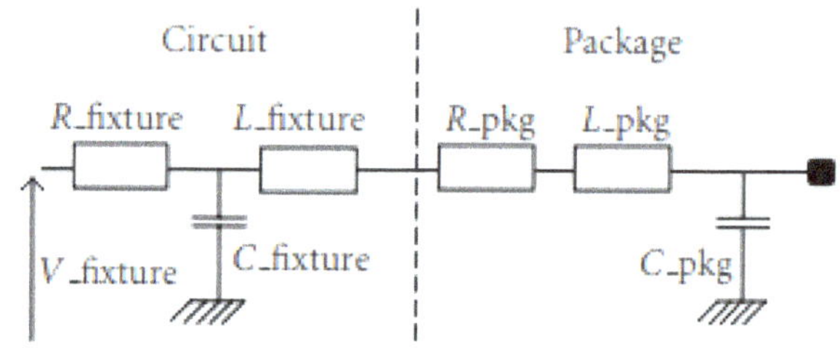

Figure 11: IBIS output model.

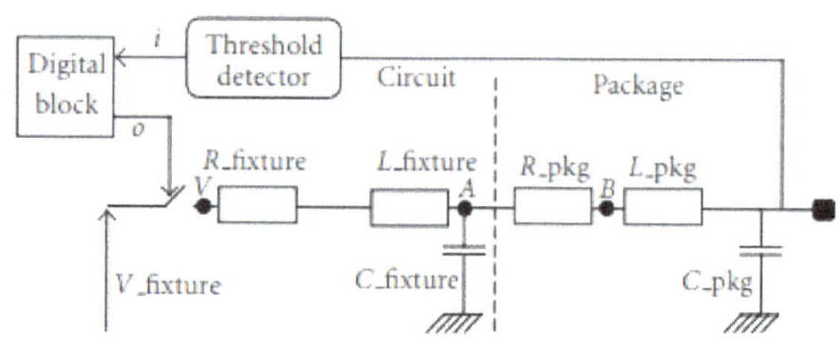

Figure 12: IBIS equivalent circuit to model in SystemC-AMS.

is to scan the IBIS file, look for those keywords, and extract useful information, in particular the number of each I/O type as well as each type's electrical characteristics, that is, the value of the R, L, C elements. The synopsis of our IBIS-to-SystemC-AMS converter is presented in Figure 13. When running, it performs a scanning phase followed by a generation phase. When it parses the IBIS file, the program first counts the number of different I/O models for the circuit as well as their number of occurrences. Then, for each I/O model, it extracts and stores the value of the R, L, C elements. To do so, the program compares the first word of each line of the IBIS file with a set of predefined keywords (e.g., "R pkg" for the package resistance). If it matches with one of the keywords, it saves the value of the element (which, in an IBIS file, is usually written next to the keyword). As an illustration, in the IBIS text file sample (Algorithm 2), the program will count 8 I/O, among them three identical (A0, A1, A2). It will also notice the R pkg, L pkg and C pkg keywords and save the values located next to these keywords (resp., 0.2 Ω, 7 nH, and 1,5 pF). When the scanning phase is completed, the program generates the SystemC-AMS file. This file includes one top level module with the correct number of I/Os. Each I/O is instantiated as a separate module inside the top level component and can be directly associated with a SystemC digital simulation model (see Figure 1): on one hand, a signal on an output pin is produced with the help of a switch component, commanded by the digital model, on the other hand, a voltage on an input pin is converted by a threshold detector to generate a boolean value that is sent to the digital model. Obviously, each I/O module models the electrical circuits presented above, with the extracted values of the R, L, C components. Algorithm 3 presents a sample of an in-out pin module created from an IBIS file specification. This tool was used to generate a SystemC-AMS model out of the IBIS description of the CAT24WC0 2 Kbit I2C

```
[Package]
| variable      typp         min          max
R_pkg           0.20         100.00 m     0.40
L_pkg           7.00 nH      4.10 nH      10.00 nH
C_pkg           1.50 pF      1.00 pF      3.00 pF
|
|************Pin Defintions************
[Pin]           signal_name     model_name
|
1               A0              InputModel1
2               A1              InputModel1
3               A2              InputModel1
4               GND             GND
5               SDA             IOModel
6               SCL             InputModel2
7               WP              InputModel3
8               VCC             POWER
```

Algorithm 2: IBIS text file sample.

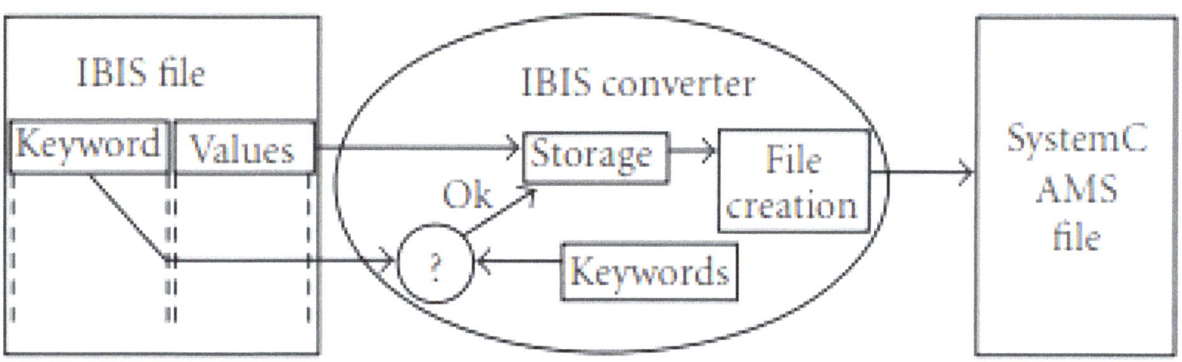

Figure 13: IBIS-to-SystemC-AMS converter tool synopsis.

EEPROM chip from catalyst semiconductors [22]. This model serves as the analog interface of both slave memory devices featured in the I2C platform.

Transmission line model

The bus lines feature a number of imperfections and can suffer from any kind of signal integrity perturbation. This section presents how the electrical model of the bus lines can be included in our modelling platform. So far, we have considered the bus lines as perfect conductors immune to any kind of interference. Still, the nature of the wires as well as their geometry can modify the behaviour of the transmitted signals and, of course, so does the external environment. These imperfections are represented with R, L, C elements that we integrate in our platform. Our line model below (Figure 14) is based on [23]. Each bus line includes the wire resistance and also an inductance that represents the wire self-inductance. The value of this inductance is given by the following formula, where l is the wire length and d the wire diameter in centimetres:

$$L = 0.002l\left(\ln\left(\frac{4l}{d}\right) - 0.75\right)\ [\mu\text{H}]. \tag{1}$$

In the situation of two (or more) adjacent lines, which is likely to appear in the case of the I^{2}C bus, a mutual coupling phenomenon appears. It is represented by a parallel

```
struct IBIS_INOUT_PIN : sc_module
{
  sc_out<bool>  i; // Input signal sent to digital block
  sc_in<bool>   o; // Output signal sent by digital block
  ...
  sca_elec_port  pin; // Inout Pin
  ...
  sca_sdf_signal<double> in_sdf; // Pin Voltage
  ...
  sca_elec_ref    gnd;    // Electrical Ground
  sca_elec_node B;        // Node between R_pkg & L_pkg
  sca_elec_node A;        // Node between R_pkg & L_fix
  sca_elec_node V;        // Node between R_fix & switch
  ...
  // Declaration of IBIS RLC elements
  sca_r *R_pkg, *R_fixture; sca_c *C_pkg, *C_fixture;
  sca_l *L_pkg; sca_vconst *V_fixture;
  sca_rswitch *sw1;    sca_vd2sdf *conv_pin;
  ...
  // Threshold converts input voltage to bool value
  THRESHOLD_DETECTOR *thd_in;
  ...
  SC_HAS_PROCESS(IBIS_INOUT_PIN);
  ...
  // Constructor: RLC values retrieved from IBIS file
  IBIS_INOUT_PIN( sc_module_name insname,
      double R_pkg_value, double C_pkg_value,
      double L_pkg_value, double C_fixture_value,
      double R_fixture_value, double V_fixture_value )
    {
      // PACKAGE ELEMENTS INSTANCIATION
        C_pkg->value = C_pkg_value;
        C_pkg->p(pin); C_pkg->n(gnd);
  ...
        L_pkg->value = L_pkg_value;
        L_pkg->p(B);    L_pkg->n(pin);
  ...
        R_pkg->value = R_pkg_value;
        R_pkg->p(A);    R_pkg->n(B);
  ...
      // PACKAGE ELEMENTS INSTANCIATION
        C_fixture->value = C_fixture_value;
        C_fixture->p(A); C_fixture->n(gnd);
  ...
        R_fixture->value = R_fixture_value;
        R_fixture->p(Vfix); R_fixture->n(B);
  ...
        V_fixture->value = V_fixture_value;
        V_fixture->p(Vfix); V_fixture->n(gnd);
  ...
      // Switch controlled by digital block
        sw1->p(V); sw1->n(gnd);
        sw1->ctrl(o); sw1->off_val = true;
  ...
      // Threshold: Pin Voltage is sent to digital
      // block as bool value
        thd_in->threshold = 1.1;
        thd_in->in (in_sdf); thd_in->out(i);
        };
      }
```

Algorithm 3: SystemC-AMS code sample generated from IBIS specification.

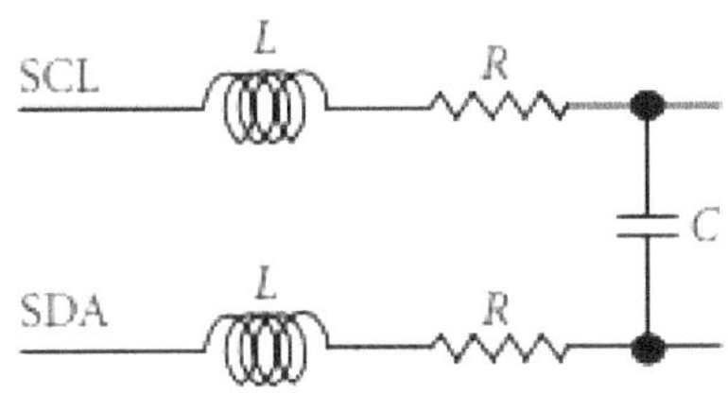

FIGURE 14: Transmission line model.

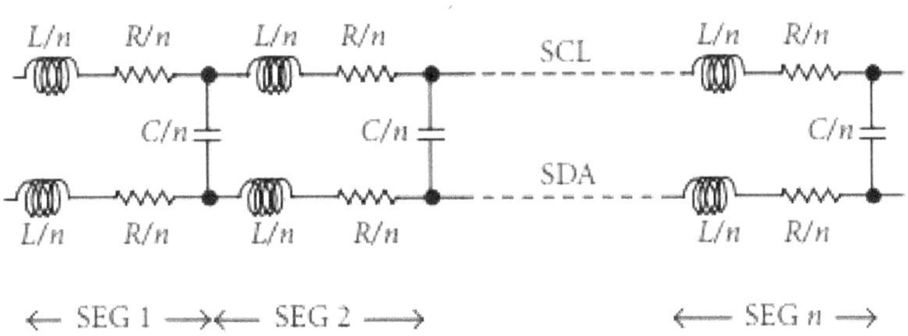

Figure 15: Transmission line model in n segments.

capacitance whose value depends on the diameter d of the wires and the distance D between them. The mutual capacitance is given by the following formula:

$$C = \frac{\pi 0.0885}{\cosh^{-1}(D/d)} \text{ [pF/cm]}. \qquad (2)$$

In a first step, we considered that the mutual inductance between the I²C wires is negligible. The length of the bus lines, for instance in an automotive context, can be quite important, up to several metres. Therefore, the global line model of Figure 14 can be divided into several segments, such as in Figure 15. With this representation, we can take into account a distance variation between two bus lines, affect only a part of the bus with a perturbation, or simply use segments with different lengths to respect the node topology in the system (Figure 16). Our SystemC-AMS line model comes in the form of one top module that instanciates n segment submodules. The length of each segment can be parameterized in the submodule constructor (Algorithm 4).

```
struct LINE_I2C:sc_module
{
  sca_elec_port scl1, scl2;    // Segment ends for SCL
  sca_elec_port sda1, sda2;    // Segment ends for SDA
  sca_elec_node lr1, lr2;      // Node between L and R

  double L_value, R_value, C_value; // Elements value

  sca_r *R_scl, *R_sda;
  sca_l *L_scl, *L_sda;
  sca_c *C_cpg;

  SC_HAS_PROCESS(LINE_I2C);

  LINE_I2C(sc_module_name insname, double length)
    {  // Length in cm
       L_value = 7.8e - 9*length;
       R_value = 0.5*length;
       C_value = 0.1e - 12*length;

     // Elements Instanciation

       L_scl->p(scl1); L_scl->n(lr1);
       L_scl->value = L_value;

       R_scl->p(lr1); R_scl->n(scl2);
       R_scl->value = R_value;

       L_sda->p(sda1); L_sda->n(lr2);
       L_sda->value = L_value;

       R_sda->p(lr2); R_sda->n(sda2);
       R_sda->value = R_value;

       C_cpg->p(scl2); C_cpg->n(sda2);
       C_cpg->value = C_value;
    };
};
```

Algorithm 4: Line segment SystemC-AMS code sample.

PLATFORM SIMULATION

We present in this section the simulation of our field-busbased platform, summarized in Figure 17. The SystemC and SystemC-AMS description allow us to cosimulate the MIPS or the 8051 embedded software with the hardware behaviour of the components, whether analog or digital. This section first gives some information about the code that is executed during simulation, then shows its impact on the digital blocks of the platform, especially the bus controllers. We focus next on the analog response on the bus, presenting the influence of the transmission line on the shape of the signal. Finally, we provide simulation performances for our mixed platform.

Embedded code

The aim of our simulation is to validate the functionality of the bus controller by testing all the protocol features.

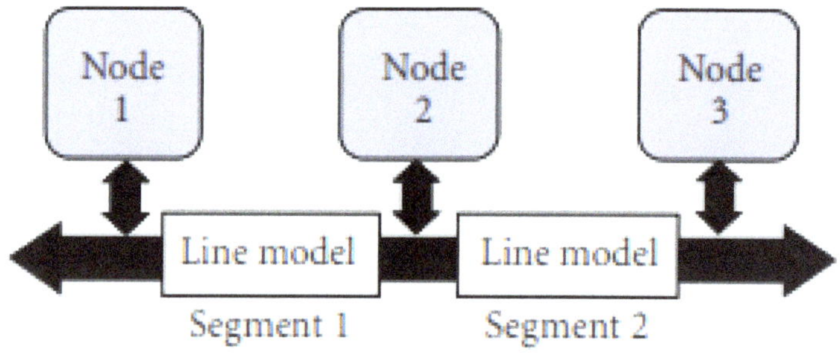

Figure 16: Line model and system topology.

Therefore, the embedded code found on the two master devices mainly consists in bus access operations. The following assembler code (Algorithm 5) is compiled and stored in the internal ROM of the 8051 microcontroller. When it runs, the program performs an I2C write operation: the data byte 4Eh is written on the B4h cell address of the slave RAM device. This write operation takes place in two stages: first the IP sends the cell address to the slave device (the A0 h value is

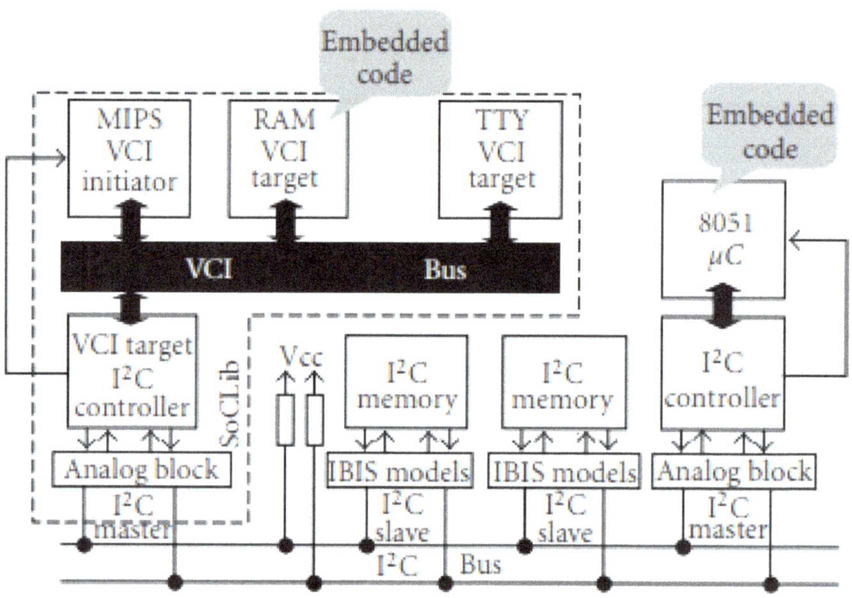

Figure 17: I2C platform.

```
        //      Write in RAM(slave address = 0x50)
        //      the value 0x4E in cell 0xF1

main: MOV       R0, #A0 // Put slave address in R0 register
      MOV       A, #B4 // Put cell address in A register
      MOVX      @RO, A // Send R0 + A to I2C IP
      MOV       R0, #A0 // Put slave address in R0 register
      MOV       A, #4E // Put data byte in A register
      MOVX      @RO, A // Send R0 + A to I2C IP
```

Algorithm 5: 8051 assembler code sample—byte writing operation.

made of the 50 h 7-bit slave address and the R/W bit set to 0), then it writes the data byte to this same slave device. Similarly, Algorithm 6 gives a sample of the MIPS code, featuring the assembler functions used to request the controller IP a read or a write on the I2C bus, as well as the interrupt subroutine which is called when a read data has been received from the I2C bus and is available for the MIPS master.

Digital simulation

When running, the codes found in the SoC and the microcontroller send requests and therefore activate their respective I2C controller. In particular, we can see in Figure 18 a sequence featuring a write operation requested by the 8051 to one of the RAM devices followed by an MIPScommanded read request of the same RAM cell address. The SystemC chronogram shows the 8051 positioning the 4 Eh byte on its data bus and the F1 h value on its address bus, resulting on a first transmission on the I2C bus, then the SoC bus controller performing a second transmission to retrieve the data from the RAM component. The chronogram shows the digital behaviour of the SDA bus line (featured on the last row), as well as the SDA commands sent by each component, indicating which device has control of the bus at any given time. We can also notice that an IRQ signal is generated when the 4Eh value is available in the MIPS I2C controller.

Analog simulation

The chronogram in Figure 19 represents the analog behaviour of the I2C bus lines during a write transmission issued by the MIPS-based SoC. It does not take into account the transmission line model. On the first row of the chronogram, the slave address (50 h) is first sent on the bus, then on the second row, after an acknowledge from this slave device, the master transmits the cell address (B4 h) it wants to write into. Finally, on the third row, after another acknowledge from the slave, the data byte (4 Eh) is sent to the slave, followed by a last acknowledge from the slave and a stop bit, signaling the end of transmission. The chronogram also displays the shape of the analog signal. The rising and falling times that can be seen depend on the analog interface of the component that has control of the bus when a particular bit is transmitted. In the case of the chronogram in Figure 19, most of the bits are issued by the 8051 bus controller interface, except for the acknowledges which are issued by the IBIS-converted interface of the RAM device. Our platform also allows us to test the multimaster arbitration of the I2C bus as it is implemented in our model. In the chronogram shown in Figure 20, both masters send a start bit at the same moment. The SDA bus line follows the two commands as long as they are identical, and when they differ, the master which sends a 0 bit (in this case the MIPS) takes control of the bus and the 8051 stops transmitting on SCL and SDA. Also of interest is the wired-AND function performed on the SDA and SCL lines of the bus. This shows that our mixed-signal controller model successfully performs electrical multimaster arbitration. Finally, Figure 21 presents the signal behaviour if we take into account the line imperfections with the model presented in Section 4. We can clearly see the effect of the wire inductance and the mutual capacitance.

Simulation performances

Table 1 presents simulation speeds for various configurations of the simulation platform presented in Section 4. The measurements were performed on a Linux workstation equipped with a Pentium M processor running at 1.73 GHz, a L2 cache of 2 Mb, and 1 Gb of SDRAM. Sampling period for the analog models is 100 nanoseconds. Our aim is to measure the increase of simulation time between a purely digital simulation in SystemC of a platform, and a mixed SystemC/SystemC-AMS simulation of the same platform. We first measure this overhead in the case of a simplified platform made of a microcontroller node and a slave RAM, first without and then with the transmission line model. Results show that for the 8051 platform, the overhead due to the AMS parts of the model is relatively important, though simulation is still very fast. In the second case that is the full 4-node platform including the transmission line model, the impact is less

```
// Interrupt Subroutine
void SwitchOnIt(int it)
{
   int i;
=
// Identify the active interrupt of highest priority
   for (i = 0; i < 8; i++) if (it&(1 << i)) break;
=
   switch (i)
   {
      case 0: break;      // SwIt 0
      case 1: break;      // SwIt 1
      case 2: read_interface(); break; // It 0
      case 3: break;      // It 1
      case 4: break;      // It 2
      case 5: break;      // It 3
      case 6: break;      // It 4
      case 7: break;      // It 5
      default: break;
   }
}
=
// Main Program
int main(void)
{
   write_byte(0x50, 0x57, 0x69);
   // Parameters = Slave Adr (stored in $4 register)
   //              Cell Adr (stored in $5 register)
   //              Data Byte (stored in $6 register)
=
   read_byte(0x51, 0xf1);
   // Parameters = Slave Adr (stored in $4 register)
   //              Cell Adr (stored in $5 register)
=
   while (1);
   return 0;
}
=
// Byte Writing ASM Code Sample
write_byte:
   la   $3,0xc0000000   // Load VCI target (I2C IP)
                        // address to $3 register
   sll  $4, $4, 9       // Slave adr in high part of register
   or   $4, $4, $5      // Cell adr in low part of register
   sh   $4, 0 ($3)      // (save half-word)
                        // send target + cell adr to IP
   sb   $6, 1 ($3)      // Send data byte to IP
   nop
   j    $31             // return from subroutine
.end write_byte
// Byte Reading ASM Code Sample
read_byte:
=
   la   $3,0xc0000000   // Load VCI target (I2C IP)
                        // address to $3 register
   sll  $4, $4, 9       // Slave adr in high part of register
   ori  $4, $4, 0x100   // Set R/W Bit to Read
   or   $4, $4, $5      // Cell adr in low part of register
   sh   $4, 0 ($3)      // save half-word ->
   nop                  // send target + cell adr to IP
   j    $31             // return from subroutine
```

Algorithm 6: MIPS code sample.

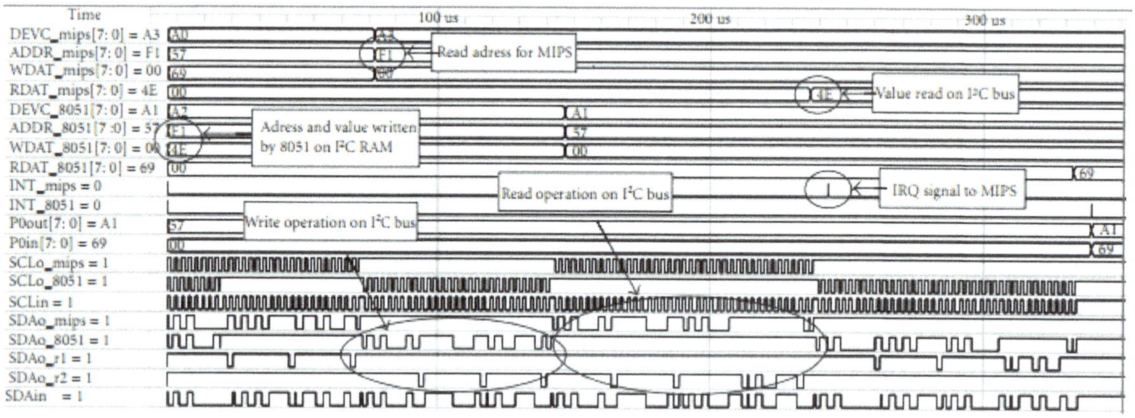

Figure 18: Simulation with 8051 and MIPS master.

Table 1: I2C Platform simulation performance in cycles/s

Application	SystemC platform	Sys.C + Sys.C-AMS equivalent plarform	Overhead
8051 node + RAM with IBIS	149 300	106 400	29%
8051 node + RAM with IBIS + line model	149 300	87 500	41%
4-node platform without line model	25 800	23 500	9%
Complete I²C platform (Figure 19)	25 800	19 500	24%

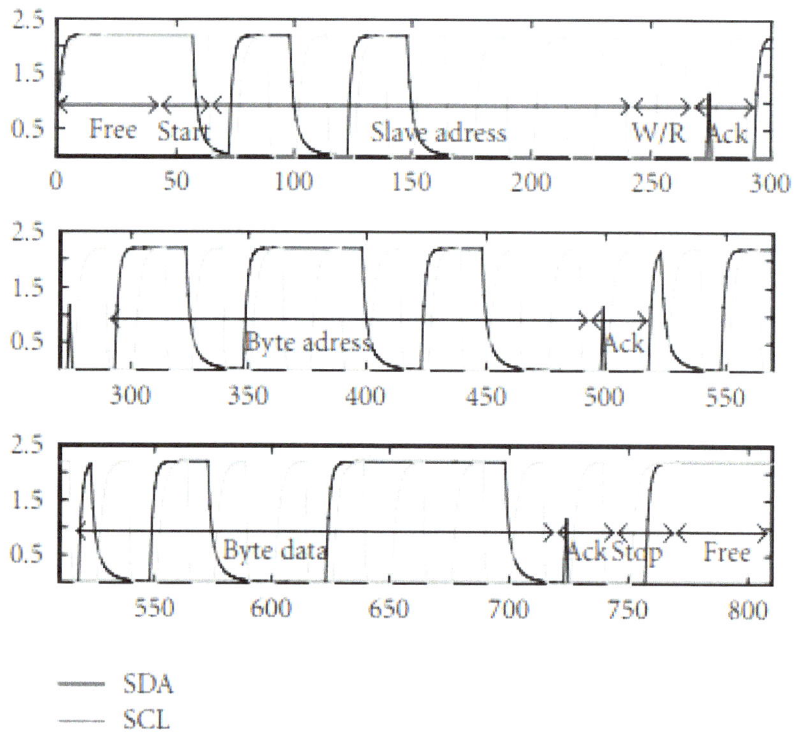

Figure 19: Analog simulation—byte writing operation.

prominent, mainly because the proportion of purely SystemC components is more important. In this latter configuration, analog simulation of the I2C bus is performed only 9% slower than a digital-only simulation if we do not include the line model, and 24% if we add a line segment model between two nodes.

CONCLUSION

In this paper, we showed how to model field-bus communications in the context of embedded systems. We described a heterogeneous model that features three main parts. Firstly,

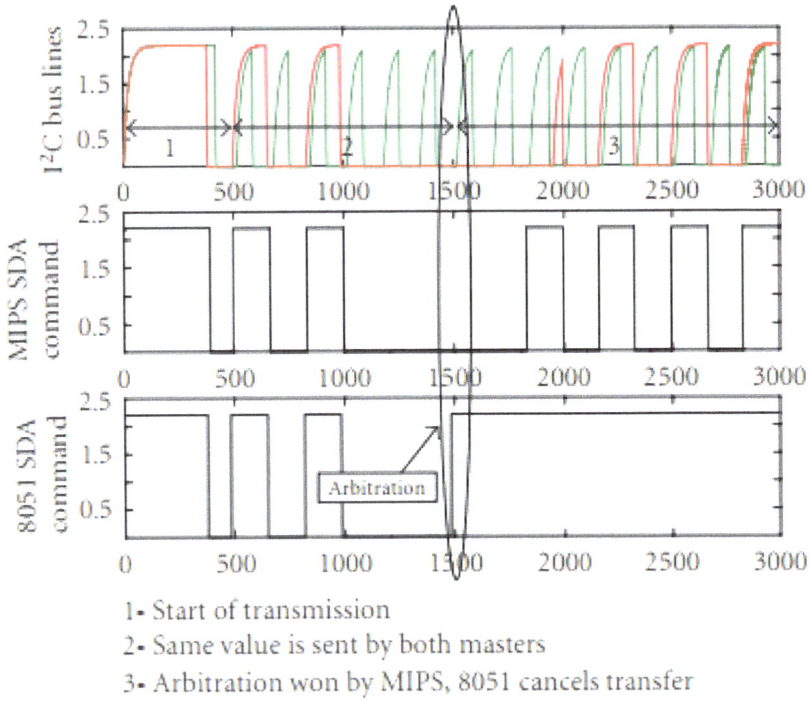

an I/O bus controller interfaces the processor cores with the field bus. Secondly, an analog part represents the interface of the I/O controller with the bus lines. To accurately model the analog behaviour of the components, we introduced an IBIS-to-SystemC-AMS converter to derive the analog description of the field bus nodes analog interface from industrial circuits. Thirdly, we modelled the bus lines as an array of segments that take into account the electrical imperfections of the bus lines. Simulations showed the successful operations of the I2C field bus. The SystemC-AMS simulation times showed different overheads over digital SystemC: 40% for a 8051 to

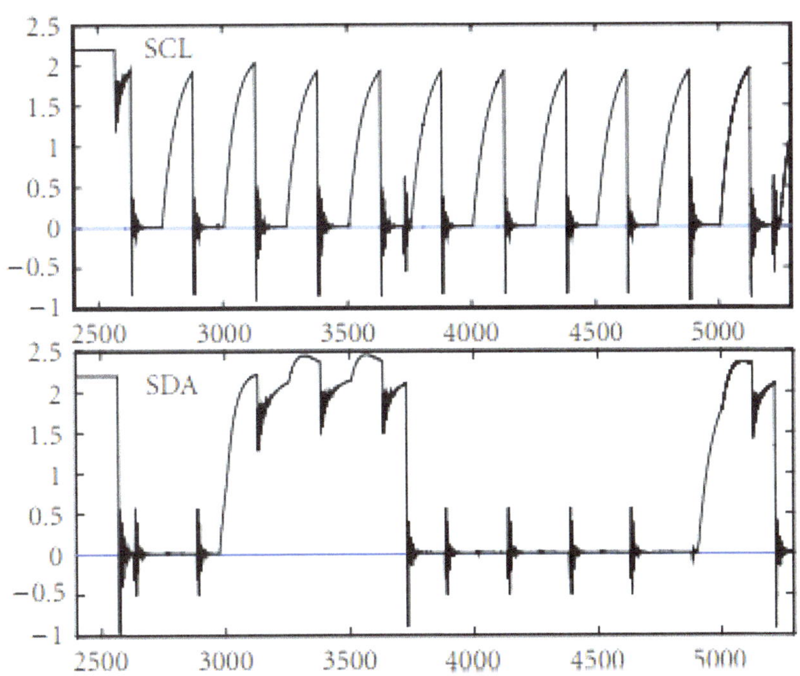

Figure 21: Bus lines simulation with transmission line model.

RAM I²C communication, and around 25% for an MIPSbased SoC plus 8051 system. This overhead is acceptable for a mixed-signal simulation. The benefits of our approach are that it includes analog and digital models in a single environment. It will therefore allow to simulate interactions between analog and digital functions that are usually discovered on the hardware prototypes. As a next step, we intend to show that our approach is generic by applying it to other field busses such as CAN. Further, this SystemC-AMS simulation platform stresses the interest of this language as a unique tool from the specification to the verification of AMS embedded systems, gathering together software execution with analog and digital behaviour.

REFERENCES

1. M. Chiodo, D. Engels, P. Giusto, et al., "A case study in computer-aided co-design of embedded controllers," Design Automation for Embedded Systems, vol. 1, no. 1-2, pp. 51–67, 1996.
2. J. Liu, X. Liu, and E. A. Lee, "Modeling distributed hybrid systems in Ptolemy II," in Proceedings of the American Control Conference (ACC '01), pp. 4984–4985, Arlington, Va, USA, June 2001.
3. Vachoux, C. Grimm, R. Kakerow, and C. Meise, "Embedded mixed-signal systems: new challenges for modeling and simulation," in Proceedings of IEEE International Symposium on Circuits and Systems (ISCAS '06), pp. 991–994, Island of Kos, Greece, May 2006.
4. J. Oudinot, J. Ravatin, and S. Scotti, "Full transceiver circuit simulation using VHDL-AMS," in Proceedings of Forum on Specification and Design Languages (FDL '02), pp. 29–33, Marseille, France, September 2002.
5. T. Grotker, S. Liao, G. Martin, and S. Swan, ¨ System Design With SystemC, Kluwer Academic Publishers, Dordrecht, The Netherlands, 2002.
6. Vachoux, C. Grimm, and K. Einwich, "SystemC-AMS requirements, design objectives and rationale," in Proceedings of Design, Automation and Test in Europe Conference and Exhibition (DATE '03), pp. 388–393, Munich, Germany, March 2003.
7. Vachoux, C. Grimm, and K. Einwich, "Extending SystemC to support mixed discrete-continuous system modeling and simulation.," in Proceedings of IEEE International Symposium on Circuits and Systems (ISCAS '05), vol. 5, pp. 5166–5169, Kobe, Japan, May 2005.
8. Vachoux, C. Grimm, and K. Einwich, "Towards analog and mixed-signal SOC design with systemC-AMS," in Proceedings of the 2nd International Workshop on Electronic Design, Test & Applications (DELTA '04), pp. 97–102, Perth, Australia, January 2004.
9. Vachoux, C. Grimm, and K. Einwich, "Analog and mixed signal modelling with SystemC-AMS," in Proceedings of IEEE International Symposium on Circuits and Systems (ISCAS '03), vol. 3, pp. 914–917, Bangkok, Thailand, May 2003.
10. Open SystemC Initiative, http://www.systemc.org/.
11. OSCI Analog/Mixed-signal Working Group (AMSWG), http://www.systemc.org/apps/group public/workgroup.php? wg abbrev=amswg.
12. O. Romain, T. Cuenin, and P. Garda, "Design and verification ´ of mixed-signal I/O IPs: an 12C bus controller," in Proceedings of IEEE International Symposium on Industrial Electronics (ISIE '04), vol. 1, pp. 77–81, IEEE Press, Ajaccio, France, May 2004.
13. M. Alassir, J. Denoulet, O. Romain, and P. Garda, "A SystemC AMS model of an 12C bus controller," in Proceedings of IEEE International Conference on Design and Test of Integrated Systems in Nanoscale Technology (DTIS '06), pp. 154–158, IEEE Press, Tunis, Tunisia, September 2006.
14. M. Alassir, J. Denoulet, O. Romain, and P. Garda, "Modeling I²C communication between SoCs with SystemC-AMS," in Proceedings of IEEE International Symposium on Industrial Electronics (ISIE '07), pp. 1412–1417, IEEE Press, Vigo, Spain, June 2007.
15. M. Alassir, J. Denoulet, G. Vasilescu, O. Romain, R. Arnaud, and P. Garda, "Modeling field bus communications for automotive applications," in Proceedings of the Forum on specification and Design Languages (FDL '07), Barcelona, Spain, September 2007.
16. Philips Semiconductors, The I²C-Bus Protocol Specification, Document Order Number: 9398 393 40011, January 2000, http://www.nxp.com/acrobat/literature/9398/39340011 21 .pdf.
17. Philips Semiconductors, PCF 8584, I2C bus controller, http://www.nxp.com/acrobat/datasheets/PCF8584 4.pdf.
18. SoCLIB, "A modelisation & simulation plat-form for system on chip," 2003, http://www.soclib.fr/.
19. VSI Alliance, Virtual Component Interface Standard—version 2 (OCB 2 2.0), http://www.vsi.org/index.htm.
20. Intel, 80C51 Single-Chip 8-bit Microcontroller datasheet, ftp://download.intel.com/MCS51/datashts/27050108.pdf.
21. IBIS (I/O Buffer Information Specification), Version 4.2, June 2006, http://vhdl.org/pub/ibis/ver4.2/ver4 2.pdf.
22. CAT24WC01/02/04/08/16, 1K/2K/4K/8K/16K-Bit Serial EEPROM datasheet, Catalyst Semiconductors, http://www .catsemi.com/datasheets/24WC.
23. G. Vasilescu, Electronic Noise and Interfering Signals: Principles and Applications, chapter 12, Springer, New York, NY, USA, 2005.

CHAPTER 14

PARALLEL BACKPROJECTION: A CASE STUDY IN HIGH-PERFORMANCE RECONFIGURABLE COMPUTING

Ben Cordes and Miriam Leeser

INTRODUCTION

In the continuing quest for computing architectures that are capable of solving more computationally complex problems, a new direction of study is *high-performance reconfigurable computing* (HPRC). HPRC can be defined as a marriage of traditional high-performance computing (HPC) techniques and reconfigurable computing (RC) devices.

HPC is a well-known set of architectural solutions for speeding up the computation of problems that can be divided neatly into pieces. Multiple general-purpose processors (GPPs) are linked together with high-speed networks and storage devices such that they can share data. Pieces of the problem are then distributed to the individual processors and computed, and the answer is assembled from the pieces. Commonly available HPC systems include Beowulf clusters and other supercomputers. Reconfigurable computing uses many of the same concepts as HPC, but at a finer grain. A special-purpose processor (SPP), often a field-programmable gate array (FPGA), is attached to a GPP and programmed to execute a useful function. Special-purpose hardware computes the answer to the problem quickly by exploiting hardware design techniques such as pipelining, the replication of small computation units, and high-bandwidth local memories.

Both of these computing architectures reduce computation time by exploiting the parallelism inherent in the application. They rely on the fact that multiple parts of the overall problem can be computed relatively independently of each other. Though HPC and RC act on different levels of parallelism, in general, applications with a high degree of parallelism are well-suited to these architectures.

The idea behind HPRC is to provide a computing architecture that takes advantage of both the *coarse-grained*parallelism exploited by clustered HPC systems and the*fine-grained* parallelism exploited by RC systems. In theory, more exploited parallelism means more speedup and faster computation times. In reality, factors such as communications bandwidth may prevent performance from improving as much as is desired.

In this paper, we examine one application that contains a very high degree of parallelism. The *backprojection* image formation algorithm for synthetic aperture radar (SAR) systems is «embarrassingly parallel», meaning that it can be broken down and parallelized on many different levels. For this reason, we chose to implement backprojection on an HPRC machine at the Air Force Research Laboratory in Rome, NY, USA, as part of an SAR processing system. We present an analysis of the algorithm and its inherent parallelism, and we describe the implementation process along with the design decisions that went into the solution.

Our contributions are as follows.

1. We implement the backprojection algorithm for SAR on an FPGA. Though backprojection has been implemented many times in the past (see Section 3), FPGA implementations of backprojection for SAR are not well represented in the literature.
2. We further parallelize this implementation by developing an HPC application that produces large SAR images on a multinode HPRC system.

Citation: Ben Cordes and Miriam Leeser, Parallel Backprojection: A Case Study in High-Performance Reconfigurable Computing, doi:10.1155/2009/727965.

The rest of this paper is organized as follows. Section 2 provides some background information on the backprojection algorithm and the HPRC system on which we implemented it. In Section 3, we discuss related research. Section 4 describes the backprojection implementation and how it fits into the overall backprojection application. The performance data and results of our design experiments are analyzed in Section 5. Finally, Section 6 draws conclusions and suggests future directions for research.

Readers who are interested in more detail about this work are directed to the master thesis on which it is based [1].

BACKGROUND

This section provides supporting information that is useful to understanding the application presented in Section 4. Section 2.1 describes backprojection and SAR, highlighting the mathematical function that we implemented in hardware. Section 2.2 presents details about the HPRC system that hosts our application.

Backprojection Algorithm

We briefly describe the backprojection algorithm in this section. Further details on the radar processing and signal processing aspects of this process can be found in [2, 3].

Backprojection is an image reconstruction algorithm that is used in a number of applications, including medical imaging (computed axial tomography, or (CAT)) and synthetic aperture radar (SAR). The implementation we describe is used in an SAR application. For both radar processing and medical imaging applications, backprojection provides a method for reconstructing an image from the data that are collected by the transceiver.

SAR data are essentially a series of time-indexed radar reflections observed by a single transceiver. At each step along the synthetic aperture, or flight path, a pulse is emitted from the source. This pulse reflects off elements in the scene to varying degrees, and is received by the transceiver. The observed response to a radar pulse is known as a "trace".

SAR data can be collected in one of two modes, "strip-map" or "spotlight". These modes describe the motion of the radar relative to the area being imaged. In the spotlight mode of SAR, the radar circles around the scene. Our application implements the strip-map mode of SAR, in which radar travels along a straight and level path.

Regardless of mode, given a known speed at which the radar pulse travels, the information from the series of time-indexed reflections can be used to identify points of high reflectivity in the target area. By processing multiple traces instead of just one, a larger radar aperture is synthesized and thus a higher-resolution image can be formed.

The backprojection image formation algorithm has two parts. First, the radar traces are filtered according to a linear time-invariant system. This filter accounts for the fact that the airplane on which the radar dish is situated does not fly in a perfectly level and perfectly straight path. Second, after filtering, the traces are "smeared" across an image plane along contours that are defined by the SAR mode; in our case, the flight path of the plane carrying the radar. Coherently, summing each of the projected images provides the final reconstructed version of the scene.

Backprojection is a highly effective method of processing SAR images. It is computationally complex, much like traditional Fourier-based image formation techniques. However, backprojection contains a high degree of parallelism, which makes it suitable for the implementation on reconfigurable devices.

The operation of backprojection takes the form of a mapping from projection data $p(t,u)$ to an image $f(x,y)$. A single pixel of the image f corresponds to an area of ground containing some number of objects which reflect radar to a certain degree. Mathematically, this relationship is written as

$$f(x,y) = \int p(i(x,y,u),u)\,du, \tag{1}$$

where $i(x,y,u)$ is an indexing function indicating, at a given u, those t that play a role in the value of the image at location (x,y). For the case of SAR imaging, the projection data $p(t,u)$ take the form of the filtered radar traces described above. Thus, the variable u corresponds to the slow-time location of the radar, and t is the fast-time index into that projection. Fast-time variables are related to the speed of radar propagation (i.e., the speed of light), while slow-time variables are related to the speed of the airplane carrying the radar. The indexing function i takes the following form for SAR:

$$i(x,y,u) = \chi(y \pm x\tan\phi)\cdot\frac{\sqrt{x^2+(y-u)^2}}{c}, \tag{2}$$

Where c is the speed of light, ϕ the beamwidth of the radar, and χ (a, b) equal to 1 for $a \le u \le b$ and 0 otherwise. x and y describe the two-dimensional offset between the radar and the physical spot on the ground corresponding to a pixel, and can thus be used in a simple distance calculation as seen in the right-hand side of (2).

In terms of implementation, we work with a discretized form of (1) in which the integral is approximated as a Riemann sum over a finite collection of projections u_k, $k \in \{1 \cdots K\}$ and is evaluated at the centers of image pixels (x_i, y_j), $i \in \{1 \cdots N\}$, $j \in \{1 \cdots K\}$. Because the evaluation of the index function at these discrete points will generally not result in a value of t which is exactly at a sample location, interpolation could be performed to increase accuracy.

HPRC Architecture

This project aimed at exploiting the full range of resources available on the heterogeneous high-performance cluster (HHPC) at the Air Force Research Laboratory in Rome, NY, USA [4]. Built by HPTi [5], the HHPC features a Beowulf cluster of 48 heterogeneous computing nodes, where each node consists of a dual 2.2 GHz Xeon PC running Linux and an Annapolis Microsystems WildStar II FPGA board.

The WildStar II features two VirtexII FPGAs and connects to the host Xeon general-purpose processors (GPPs) via the PCI bus. Each FPGA has access to 6 MB of SRAM, divided into 6 banks of 1 MB each, and a single 64 MB SDRAM bank.

The Annapolis API supports a master-slave paradigm for control and data transfer between the GPPs and the FPGAs. Applications for the FPGA can be designed either through traditional HDL-based design and synthesis tools, as we have done here, or by using Annapolis's CoreFire [6] module-based design suite.

The nodes of the HHPC are linked together in three ways. The PCs are directly connected via gigabit Ethernet as well as through Myrinet MPI cards. The WildStar II boards are also directly connected to each other through a low-voltage differential signaling (LVDS) I/O daughter card, which provides a systolic interface over which each FPGA board may talk to its nearest neighbor in a ring. Communication over Ethernet is supplied by the standard C library under Linux. Communication over Myrinet is achieved with an installation of the MPI message-passing standard, though MPI can also be directed to use Ethernet instead. Communicating through the LVDS interconnect involves writing communication modules for the FPGA manually. In this project, we relied on Myrinet to move data between nodes. This architecture represents perhaps the most direct method for adding reconfigurable resources to a supercomputing cluster. Each node architecture is similar to that of a single-node reconfigurable computing solution. Networking hardware which interfaces well to the Linux PCs is included to create the cluster network. The ability to communicate between FPGAs is included but remains difficult for the developer to employ. Other HPRC platforms, such as those developed by Cray and SRC, may employ different interconnection methods, programming methods, and communication paradigms.

RELATED WORK

Backprojection itself is a well-studied algorithm. Most researchers have focused on implementing backprojection for computed tomography (CT) medical imaging applications; backprojection for synthetic aperture radar (SAR) on FPGAs is not well-represented in the literature.

The precursor to this work is that of Coric et al. [7] Backprojection for CT uses the "spotlight" mode of imaging, in which the sensing array is rotated around the target area. (Contrast this with the "strip-map" mode described in Section 2.1.) Other implementations of backprojection for CT on FPGAs have been published [8].

CT backprojection has also been implemented on several other computing devices, including GPUs [9] and the cell broadband engine [10]. Results are generally similar (within a factor of 2) to those achieved on FPGAs.

Of the implementations of backprojection for SAR, almost none has been designed for FPGAs. Soumekh et al. have published on implementations of SAR in general and backprojection in particular [11], as well as the Soumekh reference book on the subject [2], but they do not examine the use of FPGAs for computation. Some recent work on backprojection for SAR on parallel hardware has come from Halmstad University in Sweden [12, 13]; their publications lay important groundwork but have not been implemented except in software and/or simulation.

Backprojection is not the only application that has been mapped to HPRC platforms, though signal processing is traditionally a strength of RC and so large and complex signal processing applications like backprojection are common. With the emergence of HPRC, scientific applications are also seeing significant research effort. Among these are such applications as hyperspectral dimensionality reduction [14], molecular dynamics [15, 16], and cellular automata simulations [17].

Another direction of HPRC research has been the development of libraries of small kernels that are useful as building blocks for larger applications. The Vforce framework [18] allows for portable programming of RC systems using a library of kernels. Other developments include libraries of floating-point arithmetic units [19], work on FFTs [20], and linear algebra kernels such as BLAS [21, 22].

Several survey papers [23, 24] address the trends that can be found among the reported results. The transfer of data between GPP and FPGA can significantly impact performance. The ability to determine and control the memory access patterns of the FPGA and the on-board memories is critical. Finally, sacrificing the accuracy of the results in favor of using lighter-weight operations that can be more easily implemented on an FPGA can be an effective way of increasing performance.

EXPERIMENTAL DESIGN

In this section, we describe an implementation of the backprojection image formation algorithm on a high-performance reconfigurable computer. Our implementation has been designed to provide high-speed image formation services and support output data distribution via a publish/subscribe [25] methodology. Section 4.1 describes the system on which our implementation runs. Section 4.2 explores the inherent parallelism in backprojection and describes the high-level design decisions that steered the implementation. Section 4.3 describes the portion of the implementation that runs in software, and Section 4.4 describes the hardware.

System Background

In Section 2.2, we described the HHPC system. In this section, we will explore more deeply the aspects of that system that are relevant to our experimental design.

HHPC Features

Several features of the Annapolis WildStar II FPGA boards are directly relevant to the design of our backprojection implementation. In particular, the host-to-FPGA interface, the on-board memory bandwidth, and the available features of the FPGA itself guided our design decisions.

Communication between the host GPP and the WildStar II board is over a PCI bus. The HHPC provides a PCI bus that runs at 66 MHz with 64-bit datawords. The WildStar II on-board PCI interface translates this into a 32-bit interface running at 133 MHz. By implementing the DMA data transfer mode to communicate between the GPP and the FPGA, the on-board PCI interface performs this translation invisibly and without significant loss of performance. A 133 MHz clock is also a good and achievable clock rate for FPGA hardware, so most of the hardware design can be run directly off the PCI interface clock. This simplifies the design since there are fewer clock domains (see Section 4.4.1).

The WildStar II board has six on-board SRAM memories (1 MB each) and one SDRAM memory (64 MB). It is beneficial to be able to read one datum and write one datum in the same clock cycle, so we prefer to use multiple SRAMs instead of the single larger SDRAM. The SRAMs run at 50 MHz and feature a 32-bit dataword (plus four parity bits), but they use a DDR interface. The Annapolis controller for the SRAM translates this into a 50 MHz 72-bit interface. Both features are separately important: we will need to cross from the 50 MHz memory clock domain to the 133 MHz PCI clock domain, and we will need to choose the size of our data such that they can be packed into a 72-bit memory word (see Section 4.2.4).

Finally, the Virtex2 6000 FPGA on the Wildstar II has some useful features that we use to our advantage. A large amount of on-chip memory is available in the form of BlockRAMs, which are configurable in width and depth but can hold at most 2 KB of data each. One hundred forty four of these dual-ported memories are available, each of which can be accessed independently. This makes BlockRAMs a good candidate for storing and accessing input projection data (see Sections 4.2.4 and 4.4.3.) BlockRAMs can also be configured as FIFOs, and due to their dual-ported nature, can be used to cross clock domains.

Swathbuckler Project

This project was designed to fit in as part of the Swathbuckler project [26, 27, 28], an implementation of synthetic aperture radar created by a joint program between the American, British, Canadian, and Australian defense research project agencies. It encompasses the entire SAR process including the aircraft and radar dish, signal capture and analog-to-digital conversion, filtering, and image formation hardware and software.

Our problem as posed was to increase the processing capabilities of the HHPC by increasing the performance of the portions of the application seen on the right-hand side of Figure 1. Given that a significant amount of work had gone into tuning the performance of the software implementation of the filtering process [26], it remained for us to improve the speed at which images could be formed. According to the project specification, the input data are streamed into the microprocessor main memory. In order to perform image formation on the FPGA, it is then necessary to copy data from the host to the FPGA. Likewise, the output image must be copied from the FPGA memory to the host memory so that it can be made accessible to the publish/subscribe software. These data transfer times are included in our performance measurements (see Section 5).

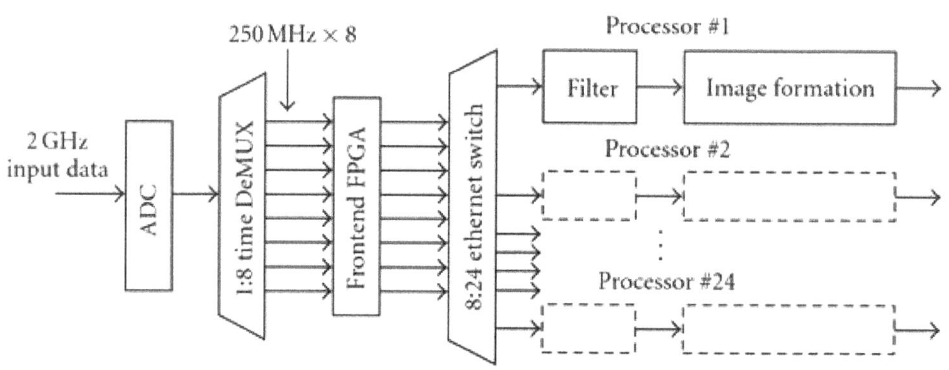

Figure 1 Block diagram of Swathbuckler system. (Adapted from [26].)

Algorithm Analysis

In this section, we dissect the backprojection algorithm with an eye toward implementing it on an HPRC machine. There are many factors that need to be taken into account when designing an HPRC application. First and foremost, an application that does not have a high degree of parallelism is generally not a good candidate. Given a suitable application, we then decide how to divide the problem along the available levels of parallelism in order to determine what part of the application will be executed on each available processor. This includes GPP/FPGA assignment as well as dividing the problem across the multiple nodes of the cluster. For the portions of the application run on the FPGAs, data arrays must be distributed among the accessible memories. Next, we look at some factors to improve the performance of the hardware implementation, namely, data formats and computation strength reduction. We conclude by examining the parameters of the data collection process that affect the computation.

Parallelism Analysis

In any reconfigurable application design, performance gains due to implementation in hardware inevitably come from the ability of reconfigurable hardware (and, indeed, hardware in general) to perform multiple operations at once. Extracting the parallelism in an application is thus critical to a high-performance implementation.

Equation (1) shows the backprojection operation in terms of projection data p(t, u)and an output image f(x, y). That equation may be interpreted to say that for a particular pixel $f(\hat{x},\hat{y})$, the final value can be found from a summation of contributions from the set of all projections p(t, u) whose corresponding radar pulse covered that ground location. The value of t for a given u is determined by the mapping function i (x, y, u) according to (2). There is a large degree of parallelism inherent in this interpretation.

1. The contribution from projection $p(\hat{u})$ to pixel $f(\hat{x},\hat{y})$, is not dependent on the contributions from all other projections $p(u)$, $u \neq \hat{u}$ to that same pixel $f(\hat{x},\hat{y})$.

2. The contribution from projection $p(\hat{u})$ to pixel $f(\hat{x}, \hat{y})$, is not dependent on the contribution from $p(\hat{u})$ to all other pixels $f(x, y)$, $x \neq \hat{x}$, $y \neq \hat{y}$.
3. The final value of a pixel is not dependent on the value of any other pixel in the target image.

It can be said, therefore, that backprojection is an "embarrassingly parallel" application, which is to say that it lacks any data dependencies. Without data dependencies, the opportunity for parallelism is vast and it is simply a matter of choosing the dimensions along which to divide the computation that best matches the system on which the algorithm will be implemented.

Dividing the Problem

There are two ways in which parallel applications are generally divided across the nodes of a cluster.

1. Split the *data*. In this case, each node performs the same computation as every other node, but on a subset of data. There may be several different ways that the data can be divided.
2. Split the *computation*. In this case, each node performs a portion of the computation on the entire dataset. Intermediate sets of data flow from one node to the next. This method is also known as task-parallel or systolic computing.

While certain supercomputer networks may make the task-parallel model attractive, our work with the HHPC indicates that its architecture is more suited to the data-parallel mode. Since internode communication is accomplished over a many-to-many network (Ethernet or Myrinet), passing data from one node to the next as implied by the task-parallel model will potentially hurt performance. A task-parallel design also implies that a new FPGA design must be created for each FPGA node in the system, greatly increasing design and verification time. Finally, the number of tasks available in this application is relatively small and would not occupy the number of nodes that are available to us.

Given that we will create a data-parallel design, there are several axes along which we considered splitting the data. One method involves dividing the input projection data $p(t,u)$ among the nodes along the u dimension. Each node would hold a portion of the projections $p(t,[u_i,u_j])$ and calculate that portion contribution to the final image. However, this implies that each node must hold a copy of the entire target image in memory, and furthermore, that all of the partial target images would need to be added together after processing before the final image could be created. This extra processing step would also require a large amount of data to pass between nodes. In addition, the size of the final image would be limited to that which would fit on a single FPGA board.

Rather than dividing the input data, the preferred method divides the output image $f(x,y)$ into pieces along the range (X) axis (see Figure 2). In theory, this requires that every projection be sent to each node; however, since only a portion of each projection will affect the slice of the final image being computed on a single node, only that portion must be sent to that node. Thus, the amount of input data being sent to each node can be reduced to $p([t_i,t_j],u)$. We refer to the portion of the final target image being computed on a single node, $f([x_i,x_j],y)$, as a "subimage".

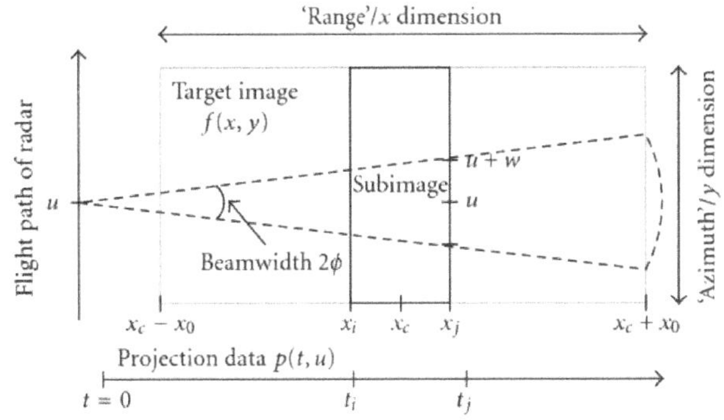

Figure 2 Division of target image across multiple nodes.

Figure 2 shows that t_j is slightly beyond the time index that corresponds to x_j . This is due to the width of the cone-shaped radar beam. The dotted line in the figure shows a single radar pulse taken at slow-time index $y = u$. The minimum distance to any part of the subimage is at the point (x_i,u), which corresponds to fast-time index t_i in the projection data. The maximum distance to any part of the subimage, however, is along the outer edge of the cone to the point $(x_j,u \pm w)$, where w is a factor calculated from the beamwidth angle of the radar and x_j . Thus, the fast-time index t_j is calculated relative to x_j and w instead of simply x_j . This also implies that the $[t_i,t_j]$ range for two adjacent nodes will overlap somewhat, or (equivalently) that some projection data will be sent to more than one node.

Since the final value of a pixel does not depend on the values of the pixels surrounding it, each FPGA needs hold only the subimage that it is responsible for computing. That portion is not affected by the results on any other FPGA, which means that the postprocessing accumulation stage can be avoided. If a larger target image is desired, subimages can be «stitched» together simply by concatenation.

In contrast to the method where input data are divided along the u dimension, the size of the final target image is not restricted by the amount of memory on a single node, and furthermore, larger images can be processed by adding nodes to the cluster. This is commonly referred to as *coarse-grained parallelism*, since the problem has been divided into large-scale independent units. Coarse-grained parallelism is directly related to the performance gains that are achieved by adapting the application from a single-node computer to a multinode cluster.

Memory Allocation

The memory devices used to store the input and output data on the FPGA board may now be determined. We need to store two large arrays of information: the target image $f(x,y)$ and the input projection data $p(t,u)$. On the Wildstar II board, there are three options: an on-board DRAM, six on-board SRAMs, and a variable number of BlockRAMs which reside inside the FPGA and can be instantiated as needed. The on-board DRAM has the highest capacity (64 MB) but is the most difficult to use and only has one read/write port. BlockRAMs are the most flexible (two read/write ports and a flexible geometry) and simple to use, but have a small (2 KB) capacity.

For the target image, we would like to be able to both read and write one target pixel per cycle. It is also important that the size of the target image stored on one node be as large as possible, so memories with larger capacity are better. Thus, we will use multiple on-board SRAMs to store the target image. By implementing a two-memory storage system, we can provide two logical ports into the target image array. During any given processing step, one SRAM acts as the source for target pixels, and the other acts as the destination for the newly computed pixel values. When the next set of projections is sent to the FPGA, the roles of the two SRAMs are reversed.

Owing to the 1 MB size of the SRAMs in which we store the target image data, we are able to save 2^{19} pixels. We choose to arrange this into a target image that is 1024 pixels in the azimuth dimension and 512 in the range dimension. Using power-of-two dimensions allows us to maximize our use of the SRAM, and keeping the range dimension small allows us to reduce the amount of projection data that must be transferred.

For the projection data, we would like to have many small memories that can each feed one of the projection adder units. BlockRAMs allow us to instantiate multiple small memories in which to hold the projection data; each memory has two available ports, meaning that two adders can be supported in parallel. Each adder reads from one SRAM and writes to another; since we can support two adders, we could potentially use four SRAMs.

Data Formats

Backprojection is generally accomplished in software using a complex (i.e., real and imaginary parts) floating-point format. However, since the result of this application is an image which requires only values from 0 to 255 (i.e., 8-bit integers), the loss of precision inherent in transforming the data to a fixed-point/ integer format is negligible. In addition, using an integer data format allows for much simpler functional units.

Given an integer data format, it remains to determine how wide the various datawords should be. We base our decision on the word width of the memories. The SRAM interface provides 72 bits of data per cycle, comprised of two physical 32-bit datawords plus four bits of parity each. The BlockRAMs are configurable, but generally can provide power-of-two sized datawords.

Since backprojection is in essence an accumulation operation, it makes sense for the output data (target image pixels) to be wider than the input data (projection samples). This reduces the likelihood of overflow error in the accumulation. We, therefore, use 36-bit complex integers (18-bit real and 18-bit imaginary) for the target image, and 32-bit complex integers for the projection data.

After backprojection, a complex magnitude operator is needed to reduce the 36-bit complex integers to a single 18-bit real integer. This operator is implemented in hardware, but the process of scaling data from 18-bit integer to 8-bit image is left to the software running on the GPP.

Computation Analysis

The computation to be performed on each node consists of three parts. The summation from (1) and the distance calculation from (2) represent the backprojection work to be done. The complex magnitude operation is similar to the distance calculation.

While adders are simple to replicate in large numbers, the hardware required to perform multiplication and square root is more costly. If we were using floating-point data formats, the number of functional units that could be instantiated would be very small, reducing the parallelism that we can exploit. With integer data types, however, these units are relatively small, fast, and easily pipelined. This allows us to maintain a high clock rate and one-result-per-cycle throughput.

Data Collection Parameters

The conditions under which the projection data are collected affect certain aspects of the backprojection computation. In particular, the spacing between samples in the $p(t,u)$ array and the spacing between pixels in the $f(x,y)$ array imply constant factors that must be accounted for during the distance-to-time index calculation (see Section 4.4.3).

For the input data, Δu indicates the distance (in meters) between samples in the azimuth dimension. This is equivalent to the distance that the plane travels between each outgoing pulse of the radar. Often, due to imperfect flight paths, this value is not regular. The data filtering that occurs prior to backprojection image formation is responsible for correcting for inaccuracies due to the actual flight path, so that a regular spacing can be assumed.

As the reflected radar data are observed by the radar receiver, they are sampled at a particular frequency ω. That frequency translates to a range distance Δt between samples equal to $c/2\omega$, where c is the speed of light. The additional factor of $1/2$ accounts for the fact that the radar pulse travels the intervening distance, is reflected, and travels the same distance back. Owing to the fact that the airplane is not flying at ground level, there is an additional angle of elevation that is included to determine a more accurate value for Δt.

For the target image (output data), Δx and Δy simply correspond to the real distance between pixels in the range and azimuth dimensions, accordingly. In general, Δx and Δy are not necessarily related to Δu, and Δt and can be chosen at will. In practice, setting $\Delta y = \Delta u$ makes the algorithm computation more regular (and thus more easily parallelizable). Likewise, setting $\Delta x = \Delta t$ reduces the need for interpolation between samples in the t dimension since most samples will line up with pixels in the range dimension. Finally, setting $\Delta x = \Delta y$ provides for square pixels and an easier-to-read aspect ratio in the output image.

The final important parameter is the minimum range from the radar to the target image, known as $R_{\min}$. This is related to the t_i parameter, and is used by the software to determine what portion of the projection data is applicable to a particular node.

Software Design

We now describe the HPRC implementation of backprojection. As with most FPGA-based applications, the work that makes up the application is divided between the host GPP and the FPGA. In this section, we will discuss the work done on the GPP; in Section 4.4, we continue with the hardware implemented on the FPGA.

The main executable running on the GPP begins by using the MPI library to spawn processes on several of the HHPC nodes. Once all MPI jobs have started, the host code configures the FPGA with the current values of the flight parameters from Section 4.2.6. In particular, the values of Δx, Δy, and R_{min} (the minimum range) are sent to the FPGA. However, in order to avoid the use of fractional numbers, all of these parameters are normalized such that $\Delta t = 1$. This allows the hardware computation to be in terms of fast-time indices in the t domain instead of ground distances.

Next, the radar data is read. In the Swathbuckler system, this input data would be streamed directly into memory and no separate «read» step would be required. Since we are not able to integrate directly with Swathbuckler, our host code reads the data from a file on the shared disk. These data are translated from complex floating-point format to integers. The host code also determines the appropriate range of t that is relevant to the subimage being calculated by this node (see Section 4.2.2).

The host code then loops over the u domain of the projection data. A chunk of the data is sent to the FPGA and processed. The host code waits until the FPGA signals processing is complete, and then transmits the next chunk of data. When all projection data have been processed, the host code requests that the final target image be sent from the FPGA. The pixels of the target image are scaled, rearranged into an image buffer, and an image file is optionally produced using a library call to the GTK+ library [29].

After processing, the target subimages are simply held in the GPP memory. In the Swathbuckler system, subimages are distributed to consumers via a publish/subscribe mechanism, so there is no need to assemble all the subimages into a larger image.

Configuration Parameters

Our backprojection implementation can be configured using several compile-time parameters in both the host code and the VHDL code that describes the hardware. In software, the values of Δx and Δy are set in the header file and compiled in. The value of R_{min} is specific to a dataset, so it is read from the file that contains the projection data.

It is also possible to set the dimensions of the subimage (1024×512 by default), though the hardware would require significant changes to support this.

The hardware VHDL code allows two parameters to be set at compile time (see Section 4.4.3). N is the number of projection adders in the design, and R is the size of the projection memories ($R \times 1024$ words). Once compiled, the value of these parameters can be read from the FPGA by the host code.

FPGA Design

The hardware that is instantiated on the FPGA boards runs the backprojection algorithm and computes the values of the pixels in the output image. A block diagram of the design is shown in Figure 3. References to blocks in this figure are printed in `monospace`.

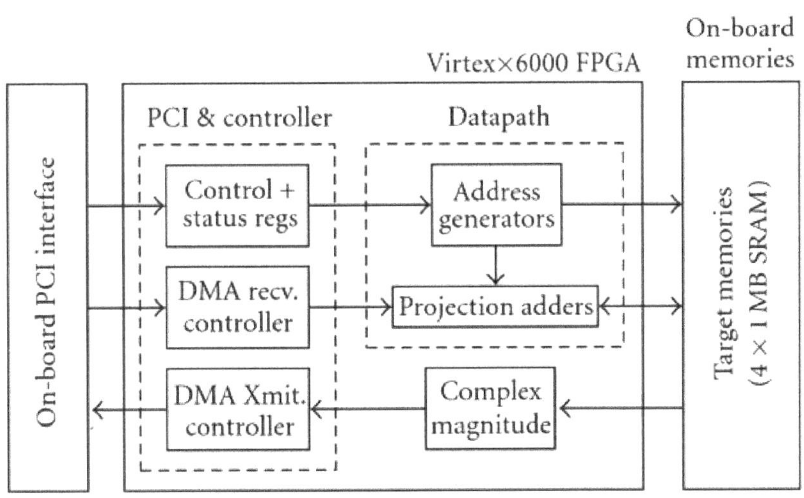

Figure 3 Block diagram of backprojection hardware unit.

Clock Domains

In general, using multiple clock domains in a design adds complexity and makes verification significantly more difficult. However, the design of the Annapolis Wildstar II board provides for one fixed-rate clock on the PCI interface, and a separate fixed-rate clock on the SRAM memories. This is a common attribute of FPGA-based systems.

To simplify the problem, we run the bulk of our design at the PCI clock rate (133 MHz). Since Annapolis VHDL modules refer to the PCI interface as the "LAD bus", we call this the *L-clock* domain. Every block in Figure 3, with the exception of the SRAMs themselves and their associated `Address Generators`, is run from the L-clock.

The SRAMs are run from the memory clock, or *M-clock*, which is constrained to run at 50 MHz. Between the `Target Memories` and the `Projection Adders` , there is some interface logic and an FIFO. This is not shown in Figure 3, but exists to cross the M-clock/L-clock domain boundary.

BlockRAM-based FIFOs, available as modules in the Xilinx CORE Generator [30] library, are used to cross clock domains. Since each of the ports on the dual-ported BlockRAMs is individually clocked, the read and write can happen in different clock domains. Control signals are automatically synchronized to the appropriate clock, that is, the «full» signal is synchronized to the write clock and the «empty» signal to the read clock. Using FIFOs whenever clock domains must be crossed provides a simple and effective solution.

Control Registers and DMA Input

The Annapolis API, like many FPGA control APIs, allows for communication between the host PC and the FPGA with two methods: "programmed" or memory-mapped I/O (PIO), which is best for reading and writing one or two words of data at a time; direct memory access (DMA), which is best for transferring large blocks of data.

The host software uses PIO to set control registers on the FPGA. Projection data is placed in a specially allocated memory buffer, and then transmitted to the FPGA via DMA. On the FPGA, the DMA Receive Controller receives the data and arranges it in the BlockRAMs inside the Projection Adders .

Datapath

The main datapath of the backprojection hardware is shown in Figure 4. It consists of five parts: the Target Memory SRAMs that hold the target image, the Distance - To - Time Index Calculator (DIC), the Projection Data BlockRAMs , theAdders to perform the accumulation operation, and the Address Generators that drive all of the memories. These devices all operate in a synchronized fashion, though there are FIFOs in several places to temporally decouple the producers and consumers of data, as indicated in Figure 4 with segmented rectangles.

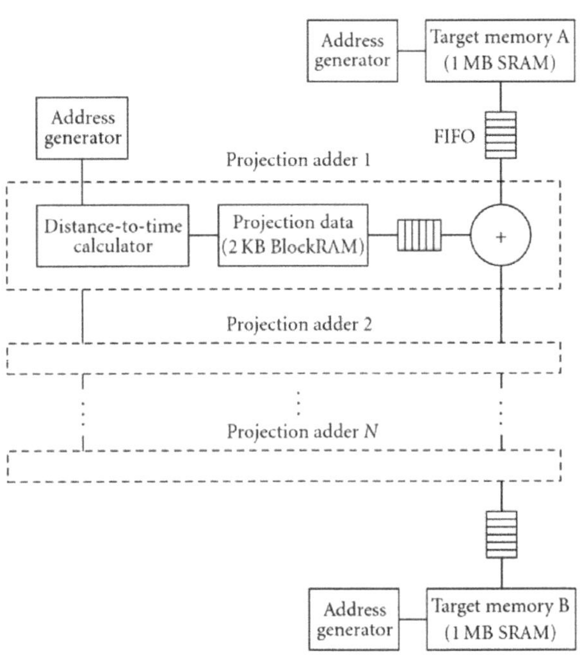

Figure 4 Block diagram of hardware datapath.

Address Generators

There are three data arrays that must be managed in this design: the input target data, the output target data, and the projection data. The pixel indices for the two target data arrays (Target Memory A and Target Memory B in Figure 4) are managed directly by separate address generators. The address generator for the Projection Data BlockRAMs also produces pixel indices; the DIC converts the pixel index into a fast-time index that is used to address the BlockRAMs.

Because a single read/write operation to the SRAMs produces/consumes two pixel values, the address generators for the SRAMs run for half as many cycles as the address generator for the BlockRAMs. However, address generators run in the clock domain relevant to the memory that they are addressing, so $n/2$ SRAM addresses take slightly longer to generate at 50 MHz than n BlockRAM addresses at 133 MHz.

Because of the use of FIFOs between the memories and the adders, the address generators for Target Memory Aand the Projection Data BlockRAMs can run freely. FIFO control signals ensure that an address generator is paused in time to prevent it from overflowing the FIFO. The address generator for Target Memory B is incremented whenever data are available from the output FIFO.

Distance-to-Time Index Calculator

The Distance - To - Time Index Calculator (DIC) implements (2), which is comprised of two parts. At first glance, each of these parts involves computation that requires large amount of hardware and/or time to calculate. However, a few simplifying assumptions make this problem easier and reduce the amount of needed hardware.

Rather than implementing a tangent function in hardware, we rely on the fact that the beamwidth ϕ of the radar is a constant. The host code performs the $\tan\phi$ function and sends the result to the FPGA, which is then used to calculate $x(a,b)$. This value is used both on a coarse-grained level to narrow the range of pixels which are examined for each processing step, and on a fine-grained level to determine whether or not a particular pixel is affected by the current projection (see Figure 2).

The right-hand side of (2) is a distance function ($\sqrt{x^2+y^2}$) and a division. The square root function is executed using an iterative shift-and-subtract algorithm. In hardware, this algorithm is implemented with a pipeline of subtractors. Two multiplication units handle the x^2 and y^2 functions. Some additional adders and subtractors are necessary to properly align the input data to the output data according to the data collection parameters discussed in Section 4.2.6. We used pipelined multipliers and division units from the Xilinx CORE Generator library; adders and subtractors are described with VHDL arithmetic operators, allowing the synthesis tools to generate the appropriate hardware.

The distance function and computation of $x(\cdot)$ occur in parallel. If the $x(\cdot)$ function determines that the pixel is outside the affected range, the adder input is forced to zero.

Projection Data Blockrams

The output of the DIC is a fast-time index into the $p(t,u)$ array. Each Projection Data BlockRAM holds the data for a particular value of u . The fast-time index t is applied to retrieve a single value of $p(t,u)$ that corresponds to the pixel that was input by the address generator. This value is stored in an FIFO, to be synchronized with the output of the `Target Memory A` FIFO.

`Projection Data` memories are configured to hold 2 k datawords by default, which should be sufficient for a 1 k range pixel image. This number is a compile-time parameter in the VHDL source and can be changed. The resource constraint is the number of available BlockRAMs.

Projection Adder

As the FIFOs from the `Projection Data` memories and the `Target Memory` are filled, the `Projection Adder` reads datawords from both FIFOs, adds them together, and passes them to the next stage in the pipeline (see Figure 4).

The design is configured with eight adder stages, meaning eight projections can be processed in one step. This number is a compile-time parameter in the VHDL source and can be changed. The resource constraint is a combination of the number of available BlockRAMs (because the `Projection Data` BlockRAMs and FIFO are duplicated) and the amount of available logic (to implement the DIC).

The number of adder stages implemented directly impacts the performance of our application. By computing the contribution of multiple projections in parallel, we exploit the *fine-grained parallelism* inherent in the backprojection algorithm. Fine-grained parallelism is directly related to the performance gains achieved by implementing the application in hardware, where many small execution units can be implemented that all run at the same time on different pieces of data.

Complex Magnitude and DMA Output

When all projections have been processed, the final target image data reside in one of the `Target Memory` SRAMs. The host code then requests that the image data be transferred via DMA to the host memory. This process occurs in three steps.

First, an `Address Generator` reads the data out of the SRAM in the correct order. Second, the data are converted from complex to real. The `Complex Magnitude` operator performs this function with a distance calculation ($\sqrt{re^2 + im^2}$). We instantiate another series of multipliers, adders, and subtractors (for the integer square root) to perform this operation. Third, the real-valued pixels are passed to the `DMA Transmit Controller`, where they are sent from the FPGA to the host memory.

EXPERIMENTAL RESULTS

After completing the design, we conducted a series of experiments to determine the performance and accuracy of the hardware. When run on a single node, a detailed profile of the execution time of both the software and hardware programs can be determined, and the effects of reconfigurable hardware design techniques can be studied. Running the same program on multiple nodes shows how well the application scales take advantage of the processing power available on HPC clusters. In this section, we describe the experiments and analyze the collected results.

Experimental Setup

Our experiments consist of running programs on the HHPC and measuring the run time of individual components as well as the overall execution time. There are two programs: one which forms images by running backprojection on the GPP (the "software" program), and one which runs it in hardware on an FPGA (the "hardware" program).

We are concerned with two factors: speed of execution and accuracy of solution. We will consider not only the execution time of the backprojection operation by itself, but also the execution time of the entire program including peripheral operations such as memory management and data scaling. In addition, we examine the full application run time.

In terms of accuracy, the software program computes its results in floating point while the hardware uses integer arithmetic. We will examine the differences between the images produced by these two programs in order to establish the error introduced by the data format conversion.

The ability of our programs to scale across multiple nodes is an additional performance metric. We measure the effectiveness of exploiting coarse-grained parallelism by comparing the relative performance of both the software program and the hardware implementation when run on one node and when run on many nodes.

Software Design

All experiments were conducted on the HHPC system as described in Section 2.2. Nodes on the HHPC run the Linux operating system, RedHat release 7.3, using kernel version 2.4.20.

Both the software program and the calling framework for the hardware implementation are written in C and produce an executable that is started from the Linux shell. The software program executes entirely on the GPP: memory buffers are established, projection data are read from disk, the backprojection algorithm is run, and output data are transformed from complex to real. The final step involves rearranging the output data and scaling it, then writing a PNG image to disk.

The hardware program begins by establishing memory buffers and initializing the FPGA. Projection data are read from disk into the GPP memory. Those data are then transferred to the FPGA, where the backprojection algorithm is run in hardware. Output data are transformed from complex to real on the FPGA, then transferred to the GPP memory. The GPP then executes the same rearrangement and scaling step as the software program.

To control the FPGA, the hardware program uses an API that is provided by Annapolis to interface to the WildStar II boards. The FPGA hardware was written in VHDL and synthesized using version 8.9 of Synplify Pro. Hardware place and route used the Xilinx ISE 9.1i suite of CAD tools.

The final step of both programs, where the target data are written as a PNG image to disk by the GPP, uses the GTK+ [29] library version 2.0.2. For the multinode experiments, version 1.2.1.7b of the MPICH [31] implementation of MPI is used to handle internode startup, communication, and synchronization.

The timing data presented in this section were collected using timing routines that were inserted into the C code. These routines use Linux system calls to display timing information. The performance of the timing routines was determined by running them several times in succession with no code in between. The overhead of the performance routines was shown to be less than 100 microseconds, so timing data are presented as accurate to the millisecond. Unless noted otherwise, applications were run five times and an average (arithmetic mean) was taken to arrive at the presented data.

We determine accuracy both qualitatively (i.e., by examining the images with the human eye) and quantitatively by computing the difference in pixel values between the two images.

Test Data

Four sets of data were used to test our programs. The datasets were produced using a MATLAB simulation of SAR taken from the Soumekh book [2]. This MATLAB script allows the parameters of the data collection process to be configured (see Section 4.2.6). When run, it generates the projection data that would be captured by an SAR system imaging that area. A separate C program takes the MATLAB output and converts it to an optimized file that can be read by the backprojection programs.

Each dataset contains four point source (i.e., 1 × 1 pixel in size) targets that are distributed randomly through the imaged area. The imaged area for each set is of a similar size, but situated at a different distance from the radar. Targets are also assigned a random reflectivity value that indicates how strongly they reflect radar signals.

Results and Analysis

In general, owing to the high degree of parallelism inherent in the backprojection algorithm, we expect a considerable performance benefit from implementation in hardware even on a single node. For the multinode program, the lack of need to transfer data between the nodes implies that the performance should scale in a linear relation to the number of nodes.

Single Node Performance

The first experiment involves running backprojection on a single node. This allows us to examine the performance improvement due to *fine-grained parallelism*, that is, the speedup that can be gained by implementing the algorithm in hardware. For this experiment, we ran the hardware and software programs on all four of the datasets. Table 1shows the timing breakdown of dataset no. 4; Table 2 shows the overall results for all four datasets. Note that dataset no. 1 is closest to the radar, and dataset no. 4 is furthest away.

Table 1 Single-node experimental performance.

Component	Software	Hardware	Ratio
Backprojection	*76.4* s	*351* ms	*217:1*
Complex magnitude	73 ms	15 ms	4.9:1
Form image (software)	39 ms	340 ms	1:8.7
Total	76.5 s	706 ms	108:1

Table 2 Single-node backprojection performance by dataset.

Dataset	Software	Hardware	BP speedup	App speedup
1	24.5 s	146 ms	167.4	49.8
2	30.4 s	169 ms	179.5	61.9
3	47.7 s	268 ms	177.5	75.5
4	76.5 s	351 ms	217.6	108.4

In Table 1, *Software* and *Hardware* refer to the run time of a particular component; *Ratio* is the ratio of software time to hardware time, showing the speedup or slowdown of the hardware program. *Backprojection* is the running of the core algorithm. *Complex Magnitude* transforms the data from complex to real integers. Finally, *Form Image*scales the data to the range [0:255] and creates the memory buffer that is used to create the PNG image.

There are a number of significant observations that can be made from the data in Table 1. Most importantly, the process of running the backprojection algorithm is greatly accelerated in hardware, running over 200x faster than our software implementation. It is important to emphasize that this includes the time required to transfer projection data from the host to the FPGA, which is not required by the software program. Many of the applications discussed in Section 3 exhibit only modest performance gains due to the considerable amount of time spent transferring data. Here, the vast degree of fine-grained parallelism in the backprojection algorithm that can be exploited by FPGA hardware allows us to achieve excellent performance compared to a serial software implementation.

The complex magnitude operator also runs about 5x faster in hardware. In this case, the transfer of the output data from the FPGA to the host is overlapped with the computation of the complex magnitude. This commonly used technique allows the data transfer time to be "hidden", preventing it from affecting overall performance.

However, the process of converting the backprojection output into an image buffer that can be converted to a PNG image (*Form Image*) runs faster when executed as part of the software program. This step is performed in software regardless of where the backprojection algorithm was executed. The difference in run time can be attributed to memory caching. When backprojection occurs in software, the result data lie in the processor cache. When backprojection occurs in hardware, the result data are copied via DMA into the processor main memory, and must be loaded into the cache before the *Form Image* step can begin.

We do not report the time required to initialize either the hardware or software program, since in the Swathbuckler system it is expected that initialization can be completed before the input data become available.

Table 2 shows the single-node performance of both programs on all four datasets. Note that the reported run times are only the times required by the backprojection operation. Thus, column four, *BP Speedup*, shows the factor of speedup (software:hardware ratio) for only the backprojection operation. Column five, *App Speedup*, shows the factor of speedup for the complete application including all of the steps shown in Table 1.

These results show that the computation time of the backprojection algorithm is data dependent. This is directly related to the minimum range of the projection data. According to Figure 2, as the subimage gets further away from the radar, the width of the radar beam is larger. This is reflected in the increased limits of the $\chi(a,b)$ term of (2), which are a function of the tangent of the beamwidth ϕ and the range. A larger range implies more pixels are impacted by each projection, resulting in an increase in processing time. The hardware and software programs scale at approximately the same rate, which is expected since they are processing the same amount of additional data at longer ranges.

More notable is the increase in application speedup; this can be explained by considering that the remainder of the application is not data dependent and stays relatively constant as the minimum range varies. Therefore, as the range increases and the amount of data to process increases, the backprojection operation takes up a larger percentage of the run time of the entire application. For software, this increase in proportion is negligible (99.5% to 99.8%), but for the hardware, it is quite large (12.6% to 25.0%). As the backprojection operator takes up more of the overall run time, the relative gains from implementing it in hardware become larger, resulting in the increasing speedup numbers seen in the table.

Single Node Accuracy

Qualitatively, the hardware images look very similar, with the hardware images perhaps slightly darker near the point source target. This is due to the quantization imposed by using integer data types. As discussed in Section 5.2.1, a longer range implies a wider radar beam. The $\chi(a,b)$ function from (2) determines how wide the beam is at a given range. When computed in fixed point for the hardware program, $y \pm x \tan\phi$ returns slightly different values than when the software program computes it in floating point. Thus, there is a slight smearing or blurring of the point source. Recall that dataset no. 4 has a longer range than the other datasets; appropriately, the smearing is most notable in that dataset.

Quantitatively, the two images can be compared pixel-by-pixel to determine the differences. For each dataset, Table 3 presents error in terms of the differences between the image produced by the software program and the image produced by the hardware program.

Table 3 Image accuracy by dataset.

Dataset	Errors		Max error		Mean error
1	4916	0.9%	18	7.0%	1.55
2	13036	2.5%	19	7.4%	1.73
3	18706	3.6%	12	4.6%	1.56
4	29093	5.6%	16	6.2%	1.64

The second column shows the number of pixels that are different between the two images. There are 1024×512 pixels in the image, so the third column shows the percent of overall image pixels that are different. The maximum and arithmetic mean error are shown in the last two columns. Recall that our output images are 256 gray scale PNG files; the magnitude of error is given by $err(x,y) = |hw(x,y) - sw(x,y)|$.

Again, errors can be attributed to the difference in the computed width of the radar beam between the software and hardware programs. For comparison, a version of each program was written that does not include the $\chi(a,b)$ function and instead assumes that every projection contributes to every pixel (i.e., an infinite beamwidth). In this case, the images are almost identical; the number of errors drops to 0.1%, and the maximum error is 1. Thus, the error is not due to quantization of processed data; the computation of the radar beamwidth is responsible.

Multinode Performance

The second experiment involves running backprojection on multiple nodes simultaneously, using the MPI library to coordinate. These results show how well the application scales due to *course-grained parallelism*, that is, the speedup that can be gained by dividing a large problems into smaller pieces and running each piece separately. For this experiment, we create an output target image that is 64 times the size of the image created by a single node. Thus, when run on one node, 64 iterations are required; for two nodes, 32 iterations are required, and so on. Table 4 shows the results for a single dataset.

Table 4 Multinode experimental performance.

Nodes		Software		Hardware		
	Mean	Standard deviation	Speedup	Mean	Standard deviation	Speedup
1	1943.2	6.15	1.0	25.0	.01	1.0
2	983.2	10.46	2.0	13.4	.02	1.9
4	496.0	4.60	3.9	7.8	.02	3.9
8	256.5	5.85	7.6	4.0	.06	6.0
16	128.4	1.28	15.1	—	—	—

For both the software and hardware programs, five trials were run. For each trial, the time required to run backprojection and form the resulting image on each node was measured, and the *maximum* time reported. Thus, the overall run time is equal to the run time of the slowest node. The arithmetic mean of the times (in seconds) from the five trials are presented, with standard deviation. The mean run time is compared to the mean run time for one node in order to show the speedup factor.

Results are not presented for a 16-node trial of the hardware program. During our testing, it was not possible to find 16 nodes of the HHPC that were all capable of running the hardware program at once. This was due to hardware errors on some nodes, and inconsistent system software installations on others.

The mostly linear distribution of the data in Table 4 shows that for the backprojection application, we have achieved a nearly ideal parallelization. This can be attributed to the lack of data passing between nodes, combined with an insignificant amount of overhead involved in running the application in parallel with MPI. The hardware program shows a similar curve, except for $N = 8$ nodes, where the speedup drops off slightly. At run times under five seconds, the MPI overhead involved in synchronizing the nodes between each processing iteration becomes significant, resulting in a slight slowdown (6x speedup compared to the ideal 8x).

The speedup provided by the hardware program is further described in Table 5. Compared to one node running the hardware program, we have already seen the nearly linear speedup. Compared to an equal number of nodes running the software program, the hardware consistently performs around 75x faster. Again, for $N = 8$, there is a slight drop off in speedup owing to the MPI overhead for short run times. Finally, we show that when compared to a single node running the software program, the combination of fine- and coarse-grained parallelism results in a very large performance gain.

Table 5 Speedup factors for hardware program.

Nodes	Ratio compared to		
	1 hardware	N software	1 software
1	1.0	77.8	77.8
2	1.9	75.8	149.8
4	3.9	76.5	299.8
8	6.0	61.1	463.0

DISCUSSION

The results from Section 5.2.3 show that excellent speedup can be achieved by implementing the backprojection algorithm on an HPRC machine. As HPRC architectures improve and more applications are developed for them, designers will continue to search for ways to carve out more and more performance. Based on the lessons learned in Section 3 and our work on backprojection, in this section we suggest some directions for future research.

Future Backprojection Work

This project was developed with an eye toward implementation as a part of the Swathbuckler SAR system (see Section 4.1.2). Owing to the classified nature of that project, additional work beyond the scope of this project is required to integrate our backprojection implementation into that project. To determine the success of this aspect of the project, we would need to compare backprojection to the current Swathbuckler image formation algorithm, both in terms of run time as well as image quality.

Despite excellent speedup results, there are further avenues for improvement of our hardware. The Wildstar II boards feature two identical FPGAs, so it may be possible to process two images at once. If the data transfer to one FPGA can be overlapped with computation on the other, significant speedup is possible. It may also be possible for each FPGA to create a larger target image using more of the on-board SRAMs.

An interesting study could be performed by porting backprojection to several other HPRC systems, some of which can be targeted by high-level design languages. This would be the first step toward developing a benchmark suite for testing HPRC systems; however, without significant tool support-to-support application portability between HPRC platforms, this process would be daunting.

HPRC Systems and Applications

One common theme among the FPGA applications mentioned in Section 3 is data transfer. Applications that require a large amount of data to be moved between the host and the FPGA can eliminate most of the gains provided by increased parallelism. Backprojection does not suffer from this problem because the amount of parallelism exploited is so high that the data transfer is a relatively small portion of the run time, and some of the data transfers can be overlapped with computation. These are common and well-known techniques in HPRC application design.

This leads us to two conclusions. First, when considering porting an application to an HPRC system, it is important to consider whether the amount of available parallelism is sufficient to provide good speedup. Tools that can analyze an application to aid designers in making this decision are not generally available.

Second, it is crucial for the speed of the data transfers to be as high as possible. Early HPRC systems such as the HHPC use common bus architectures like PCI, which do not provide very high bandwidth. This limits the effectiveness of many applications. More recent systems such as the SRC-7 have included significantly higher bandwidth interconnect, leading to improved data transfer performance and increasing the number of applications that can be successfully ported. Designers of future HPRC systems must continue to focus on ways to improve the speed of these data transfers.

It is also noteworthy that the backprojection application presented here was developed using hand-coded VHDL, with some functional units from the Xilinx CoreGen library [30]. Writing applications in an HDL provides the highest amount of flexibility and customization, which generally implies the highest amount of exploited parallelism. However, HDL development time tends to be prohibitively high. Recent research has focused on creating programming languages and tools that can be used to increase programmer productivity, but applications developed with these tools have not provided speedups comparable to those of hand-coded HDL applications. The HPRC community would benefit from the continued improvement of development tools such as these.

Finally, each HPRC system has its own programming method that is generally incompatible with other systems. The standardization of programming interfaces would make the development of design tools easier, and would also increase application developer productivity when moving from one machine to the next. Alternately, tools to support portability of HPRC applications such as the VForce project [18] would also help HPRC developers.

CONCLUSIONS

In conclusion, we have shown that backprojection is an excellent choice for porting to an HPRC system. Through the design and implementation of this application, we have explored the benefits and difficulties of HPRC systems in general, and identified several important features of both these systems and applications that are candidates for porting. Backprojection is an example of the class of problems that demand larger amounts of computational resources than can be provided by desktop or single-node computers. As HPRC systems and tools mature, they will continue to help in meeting this demand and making new categories of problems tractable.

DECLARATIONS

Acknowledgments

This work was supported in part by the Center for Subsurface Sensing and Imaging Systems (CenSSIS) under the Engineering Research Centers Program of the National Science Foundation (Award no. EEC-9986821) and by the DOD High Performance Computer Modernization Program. The authors would also like to thank Dr. Richard Linderman and Prof. Eric Miller for their input to this project as well as Xilinx and Synplicity Corporations for their generous donations.

REFERENCES

1. Cordes B: *Parallel backprojection: a case study in high-performance reconfigurable computing, M.S. thesis.* Department of Electrical and Computer Engineering, Northeastern University, Boston, Mass, USA; 2008.
2. Soumekh M: *Synthetic Aperture Radar Signal Processing with MATLAB Algorithms.* John Wiley & Sons, New York, NY, USA; 1999.
3. Kak AC, Slaney M: *Principles of Computerized Tomographic Imaging.* IEEE Press, New York, NY, USA; 1988.
4. Ross VW: Heterogeneous high performance computer. *Proceedings of the High Performance Computing Modernization Program Users Group Conference (HPCMP ‹05), June 2005, Nashville, Tenn, USA* 304-307.
5. High Performance Technologies Inc *Cluster Computing,* January 2008, http://www.hpti.com/ Cluster Computing, January 2008,
6. Annapolis Microsystems, Inc *CoreFire FPGA Design Suite,* January 2008, http://www.annapmicro.com/corefire.htmlCoreFire FPGA Design Suite, January 2008,
7. Coric S, Leeser M, Miller E, Trepanier M: Parallel-beam backprojection: an FPGA implementation optimized for medical imaging. *Proceedings of the 10th ACM/SIGDA International Sysmposium on Field-Programmable Gate Arrays (FPGA ‹02), February 2002, Monterey, Calif, USA* 217-226.
8. Gac N, Mancini S, Desvignes M: Hardware/software 2D-3D backprojection on a SoPC platform. *Proceedings of the ACM Symposium on Applied Computing (SAC ‹06), April 2006, Dijon, France* 222-228.
9. Xue X, Cheryauka A, Tubbs D: Acceleration of fluoro-CT reconstruction for a mobile C-arm on GPU and FPGA hardware: a simulation study. In *Medical Imaging 2006: Physics of Medical Imaging, February 2006, San Diego, Calif, USA, Proceedings of SPIE* Edited by: Flynn MJ, Hsieh J. 6142: 1494-1501.
10. Bockenbach O, Knaup M, Kachelrieß M: Implementation of a cone-beam backprojection algorithm on the cell broadband engine processor. *Medical Imaging 2007:*

Physics of Medical Imaging, February 2007, San Diego, Calif, USA, Proceedings of SPIE 6510: 1-10.

11. Nguyen L, Ressler M, Wong D, Soumekh M: Enhancement of backprojection SAR imagery using digital spotlighting preprocessing. *Proceedings of the IEEE Radar Conference, April 2004, Philadelphia, Pa, USA* 53-58.
12. Hast A, Johansson L: *Fast factorized back-projection in an FPGA, M.S. thesis*. Halmstad University, Halmstad, Sweden; 2006. http://hdl.handle.net/2082/576
13. Ahlander A, Hellsten H, Lind K, Lindgren J, Svensson B: Architectural challenges in memory-intensive, real-time image forming. *Proceedings of the 36th International Conference on Parallel Processing (ICPP ‹07), September 2007, Xian, China* 35.
14. El-Ghazawi E, El-Araby E, Agarwal A, LeMoigne J, Gaj K: Wavelet spectral dimension reduction of hyperspectral imagery on a reconfigurable computer. *Proceedings of the International Conference on Military and Aerospace Programmable Logic Devices (MAPLD ‹04), September 2004, Washington, DC, USA*
15. Alam SR, Agarwal PK, Smith MC, Vetter JS, Caliga D: Using FPGA devices to accelerate biomolecular simulations.*Computer* 2007,40(3):66-73.
16. Meredith JS, Alam SR, Vetter JS: Analysis of a computational biology simulation technique on emerging processing architectures. *Proceedings of the 21st International Parallel and Distributed Processing Symposium (IPDPS ‹07), March 2007, Long Beach, Calif, USA* 1-8.
17. Tripp JL, Gokhale MB, Hansson AA: A case study of hardware/software partitioning of traffic simulation on the cray XD1. *IEEE Transactions on Very Large Scale Integration (VLSI) Systems* 2008,16(1):66-74.
18. Moore N, Conti A, Leeser M, King LS: Vforce: an extensible framework for reconfigurable supercomputing. *Computer*2007,40(3):39-49.
19. Wang X, Braganza S, Leeser M: Advanced components in the variable precision floating-point library. *Proceedings of the 14th Annual IEEE Symposium on Field-Programmable Custom Computing Machines (FCCM ‹06), April 2006, Napa, Calif, USA* 249-258.
20. Underwood KD, Hemmert KS, Ulmer C: Architectures and APIs: assessing requirements for delivering FPGA performance to applications. *Proceedings of the ACM/IEEE Conference on Supercomputing (SC ‹06), November 2006, Tampa, Fla, USA*
21. Smith MC, Vetter JS, Alam SR: Scientific computing beyond CPUs: FPGA implementations of common scientific kernels. *Proceedings of the 8th International Conference on Military and Aerospace Programmable Logic Devices (MAPLD ‹05), September 2005, Washington, DC, USA*
22. Zhuo L, Prasanna VK: High performance linear algebra operations on reconfigurable systems. In *Proceedings of the ACM/IEEE Conference on Supercomputing (SC ‹05), November 2005, Seatle, Wash, USA*. IEEE Computer Society; 2.
23. Gokhale M, Rickett C, Tripp JL, Hsu C, Scrofano R: Promises and pitfalls of reconfigurable supercomputing.*Proceedings of the International Conference on Engineering of Reconfigurable Systems and Algorithms (ERSA '06), June 2006, Las Vegas, Nev, USA* 11-20.
24. Herbordt MC, VanCourt T, Gu Y, *et al.*: Achieving high performance with FPGA-based computing. *Computer*2007,40(3):50-57.
25. Eugster PTh, Felber PA, Guerraoui R, Kermarrec A-M: The many faces of publish/subscribe. *ACM Computing Surveys*2003,35(2):114-131. 10.1145/857076.857078
26. Rouse S, Bosworth D, Jackson A: Swathbuckler wide area SAR processing front end. *Proceedings of the IEEE Radar Conference, April 2006, New York, NY, USA* 1-6.
27. Linderman RW: Swathbuckler: wide swath SAR system architecture. *Proceedings of the IEEE Radar Conference, April 2006, Verona, NY, USA* 465-470.
28. Tucker S, Vienneau R, Corner J, Linderman RW: Swathbuckler: HPC processing and information exploitation.*Proceedings of the IEEE Radar Conference, April 2006, New York, NY, USA* 710-717.
29. *GTK+ Project*, March 2008, http://www.gtk.org/
30. Xilinx, Inc *CORE Generator*, March 2008,http://www.xilinx.com/products/design_tools/logic_design/design_entry/coregenerator.htm CORE Generator, March 2008,
31. Argonne National Laboratories *MPICH*, March 2008, http://www.mcs.anl.gov/research/projects/mpich2/ MPICH, March 2008,

CHAPTER 15

DESIGN OF PROCESSORS WITH RECONFIGURABLE MICROARCHITECTURE

Andrey Mokhov, Maxim Rykunov, Danil Sokolov and Alex Yakovlev

Keywords: Microarchitecture, microprocessor, reconfigurability, power proportionality, Conditional Partial Order Graphs

INTRODUCTION

Since 2007, our society has used more energy for browsing the Internet than for air travel [1]. It is also predicted that the energy (and environmental) footprint of computation and data traffic will steadily increase in the near future: Data centres will grow and so will the network infrastructure together with the number of terminal nodes of the global information network such as computers, phones, gadgets and other connected cyber-physical devices (so called Internet of Things) [1]. Energy-efficiency of components at all levels of the computation hierarchy is thus becoming a major concern for the microelectronics industry. A serious factor impeding the progress in addressing this concern is a wide gap between the ways of how energy efficiency is approached by hardware and software engineers, and this gap is emphasised by the lack of common understanding between the two communities.

In this paper we discuss an approach to bridging this gap by developing a shared design criterion, called power-proportionality, on the basis of which both electronics and programming solutions can be judged. A computing system, for it to be considered power-proportional, has to keep power consumption and computation load proportional to each other. That is, an idle system would ideally consume no power, whereas given a small energy budget the system would adapt itself by reducing its computation flow and lowering the delivered Quality-of-Service (QoS), and still remain functional; seeSection 2 for a detailed discussion. The state-of-the-art systems have generally poor power-proportionality; for example, the servers used in data centres typically consume 50%–60% of peak power under 10% of peak load [2]. We hypothesise that modern systems exhibit poor power proportionality because they are designed to operate in a narrow scope of conditions, typically optimised for either high performance or low power consumption. This approach is inherently flawed because all the design effort is focused on one particular operation mode disregarding others. Another issue that makes it difficult to handle unpredictable environmental conditions is the absence of a cross-layer feedback-control loop between energy supply components (in hardware) and an operating system or an active application (in software), see Figure 1: software is unable to steer the computation with the aim of maximising QoS by prioritising tasks, trading accuracy for energy, and, in general, by strategic planning of energy resource utilisation.

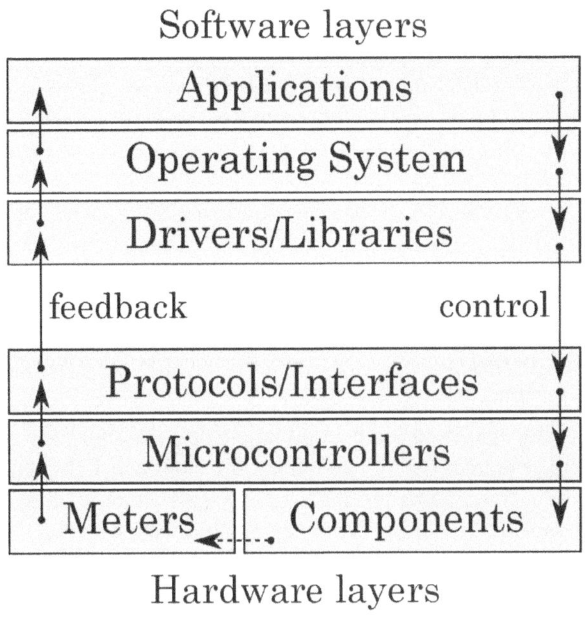

Figure 1. Cross-layer power proportionality.

Citation: Andrey Mokhov, Maxim Rykunov, Danil Sokolov and Alex Yakovlev, Design of Processors with Reconfigurable Microarchitecture, doi:10.3390/jlpea4010026.

In this paper we discuss cross-layer hardware-software mechanisms for online energy monitoring and reconfiguration. The particular focus is on methodology for design and implementation of power-proportional microprocessors, capable of adapting to varying operating conditions (such as low or even unstable voltage level) and application requirements in runtime. We use a popular Intel 8051 microprocessor as our running example to demonstrate the soundness of the approach. The obtained reconfigurable implementation compares favourably to existing specialised (non-reconfigurable) Intel 8051 designs. Our approach is different from conventional power optimisation techniques, such as [3], because we extend the microarchitecture of the processor and allow its runtime reconfiguration, thereby facilitating the cross-layer feedback-control loop shown in Figure 1.

The paper is organised as follows. Section 2 discusses the challenges in design of reconfigurable processors and introduces the concept of power-proportional computing. Section 3 gives a brief background on the Conditional Partial Order Graph (CPOG) model which we use for the specification and synthesis of processor microarchitectures. We then focus specifically on design of reconfigurable microarchitectures in Section 4. This is followed by a case study on design and implementation of a reconfigurable and power-proportional version of the Intel 8051 core in Section 5.

RECONFIGURABILITY AND POWER PROPORTIONALITY

How can one build a reconfigurable and adaptable processor? The first thought could be to implement it in the field-programmable gate array (FPGA) technology allowing static and, sometimes, dynamic reconfiguration. Such a processor would be capable of adjusting its internal structure or behaviour by rewiring interconnections between its components or even by changing functionality at the level of individual gates. This technology can provide a very fine-grain control over a system at runtime, however, the associated overheads are extremely high. In particular, in terms of energy consumption, FPGAs are typically 10–15 times more expensive than application-specific integrated circuits (ASICs); see a collection of recent FGPA vs. ASIC interviews with industrial experts in [4].

Since the fine-grain reconfigurability offered by FPGAs is overly costly in the context of energy-constrained systems, one shall consider the coarse-grain reconfigurable architectures in the ASIC realm. They significantly lower the overheads by dropping reconfigurability in datapath components and utilising their custom designed versions instead. The control logic and interconnect fabric, however, retain the capability to reconfiguration. In this setting, the key design and implementation challenge is to formally describe and synthesise the controller whose task is to coordinate hard system resources (the datapath components and interconnects) according to the runtime information on availability of soft system resources (energy and time); the latter can also include information on hardware faults in a system, thereby allowing the controller to bypass faulty components whenever possible.

Figure 2 illustrates the concept of power-proportionality and its duality. On the leftmost chart one can see dependency of power consumption on service demand. Ideally, the former should grow proportionally to the latter: when there is no service demand, the system should consume (near) zero power, while any increase in service demand should cause a proportional increase in power consumption until the point of saturation, when the system can no longer serve all the incoming requests. Our design goal is to reduce the interval of wasted energy as much as possible. On the right hand side ofFigure 2 one can see the dual chart illustrating dependency of service provision on power supply. Indeed, an ideal computation system should provide service proportionally to the inflow of energy. From this perspective, our design goal is to build computation systems that can start delivering certain level of QoS on as low power supply as possible, thus minimising the interval of unutilised energy.

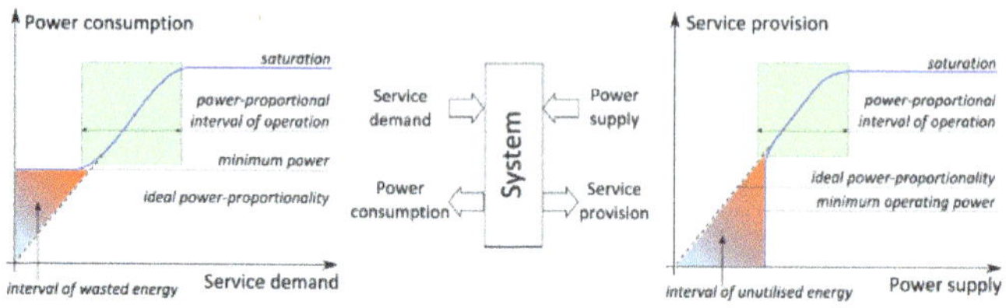

Figure 2. Power proportionality: power supply/consumption vs. service demand/provision.

The duality of power-proportionality creates a four dimensional space of power-service interplay: a computation system can be considered a black box which has two inputs (service demand and power supply) and two outputs (service provision and power consumption) as illustrated in the centre of Figure 2. The combined design goal is therefore to achieve proportionality between the inputs and outputs of the system. In this paper we show how to design a processor microarchitecture with this goal in mind.

CONDITIONAL PARTIAL ORDER GRAPHS

The conventional approach to control logic specification and synthesis is to employ Finite State Machines (FSMs) [5,6] or interpreted Petri Nets (PNs) [7] as an underlying modelling formalism. Within this approach the designer explicitly describes the controller's behaviour for each combination of available resources and operating conditions. The number of such combinations and corresponding behaviours grows extremely fast with the size and degree of adaptability of the system. This leads not only to the state space explosion problem, but also to explosion of the specification size [8], thus slowing down the synthesis tools, reducing productivity, and increasing the overall cost of ASIC development.

Our approach is based on the crucial observation that the controller's behaviours are strongly related to each other in different operating conditions. Indeed, when a system configuration is changed incrementally, e.g., a datapath component goes offline and another one is used in its place, the overall behaviour of the controller is affected in the same incremental manner, therefore one would want to have a joint behaviour specification for these two configurations.

It has been demonstrated that the FSM and PN formalisms are not well-suited to describing families of many related behaviours [8] and the design methodologies based on them have poor scalability in the context of reconfigurable systems. As an appropriate alternative, the Conditional Partial Order Graph model was introduced by Mokhov et al. [8,9]. The model was devised to allow implicit description of families of related behaviours in a compact form as will be demonstrated inSection 3 and Section 4. The preliminary investigations have shown that the model provides a scalable specification and synthesis framework, capable of producing reconfigurable circuits with only 20% of overheads in terms of latency and power in comparison to their non-reconfigurable counterparts. In this paper we further develop the approach and apply it to a more sophisticated real-life example, namely, the Intel 8051 microprocessor, a popular benchmark for power optimisation methods and new microarchitectural solutions, e.g., [3,10].

Formally, a Conditional Partial Order Graph [8] (further referred to as CPOG or graph for short) is a quintuple H=(V,E,X,ρ,φ)H=(V,E,X,ρ,φ) where:

- V is a set of vertices which correspond to events (or atomic actions) in a modelled system;
- VE⊆V×V is a set of arcs representing dependencies between the events;
- Operational vector X is a set of Boolean variables. An opcode is an assignment $(x_1, x_2, \cdots, x_{|X|}) \in \{0,1\}^{|X|}$ of these variables. An opcode selects a particular partial order from those contained in the graph;
- ρ∈F(X) is a restriction function, where F(X) is the set of all Boolean functions over variables in X. ρ defines theoperational domain of the graph: X can be assigned only those opcodes $(x_1,x_2,\cdots,x|X|)$ which satisfy the restriction function, i.e., $\rho(x_1,x_2,\cdots,x|X|)=1\rho(x_1,x_2,\cdots,x|X|)=1$;
- Function φ:(V∪E)→F(X) assigns a Boolean condition φ(z)∈F(X) to every vertex and arc z∈V∪E in the graph. Let us also define φ(z)=df0 for z∉V∪E for convenience.

CPOGs are represented graphically by drawing a labelled circle ○ for every vertex and drawing a labelled arrow → for every arc. The label of a vertex v consists of the vertex name, a colon and the vertex condition φ(v)φ(v), while every arc e is labelled with the corresponding arc condition φ(e) φ(e). The restriction function ρ is depicted in a box next to the graph; operational variables X can therefore be observed as parameters of ρ. Figure 3a shows an example of a CPOG with |V|=5 vertices and |E|=7 arcs. There is a single operational variable x; the restriction function is ρ(x)=1ρ(x)=1, hence both opcodes x=0 and x=1 are allowed. Vertices {a,b,d} have constant φ=1 conditions and are called unconditional, while vertices {c,e} are conditional and have conditions φ(c)=x and φ(e)=$\overline{x}$ respectively. Arcs also fall into two classes: unconditional(arc c→dc→d) and conditional (all the rest). As CPOGs tend to have many unconditional vertices and arcs we use a simplified notation in which conditions equal to 1 are not depicted in the graph, see Figure 3b.

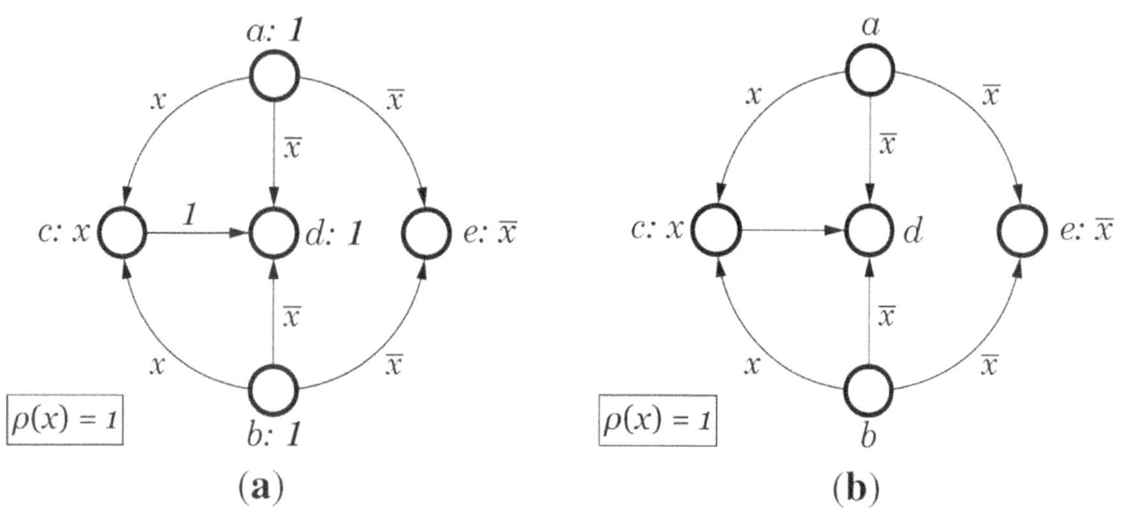

Figure 3. Graphical representation of Conditional Partial Order Graphs (CPOGs): (**a**) Full notation; (**b**) Simplified notation.

The purpose of conditions φ is to "switch off" some vertices and/or arcs in a CPOG according to a given opcode, thereby producing different CPOG projections. An example of a graph and its two projections is presented in Figure 4. The leftmost projection is obtained by keeping in the graph only those vertices and arcs whose conditions evaluate to 1 after substitution of variable x with 1 (such projections are conventionally denoted by H|x=1). Hence, vertex e disappears (shown as a dashed circle ◌), because its condition evaluates to 0: φ(e)=$\overline{x}$=$\overline{1}$=0. Arcs {a→d,a→e,b→d,b→e} disappear for the same reason; they are shown as dashed arrows ⇢⇢. The rightmost projection is obtained in the same way with the only difference that variable x is set to 0; it is denoted by H|x=0, respectively. Note that although the condition of arc c→d evaluates to 1 (in fact it is constant 1) the arc is still excluded from the resultant graph because one of the vertices it connects, viz. vertex c, is excluded and naturally an arc cannot appear in a graph without one of its vertices. Each of the obtained projections can be regarded as specification of a particular behavioural scenario of the modelled system, e.g., as specification of a processor instruction. Potentially, a CPOG H=(V,E,X,ρ,φ) can specify an exponential number of different instructions (each composed from atomic actions in V) according to one of 2|X| different possible opcodes. We can now establish a formal correspondence between CPOGs and instruction sets.

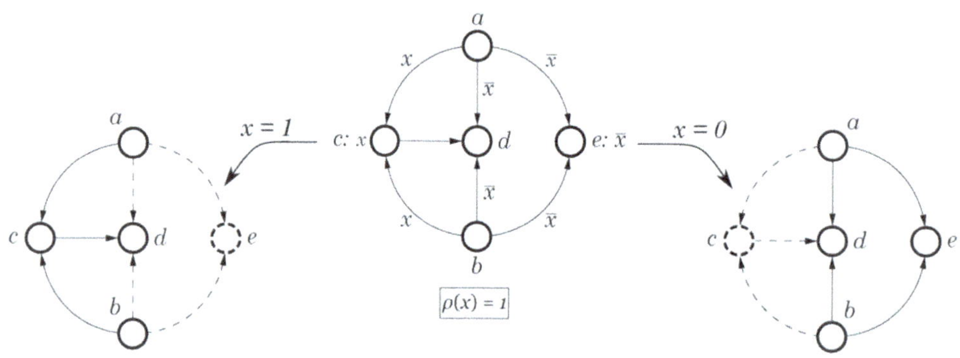

Figure 4. CPOG projections: H|x=1 (left) and H|x=0 (right).

An instruction is a pair I=(ψ,P), where ψ is the instruction opcode (a vector of Boolean values), and P is a partial order that specifies the instruction behaviour or microprogram [11]. An instruction set, denoted by ISIS, is a set of instructions with unique opcodes, that is, for any IS={I1,I2,⋯,In} such that Ik=(ψk,Pk), all opcodes ψk must be different.

Given a CPOG H=(V,E,X,ρ,φ) there is a natural correspondence between its projections and instructions: an opcode ψ=(x1,x2,⋯,x|X|)ψ=(x1,x2,⋯,x|X|) induces a partial order H|ψ, and paired together they form an instruction Iψ=(ψ,H|ψ) according to the above definition. This leads to the following formal link between CPOGs and instruction sets. A CPOG H=(V,E,X,ρ,φ) is a specification of an instruction set IS(H) defined as a union of instructions (ψ,H|ψ) which are allowed by the restriction function ρ:

$$\mathsf{IS}(H) \overset{\text{df}}{=} \{(\psi, H|_{\psi}),\ \rho(\psi) = 1\} \tag{1}$$

Note that this definition automatically guarantees that no two instructions have the same opcode.

DESIGN OF RECONFIGURABLE MICROARCHITECTURES

In this section we demonstrate how to use the CPOG model to build large instruction sets from basic building elements, how to introduce reconfigurability, and finally how to automatically derive an implementation for the specified microarchitecture.

Building Instruction Sets

CPOGs can be deconstructed by means of projections (Section 3). The opposite operation, that is constructing a CPOG out of parts, is called composition. This operation is essential for building large instruction sets from smaller ones, thus enabling modularisation, reuse, and incremental development.

Let instruction sets IS_1 and IS_2 be specified with graphs $H_1 = (V_1, E_1, X, \rho_1, \varphi_1)$ and $H_2 = (V_2, E_2, X, \rho_2, \varphi_2)$, respectively, as in Equation (1). Then their composition has the CPOG specification $H = (V_1 \cup V_2, E_1 \cup E_2, X, \rho_1 + \rho_2, \varphi)$, where the vertex/arc conditions φ are defined as

$$\forall z \in V_1 \cup V_2 \cup E_1 \cup E_2,\ \phi(z) \overset{\text{df}}{=} \rho_1\phi_1(z) + \rho_2\phi_2(z)$$

We call H the CPOG composition of H_1 and H_2 and denote this operation as $H = H_1 \cup H_2$. Composition is defined only if $\rho_1 \cdot \rho_2 = 0$, as otherwise $IS(H_1)$ and $IS(H_2)$ contain instructions with the same opcode making the result ambiguous. It is possible to formally prove that ∪ is a commutative and associative operation and that $IS(H) = IS(H_1) \cup IS(H_2)$ using the algebraic methods presented in [8], which leads to the following important result: $IS(H_1 \cup H_2) = IS(H_1) \cup IS(H_2)$. Once an instruction set is built, it usually undergoes a set of transformations and optimisations which include modification of the set of computational units, changing opcodes of instructions or groups of instructions, modifying the instruction pipeline, etc. Most of these transformations are outside the scope of this work and are defined in [9]. In the following subsections we introduce two new transformaitons.

Once an instruction set is built, it usually undergoes a set of transformations and optimisations which include modification of the set of computational units, changing opcodes of instructions or groups of instructions, modifying the instruction pipeline, etc. Most of these transformations are outside the scope of this work and are defined in [9]. In the following subsections we introduce two new transformaitons.

Mode-Specific Components

There are many ways in which we can implement a particular datapath component (e.g., a multiplier) and they all differ in their performance and energy-efficiency characteristics. If the energy budget of the system varies in runtime it may be beneficial to include more than one implementation in the processor: in this way, it would be possible to choose the most appropriate one in runtime according to the availability of energy and current application requirements. The CPOG model provides a convenient way of introducing such mode-specific component implementations into the processor microarchitecture.

Let H = (V, E, X, ρ, φ) be the specification of an instruction set and v ∈ V correspond to a datapath component whose implementation we would like to make mode-specific. This can be achieved by the CPOG transformation called vertex splitting. The operation transforms H in the following way:

- Vertex v is replaced by two vertices v_0 and v_1, corresponding to two mode-specific implementations of the component. In other words, vertex v is split into two new ones $\{v_0, v_1\}$;
- Arcs (u, v) are replaced with two arcs (u, v_0) and (u, v_1) and, similarly, arcs (v, u) are replaced with two arcs (v_0, u) and (v_1, u). Arc conditions are preserved;
- Set of variables X is extended with a new mode variable m that is used to determine which of the

two implementations is active. Mode variables are changeable in runtime from software, thus facilitating the cross-layer feedback control illustrated in Figure 1;

- Function ρ is unchanged, thus instruction opcodes are not affected by this transformation and the new microarchitecture has no penalties in terms of compactness of the instruction encoding;
- The new vertices get conditions $\varphi(v_0) = \overline{m}\varphi(v)$ and $\varphi(v_1) = m\varphi(v)$, where $\varphi(v)$ is the condition that was assigned to the removed vertex v.

The vertex splitting transformation is illustrated in Figure 5. This transformation increases area and leakage current, but the latter can be mitigated by power gating using the mode variable m to disconnect inactive components from the power supply. As a side benefit, vertex splitting improves the system's fault tolerance: if one of the two components fails, we can disable it and use the other component regardless of the current performance vs. energy-efficiency strategy. In this way the system will be able to maintain the correct functionality but the degree of power-proportionality will be decreased.

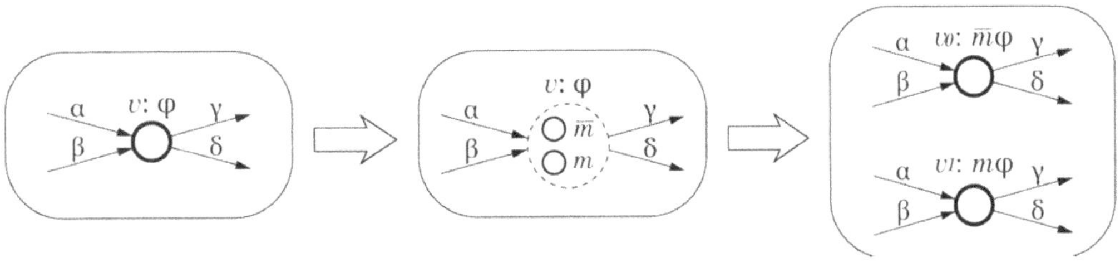

Figure 5. Vertex splitting.

Note that this transformation can also be defined in terms of CPOG composition of two mode-specific instruction sets: $\overline{m}H_0 \cup mH_1$, where H_0 (resp., H_1) corresponds to CPOG H with vertex v replaced with vertex v0 (respectively, v_1). The resulting CPOG specifies the same instruction set but can be derived in a simpler and more compositional manner.

Concurrency Management

It is often the case that an instruction can be executed with different degrees of concurrency. For example, we can typically perform an ALU operation either concurrently with fetching the next instruction opcode or in sequence with it.

Consider two actions a and b. If they have no mutual dependencies in terms of functionality then they can seemingly be executed concurrently. However, they both consume a certain amount of power and this fact should not be ignored when designing power-aware systems. If concurrent execution of a and b exceeds the available power budget, the supply voltage may drop significantly which can potentially lead to the failure of the whole system. On the other hand, executing a and bonly in sequence may be too conservative and will result in a slower system. How can we deal with this situation?

We propose to provide support for both execution scenarios in the microarchitecture and choose the most appropriate one according to the current power budget in runtime. Using the CPOG specification model this can be achieved by making the arc between vertices a and b conditional: $\varphi(a \rightarrow b) = m$. Now If the mode variable m is equal to 0 then components a and b will be executed concurrently (the arc disappears when m=0); otherwise, when m=1m=1 execution of a and b will be scheduled in sequence, hence limiting the peak power consumption.

Synthesis of Microcontroller

As soon as all the intended manipulations with the instruction set are performed, we can proceed to the stage of mapping the resultant CPOG into Boolean equations to produce a physical implementation of the specified microcontroller. In order to descend from the abstract level of atomic actions to the physical level of digital circuits the following signal-level refinements are necessary.

To interface with an asynchronous datapath component a it is possible to use the standard request-acknowledgement handshake (req_a,ack_a), as shown in Figure 6. In case of a synchronous component b the request signal is used to start the computation but, as there is no completion detection, the acknowledgement signal has to be generated using a matched delay [12]. Also, there are cases when a matched delay has to be replaced with a counter connected to the clock signal to provide an accurate multi-cycle delay—see the interface of component c in the same figure. Note that we do not explicitly show synchronisers [13] in the diagram; it is assumed that components b and c are equipped with the necessary synchronisation mechanisms to accept asynchronous requests from the microcontroller.

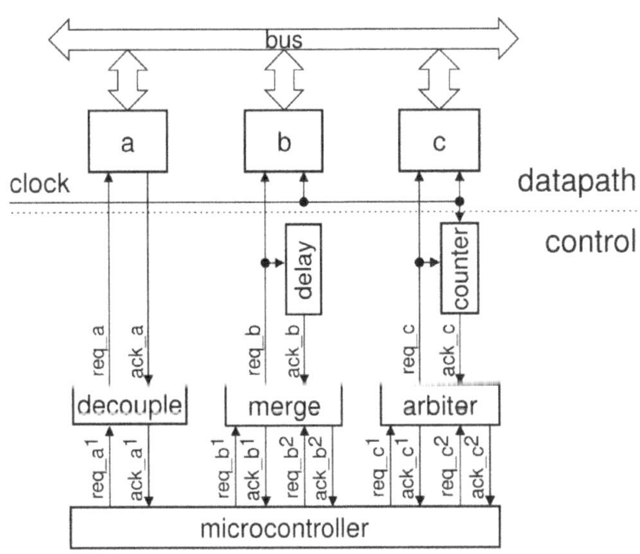

Figure 6. Datapath interface architecture.

To explicitly specify handshake signals it is possible to perform a graph transformation explained in Figure 7. Every atomic action a1a1 is split into a pair of events req_

req_a^1+ and ack_a^1+ standing for rising transitions of the corresponding handshake signals. If there are two occurrences of an atomic action, e.g., $b_1$and b_2, then both vertices are split (we use superscripts to distinguish different occurrences of the same event), etc. Semantically, when an atomic action a_1 is ready for execution, the controller should issue the request signal req_a_1 to component a; then the high value of the acknowledgement signal ack_a1ack_a1 will indicate completion of a.

Notice that the microcontroller does not reset handshakes until all of them are complete. This leads to a potential problem: a component cannot be released until the instruction execution is finished. To deal with this we decouple the microcontroller from the component, see box "decouple" in Figure 6 and its gate-level implementation in Figure 8a. Also, when a component b is used twice in an instruction we combine two handshakes $(req\_b^{1,2}, ack\_b^{1,2})$ into one using the merge controller, see Figure 8b. Merge controllers can only be used if the requests are mutually exclusive. If this is not the case, as e.g., for concurrent actions c1c1 and c2c2, then we have to set an arbiter guarding access to the component. Its implementation consists of the merge controller and the mutual exclusion (ME) element [13], see Figure 8c.

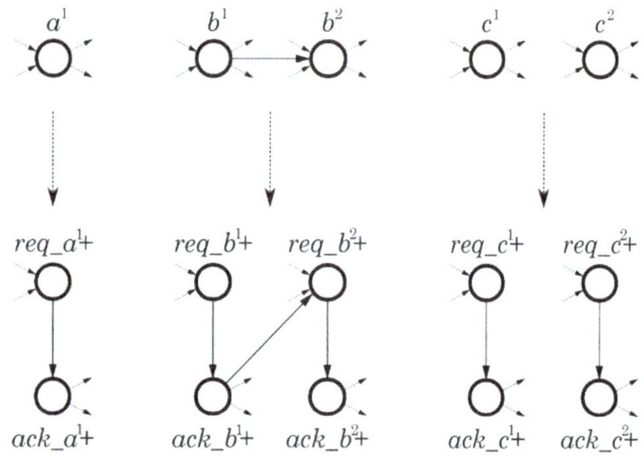

Figure 7. Signal-level refinement.

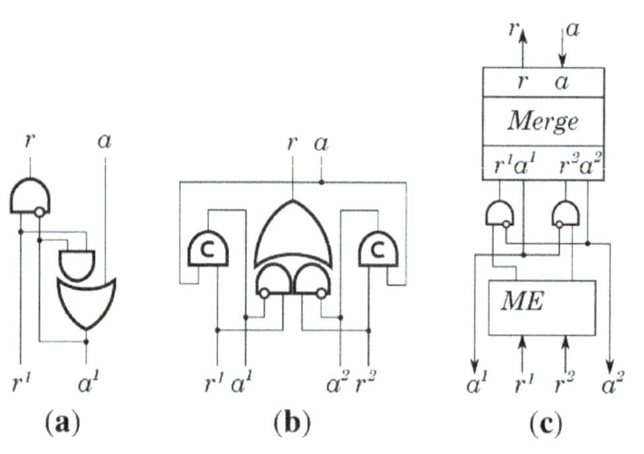

Figure 8. Handshake controllers: (**a**) Decouple; (**b**) Merge and (**c**) Arbiter.

Finally, the refined graph can be mapped into Boolean equations. An event associated with vertex v∈V is enabled to fire (req_v+ is excited) when all the preceding events u∈V have already fired (ack_u have been received) [8]:

$$req\_v = \phi(v) \cdot \prod_{u \in V} \left(\phi(u) \cdot \phi(u \to v) \Rightarrow ack\_u\right)$$

where a⇒b stands for Boolean implication indicating "b if a" relation. Mapping is a simple structural operation, however the obtained equations may be suboptimal and should undergo the conventional logic minimisation [8,11] and technology mapping [7] procedures.

It is interesting to note that the size of the microcontroller does not depend on the number of instructions directly. There are Θ(|V|2) conditions φ in all the resultant equations; the average size of these conditions is difficult to estimate, but in practice we found that the overall size of the microcontroller never grows beyond Θ(|V|2).

RECONFIGURABLE 8051 MICROCONTROLLER

In this section we describe our power-proportional implementation of the Intel 8051 microprocessor with reconfigurable microarchitecture designed using the CPOG model.

The instruction set of Intel 8051 contains 255 instructions and our implementation supports all of them. We generally follow the standard 8051 architecture with two on-chip RAMs; the program is stored in a reprogrammable off-chip ROM. However several important changes were made:

- Our implementation is asynchronous, which gives certain advantages over clocked designs in case of varying operating conditions [10]. The communication between the control logic and the datapath units is arranged by means of request and acknowledgement control signals;
- The 16-bit datapath was modified. The computation units were split into pairs as described in Section 4.2 to achieve power-proportionality; each unit in a pair was optimised either for high performance or energy-efficiency. Table 1describes the computational units. By using adjustable delay lines we achieved robust operation in a wide range of supply voltages (see Section 5.2);
- Instructions are executed according to application/environmental requirements, and taking into account the functional correctness of components, thus addressing the issues of fault tolerance.

Figure 9 shows the top-level view of the microprocessor architecture. The shaded block corresponds to the central control unit (the main design objective), the rest of the blocks correspond to datapath units. The top-level CPOG specification of the microprocessor is shown in Figure 10. Note that the main datapath block, the Arithmetic Logic Unit (ALU), is a complicated subsystem with its own control unit and multi-mode operation; its microarchitecture was designed similarly to the top-level one.

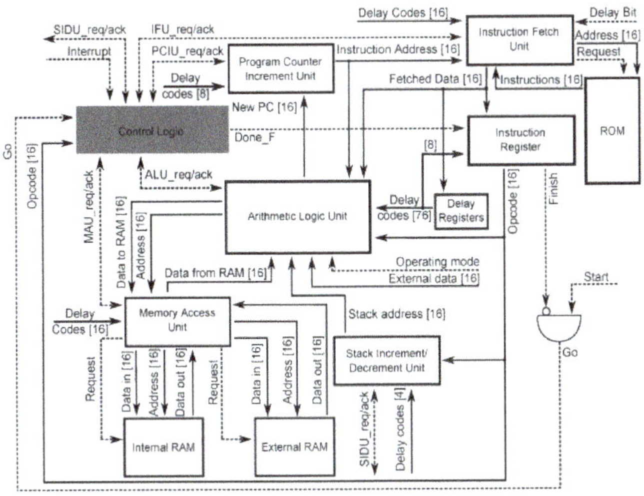

Figure 9. Architecture of the designed microprocessor (dashed lines represent a 1 bit wide control signal line, the bit-width of other connections is shown in brackets).

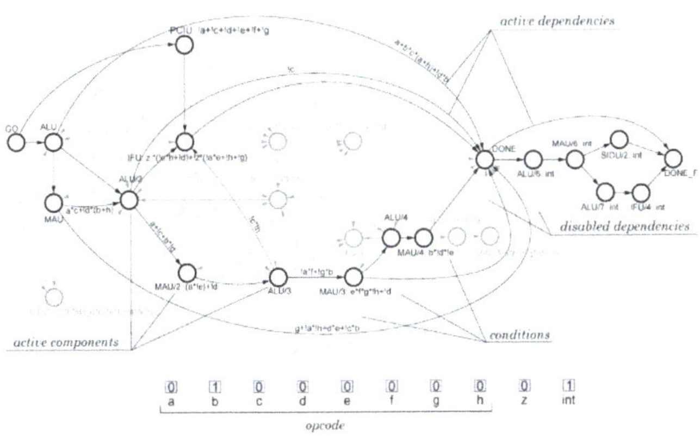

Figure 10. CPOG model of Intel 8051 control logic.

Design Flow

The first stage in microprocessor design is the specification of instructions. Each instruction corresponds to a schedule of primitive computation steps such as data transfer, arithmetic operation, memory access, etc., performed by datapath components. Overall, the proposed microprocessor design flow consists of the following steps:

Extraction of datapath components. We refine the architecture to the level of computational units and behavioural scenarios. There are several ways in which this can be done: (i) direct refinement from software; (ii) specification with Architecture Description Languages [14,15] and (iii) formalisation of existing instruction specifications (e.g., ISA manual). In our case we had a complete description of the microprocessor, its instruction set, and architecture details [16], therefore the last option was the most suitable. Table 1 describes five computational units extracted from the 8051 instruction set.

Partial order specification of instructions. The behaviour of each instruction is captured by a partial order (PO) in which the required computation units need to be activated. Different instructions often have the same PO representations; we grouped all 8051 instructions into 37 classes according to their POs, with the largest class containing 60 instructions.

Figure 11 shows the PO representation of one of the classes, which corresponds to a group of 17 instructions such as, for example, ADD A, immediate that performs addition of the accumulator and an immediate constant. We first read data from the accumulator (ALU followed by MAU) while concurrently fetching an immediate constant (PCIU followed by IFU). Then we perform addition (ALU/2) concurrently with another increment of PC, and finally write the result back into accumulator (MAU) and fetch the next instruction (IFU/2). Many other instructions have the same behaviour: ADDC A, immediate; SUBB A, immediate; ORL A, immediate; ANL A, immediate, etc.

Table 1. List of datapath units.

Name	Description
PCIU	Program Counter Increment Unit is responsible for incrementing the Program Counter—a special register pointing to the address of the next instruction in the program memory.
IFU	Instruction Fetch Unit loads the operational code of the current instruction from the program memory using the address stored in Program Counter.
MAU	Given an address pointer and data, the Memory Access Unit can access the internal and external RAM.
ALU	Arithmetic Logic Unit is the main computation unit responsible for performing arithmetic and logic operations in the microprocessor. To make it adjustable towards a wide range of operating conditions and application requirements, we designed two sets of ALU datapath components, each optimised to work in a specific mode: one optimised for performance and the other one for energy efficiency.
SIDU	Stack pointer Increment/Decrement Unit, as the name suggests, is in charge of incrementing and decrementing the stack pointer.

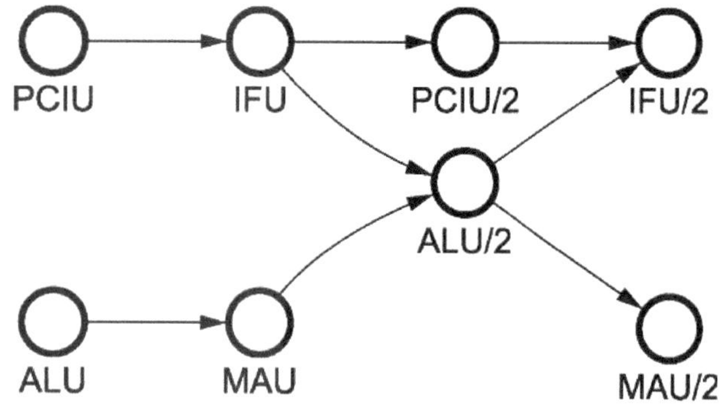

Figure 11. Partial order (PO) for ADD instruction.

Encoding of partial orders. In order to distinguish between the POs, we need to encode them. The codes can either be assigned arbitrarily or can be provided as part of system specification. Importantly, the size and latency of the final microcontroller circuit depends significantly on the chosen encoding of the scenarios. There are several types of encoding that can be used, such as one hot, matrix, balanced, and Huffman encoding [17]. A new technique for optimal encoding was recently introduced in [18], but it was unable to process our set of POs due to high complexity of the optimal encoding algorithm, hence we used Huffman encoding in our design.

CPOG synthesis. Once all instructions are represented as POs and encoded, we can synthesise a CPOG containing all of them as described in Section 4. We used the Workcraft framework [19] which automates the specification and synthesis stages. The resulting CPOG is shown in Figure 10.

Mapping. The final stage is mapping of the CPOG representation into a set of logic gates to produce a physical implementation (a gate-level netlist) of the specified microcontroller. Finally, we can translate the obtained Boolean equations into VHDL, Verilog or other HDL language to use in standard EDA flow.

Adaptability Features

To enable power-proportionality in a wide range of power supply and service demand conditions, we designed two sets of computation units: one optimised for performance and the other one optimised for energy efficiency. Depending on whether there is a shortage of energy or on any other restrictions imposed by an active application, we can choose the most appropriate datapath unit to be used during an instruction execution. The decision can be made at different levels of system control: software level, sensors, external signals, etc.

Our implementation of self-timed datapath components is based on bundled-data approach, where each component is accompanied with a matching delay line to signal completion of its computation. In order to correctly function in a wide range of operating conditions (e.g., supply voltage or temperature), a component needs to adjust the latency of its completion signal. We addressed this issue by use of adjustable delay lines, whose latency is selectable by the delay code at runtime, as shown in Figure 12.

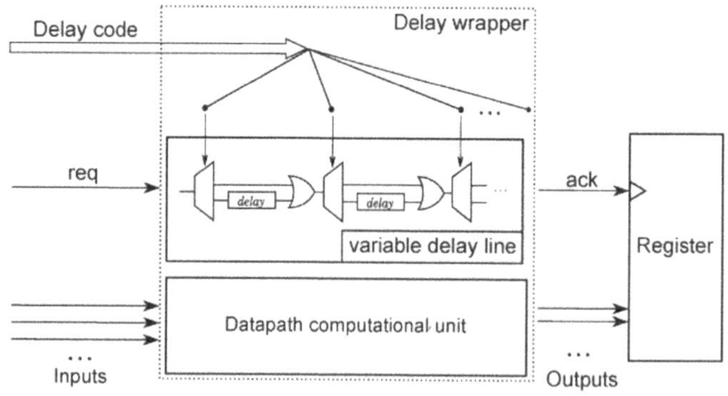

Figure 12. Adjustable delay lines.

Experiments

We implemented the microprocessor in STMicroelectronics 130nm CMOS technology process and conducted experiments to evaluate the level of adaptability we can achieve using the presented approach. In this section we report results of our measurements. The nominal supply voltage of the chip is 1.2 V; we varied the supply voltage in the range from 0.22 V to 1.5 V and observed the following:

- **0.89 V to 1.5 V**: the chip works in full capability mode;
- **0.74 V to 0.89 V**: at 0.89 V the RAM starts to fail; the chip operates using internal registers only;
- **0.22 V to 0.74 V**: at 0.74 V the Program Counter unit starts to fail, so the chip gets stuck at executing the same instruction forever. The control logic synthesised using the CPOG model continues to operate correctly down to 0.22 V.

It should be noted that within the above-indicated range of 0.22–0.74 V, somewhere below 0.5 V, most of the microprocessor's hardware is subjected to work in the near-threshold and even sub-threshold regime. Our implementation was based on standard STM cells, therefore we were not surprised by the fact that some registers like the Program Counter began to fail in this range. However, the important knowledge that we gleaned from testing our processor was that that the control circuitry, even though built of standard cells, was still operational in the sub-threshold regime. This shed light on how we can build future reconfigurable and power-proportional systems that would be able to operate in parts under different supply voltage levels.

We can conclude that our chip can work in dual-memory mode in the voltage range from 0.89 V to 1.5 V. This mode is to be used for applications requiring high-performance and large memory. The single-memory mode available in the voltage range from 0.74 V to 0.89 V can be used for low-power and scratchpad memory applications.

In terms of performance, we measured the range from 67 millions of instructions per second at the nominal voltage of 1.2 V down to 2700 instructions per second at 0.25 V. Note that since the Program Counter fails below 0.74 V the processor starts executing the same instruction forever. Figure 13 shows measurements of latency and power consumption of the ADD instruction run in the voltage range from 0.88 V to 1.2 V. In our experiments other instructions exhibited similar latency/power characteristics.

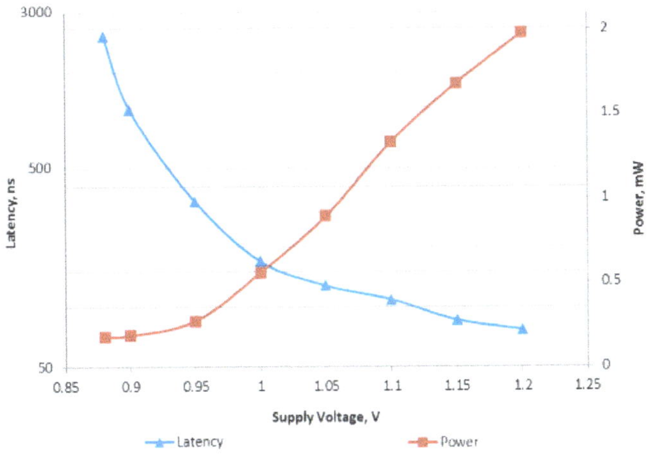

Figure 13. Measurements for ADD instruction.

To compare our implementation with existing Intel 8051 designs we run the Dhrystone V2.1 testbenches [20], see the summary of obtained results in Table 2. Sync 80C51, H8051 and DS89C420 are synchronous designs; the first two employ non-pipeline architectures and the last one has a pipelined architecture. Async 80C51 is an asynchronous counterpart of the Sync 80C51 design. Lutonium is another asynchronous implementation which utilises a highly parallel processing with a deep pipeline architecture; it has high MIPS rate and power consumption. Nanyang A8051 is a self-timed, ultra low power (and hence low performance) implementation. After normalising the results with respect to the technology difference, our implementation is clearly placed between ultra low power design (runs twice as fast) and the highly parallel version (consumes 20 times less power).

Table 2. Performance comparison with other 8051 versions.

Processor	Technology	11 cm MIPS	Average power, mW	MIPS per W	Energy, pJ per instruction
Sync 80C51 [21]	3.3 V, 350 nm	4	40	100	10000
Async80C51 [21]	3.3 V, 350 nm	4	9	44	2250
H8051 [22]	3.3 V, 350 nm	4	44.7	89.5	11175
Lutonium [23]	1.8 V, 180 nm	200	100	1800	500
DS89C420 [24]	1.1 V, 350 nm	11	18.52	600	1684
Nanyang A8051 [25]	1.1 V, 350 nm	0.6	0.07	8000	130
Lutonium [23]	1.1 V, 180 nm	100	20.7	4830	207
Proposed 8051	**1.2 V, 130 nm**	**1.5**	**0.74**	**2027**	**493**

CONCLUSIONS

We presented an approach to design of processors with reconfigurable microarchitecture for power-proportional computation systems. The approach was tested by designing and implementing the Intel 8051 microprocessor, whose control core is capable of operating in the range from 0.2 V to 1.5 V of supply voltages. Unfortunately due unreliability of the Program Counter the processor is only usable starting from 0.74 V. Our future research is to push the boundaries of power-proportionality even further by using the self-timed SRAM capable of operating at low voltages [26], as well as by implementing the critical control and datapath components, such as Program Counter, using high-reliability logic with low fan-in gates and, possibly, with dual-rail control logic for added resilience against single event upsets.

ACKNOWLEDGEMENTS

This work was supported by EPSRC grants EP/I038357/1 (eFuturesXD, project PowerProp) and EP/K034448/1 (PRiME).

Conflicts of Interest

The authors declare no conflict of interest.

REFERENCES

1. The Zettabyte Era. In Cisco Visual Networking Index: Forecast and Methodology, 2010–2015; Cisco Systems Inc: San Jose, CA, USA, 2011.
2. Meisner, D.; Sadler, C.M.; Barroso, L.A.; Weber, W.D.; Wenisch, T.F. Power Management of Online Data-Intensive Services. In Proceedings of the 38th Annual International Symposium on Computer Architecture, San Jose, CA, USA, 4–8 June 2011; pp. 319–330.
3. Saponara, S.; Fanucci, L.; Terreni, P. Architectural-level power optimization of microcontroller cores in embedded systems. IEEE Trans. Ind. Electron. **2007**, 54, 680–683.
4. Trending Back To ASICs. Available online: http://semiengineering.com/trending-asics/ (accessed on 30 October 2013).
5. Nowick, S. Automatic Synthesis of Burst-Mode Asynchronous Controllers. Ph.D. Thesis, Stanford University, Stanford, CA, USA, March 1993.
6. Baranov, S.I. Logic Synthesis for Control Automata; Kluwer Academic Publishers: Dordrecht, The Netherlands, 1994.
7. Cortadella, J.; Kishinevsky, M.; Kondratyev, A.; Lavagno, L.; Yakovlev, A. Logic Synthesis of Asynchronous Controllers and Interfaces; Advanced Microelectronics; Springer-Verlag: New York, NY, USA, 2002.
8. Mokhov, A.; Yakovlev, A. Conditional partial order graphs: Model, synthesis and application. IEEE Trans. Comput.**2010**, 59, 1480–1493.
9. Mokhov, A.; Iliasov, A.; Sokolov, D.; Rykunov, M.; Yakovlev, A.; Romanovsky, A. Synthesis of processor instruction sets from high-level ISA specifications. IEEE Trans. Comput. **2013**. PP, 1.

10. Chang, K.L.; Chang, J.S.; Gwee, B.H.; Chong, K.S. Synchronous-logic and asynchronous-logic 8051 microcontroller cores for realizing the internet of things: A comparative study on dynamic voltage scaling and variation effects. IEEE J. Emerg. Sel. Top. Circuits Syst. **2013**, 3, 23–34.
11. De Micheli, G. Synthesis and Optimization of Digital Circuits; McGraw-Hill Higher: Stanford, CA, USA, 1994.
12. Sparsoe, J.; Furber, S. Principles of Asynchronous Circuit Design: A Systems Perspective.; Kluwer Academic Publishers: Dordrecht, The Netherlands, 2001.
13. Kinniment, D.J. Synchronization and Arbitration in Digital Systems; John Wiley and Sons: Cornwall, UK, 2008; ISBN: 978-0-470-51082-7.
14. Hoffmann, A.; Schliebusch, O.; Nohl, A.; Braun, G.; Wahlen, O.; Meyr, H. A Methodology for the Design of Application Specific Instruction Set Processors (ASIP) Using the Machine Description Language LISA. In Proceedings of the ACM/IEEE International Conference on Computer-Aided Design, San Jose, CA, USA, 4–8 November 2001; pp. 625–630.
15. Fauth, A.; Freericks, M. Describing Instruction Set Processors Using nML. In Proceedings of the 1995 European Conference on Design and Test, Paris, France, 6–9 March 1995; p. 503.
16. Intel 8051 Instruction Set. Available online: http://www.keil.com/support/man/docs/is51 (accessed on 30 October 2013).
17. Huffman, D.A. A method for the construction of minimum-redundancy codes. Proc. IRE **1952**, 40, 1098–1101.
18. Mokhov, A.; Alekseyev, A.; Yakovlev, A. Encoding of processor instruction sets with explicit concurrency control. IET Comput. Digit. Tech. **2011**, 5, 427–439.
19. The Workcraft Framework Homepage. Available online: http://www.workcraft.org (accessed on 30 October 2013).
20. Price, W.J. A benchmark tutorial. IEEE Micro **1989**, 9, 28–43.
21. Van Gageldonk, H.; van Berkel, K.; Peeters, A.; Baumann, D.; Gloor, D.; Stegmann, G. An Asynchronous Low-Power 80C51 Microcontroller. In Proceedings of the Fourth International Symposium on Advanced Research in Asynchronous Circuits and Systems, San Deigo, CA, USA, 30 March–2 April 1998; pp. 96–107.
22. HMS99C52 Datasheet; Hynix Co.: Gyeonggi, South Korea, 2003; Available online: http://www.gaw.ru/pdf/Hynix/mcs51/HMS99C52.pdf (accessed on 30 October 2013).
23. Martin, A.J.; Nystrom, M.; Papadantonakis, K.; Penzes, P.I.; Prakash, J.T.; Tura, A. The Lutonium: A Sub-Nanojoule Asynchronous 8051 Microcontroller. In Proceedings of the Ninth International Symposium on Asynchronous Circuits and Systems, Vancouver, BC, Canada, 12–16 May 2003; pp. 14–23.
24. DS89C420 Microcontroller; Maxim Integrated Products: San Jose, CA, USA, 2000.
25. Chang, K.L.; Gwee, B.H. A Low-Energy Low-Voltage Asynchronous 8051 Microcontroller Cores. In Proceedings of the IEEE International Symposium on Circuits and Systems (ISCAS), Island of Kos, Greece, 21–24 May 2006.
26. Baz, A.; Shang, D.; Xia, F.; Yakovlev, A. Self-timed SRAM for Energy Harvesting Systems. In Proceedings of the 20th International Workshop on Power and Timing Modeling, Grenoble, France, 7–10 September 2010; pp. 105–115.

CHAPTER 16 MICROPROCESSOR PROTECTION DEVICES

Vladimir Gurevich

Keywords: Microprocessor-based protective devices, Relay damages, Relay protection reliability.

INTRODUCTION

Over a long period of time, electromechanical protective relays of the past generation were meeting completely all requirements set for the protection devices of electrical power equipment. In the latest microprocessor-based devices the function of relay protection has been united with functions of other devices: communication and data transmission devices, fault recorders, substation logic units, etc. Such multipurpose complexes are compared now to the individual one-functional electromechanical relays which have been functioning for decades and which are now considered as rather worn out. The advantages of microprocessor relays over electromechanical ones are quite obvious. What has been overlooked, though, is that these devices perform completely different functions which simply cannot be compared with each other [1]. Specialists of the world-leading manufactures have published a considerable number of half-advertising articles stressing only the positive qualities of microprocessor-based protective relays (MBR). The literature provides but a few publications devoted solely to the analysis of problems related to the transition to MBR although there are many more others. Though the problems related to MBR expansion are obvious, the fact that their distribution is increasing and that they are fully replacing electromechanical relays is inevitable owing to the fact that almost all world-leading manufacturers have stopped producing electromechanical protective relays. This is not due to the fact that electromechanical relays have insuperable disadvantages of any great importance (they have just worn out and have not been improved for the past 30 to 40 years), but owing to the enormous profit (gained by MBR producers) which is beyond comparison in relation to the manufacture of electromechanical protective relays [1]. As the future of relay protection is inevitably connected with MBR (at least, for complex protection functions), the forecast of ways of development of this equipment is of definite interest.

Citation: Vladimir Gurevich, Microprocessor Protection Devices, http://www.doiserbia.nb.rs/img/doi/1451-4869/2008/1451-48690802325G.pdf.

MODERN LINES IN DESIGNING MICROPROCESSOR PROTECTION DEVICES

If one considers the «Factor of Unit Functionality» (FUF) parameter – the number of functions in a single MBR – it can be observed that this parameter grows year in year out. The MBR's physical size remains constant (and even decreases), but its functionality has continuously grown diverse. What is the reason of such a growth? In our opinion, there is no objective necessity for increasing the FUF, it is only a matter of the competitive activity among MBR producers and their aspiration to excel each other in technological level of produced devices. As the FUF is directly related to the technical and technological level of the equipment, increasing the FUF is mainly directly associated with MBR technological level and capabilities of the manufacturer. Thus, in our opinion, increasing the FUF in MBR is no more than the means of competitive activity. Is this tendency useful and does it really results in increasing MBR quality? At first sight, the answer is positive, since, as it has already been mentioned above, FUF increase is provided by the use of more progressive materials, elements and technologies, and it is directly associated with MBR technological level increase. However, in practice, nothing appears to be as simple. As progress in the sphere of new materials and technologies is actually not as prompt as MBR producers would like it to be, they use all available methods to achieve their aims, i.e. to increase the FUF. Furthermore, they constantly tend to make software more and more complicated, and the interface more intricate, including thereby huge amounts of functions not used in practice. All this makes work with MBR considerably more complicated and error probability is raised owing to the «human factor» [1]. Some frequently promoted «outstanding abilities» of MBR, with increased FUF, such as full internal self-diagnostics, in practice appear to be no more than an advertising gimmick created to justify MBR complexity and cost increase. How can we speak about the efficacy of self-diagnostics if it is not able to discover the replacement of a whole printed circuit board of the MBR, not to mention damages (and not only in a peripheral board, but also in the main board with the central processor). For example, such damages that disable MBR to contact the computer, or refuse to accept and remember absolutely correct setting changes, or fails to work in mode of measurement of input analog values (currents and voltages), destroying elements in the input and output circuits, and many others which have not been discovered by the self-diagnostics system. And what can one say about the MBR's so called 'ability' to constantly control even the serviceability of output electromechanical relays other than to be an advertising gimmick? It would be very interesting for us to learn how is it is possible to control the serviceability of the electromechanical relay all the time without causing it to operate.

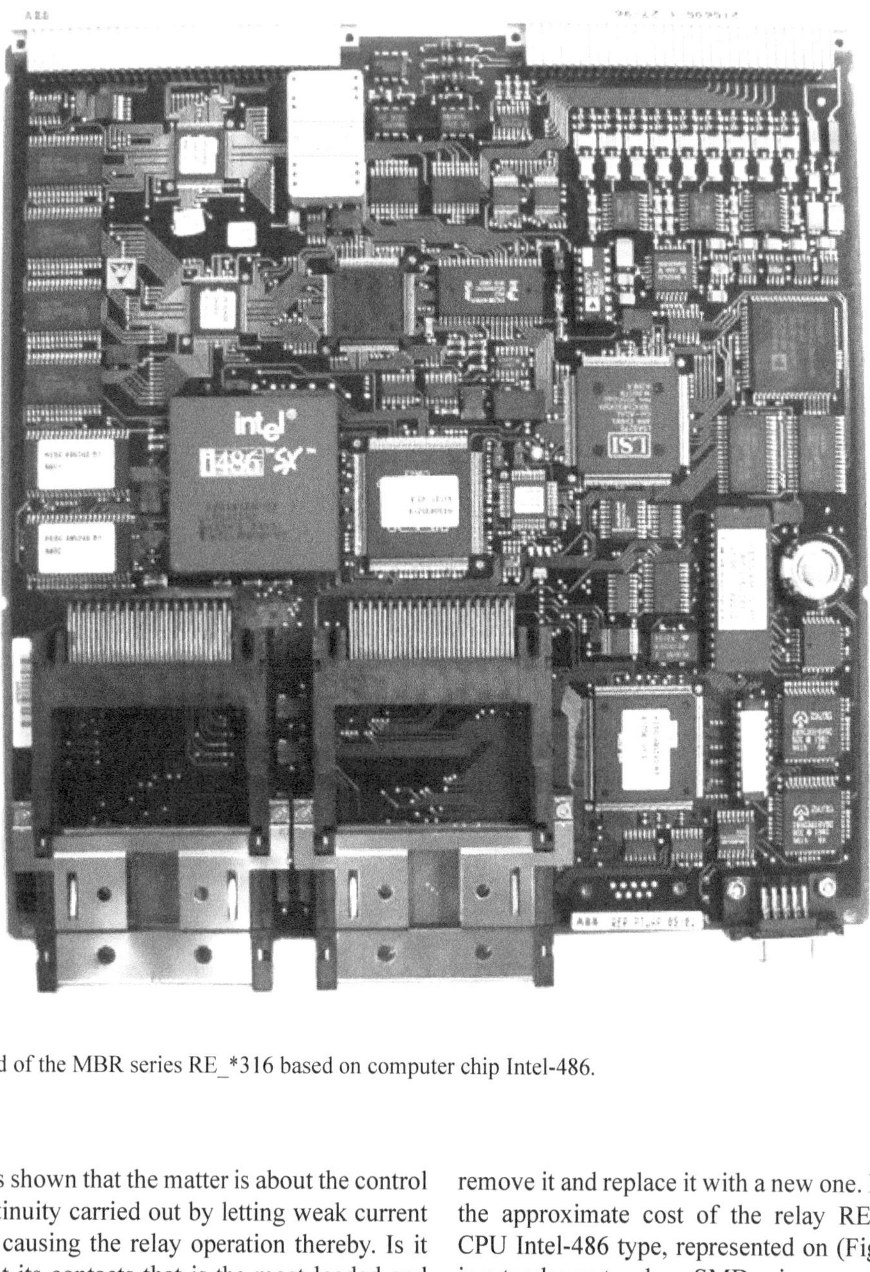

Figure. 1 – Main board of the MBR series RE_*316 based on computer chip Intel-486.

Our research has shown that the matter is about the control of a relay coil's continuity carried out by letting weak current flow through it, not causing the relay operation thereby. Is it the relay coil but not its contacts that is the most loaded and most unreliable element? Certainly not, but it is much easier to organize a useless control of the coil than the control of the condition of the contacts. What is controlled is apparently absolutely unimportant for a new advertising slogan.

The MBR printed circuit board (PCB) is more commonly implemented by using SMD (Surface Mounting Devices) technology; as the density of the SMD micro-components on the board is so high, Fig. 1, it is absolutely hopeless even to discuss troubleshooting and repairing such a PCB. One can only remove it and replace it with a new one. Let the reader imagine the approximate cost of the relay REL-316 PCB with the CPU Intel-486 type, represented on (Fig. 1). Nowadays there is a tendency to place SMD micro-components together with large power elements, electrically connected with each other on one PCB (Fig. 2). With regard to the fact that manufacture companies never give the consumer detailed circuit diagrams of the PCBs, searching for malfunctions in them and repairing them turn out also very difficult. One often has to remove it , despite the number of large discrete components usually found on them. The tendency of increasing the density installation is peculiar to PCBs on the basis of usual discrete elements.

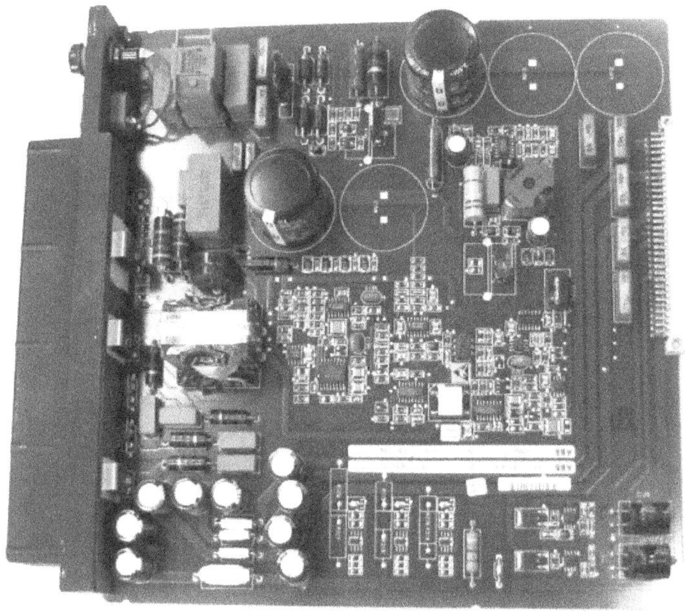

Figure. 2 – The PCB of the switching power supply of the MBR, combining SMD microcomponents with large discrete elements of usual installation

As for power supplies functioning at relatively high voltages and powers (Fig. 3), installations such as these result in a dangerous approach to printed conductors having different potentials on the board and increase the possibility of electrical breakdown of dusty PCBs with the increase of humidity. In addition, dense installation makes thermal conditions of electronic components work more difficult, which leads to a considerable shortening of their operating life. It is not at all simple and cheap to search for malfunctions and repair PCBs due to dense installation such as these, and one should observe substantial expenditure of time this entails.

Figure. 3 – The board of the MBR uniting different functional units and fully made on electronic elements of usual, but high-density installation.

A serious problem is represented by the electrolytic capacitors widely used in MBR power supplies. After 7 – 10 years of operation in switching highfrequency power supplies, even high-quality capacitors produced by well-known Japanese companies, begin to leak, emitting a chemically aggressive electrolyte. Even with this, the power supply functionality continues until adjacent micro components are destroyed or copper tracks of printed circuit dissolved. Even the layer of special firm lacquer called "mask" cannot protect them (Fig. 4). Nobody would dare guarantee the good condition of the unit even after attempts have been made to rectify the damage and to put such crucial device as MBR into operation again after damages of this kind have been made.

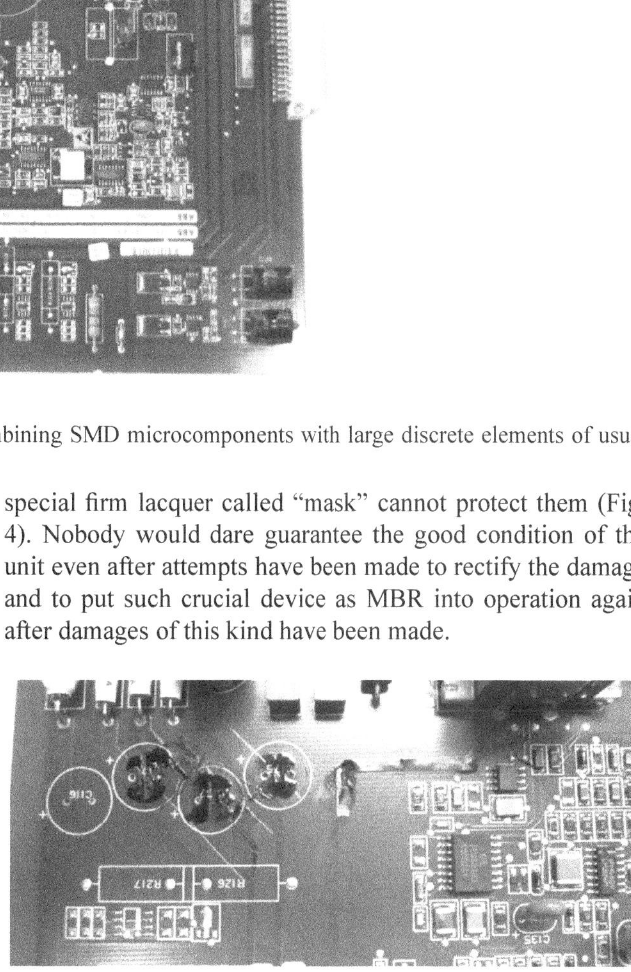

Figure. 4 – Fragment of MBR PCB with damages caused by electrolyte leakage from capacitors.

- Only the desire for decreasing the MBR dimensions can explain the fact that almost all the world-leading producers of MBRs use miniature electromagnetic relays for the direct switching of trip coils of high-voltage circuit breakers and to control rather powerful external intermediate relays. As it is shown in [2-4], the technical specifications of a vast majority of such relays fall short of their real operating conditions in the MBR. Naturally, this reduces protection reliability.
- In some types of MBR produced by world-leading companies, small connectors (slots) are found even in 250V voltage circuits not intended for functioning at such voltages. In order to decrease the risk of electrical breakdown between close outputs of miniature relays and connectors, one sometimes makes special cuts in PCB of MBR between the outputs (Fig. 5).
- Constantly increasing the density of installation considerably affects electromagnetic compatibility, especially in the atmosphere of the increased risk of

deliberate remote impact on the MBR by very powerful directional electromagnetic pulses [5]. Special powerful pulse, ultrahigh-frequency generators (Fig. 6) allow disabling of the existing MBRs instantly from a distance of several dozens of meters to half a kilometer even from a driving car, let alone military target killers like electromagnetic bombs and warheads which anyone can freely buy at arms market today.

- It is absolutely obvious by now that constantly increasing the concentration of functions of relay protection in one device along with simultaneous increase of loading factors of electronic components does not at all favour the growth of MBR reliability. In this case, only one component serving any general purpose for the whole MBR unit may fail (central microprocessor, memory, power supply, etc.) leading thus to false switching off of an important power object or serious damage due to the fact that MBR has failed to respond in an emergency situation.

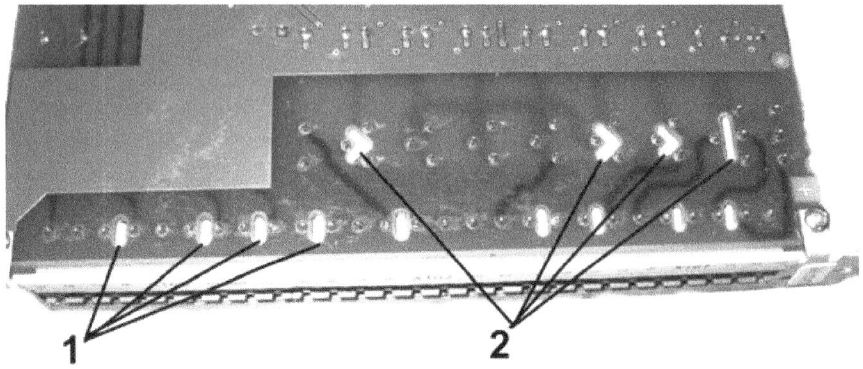

Figure. 5 – Fragment of MBR PCB with cuts made between close outputs of small connectors (1) and miniature output relays (2) with the view of preventing of electrical breakdown.

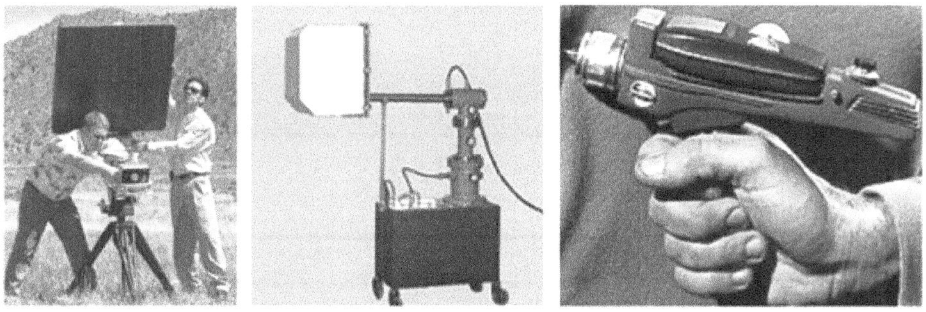

Figure. 6 – Special powerful pulse ultrahigh-frequency generators used to make impact on electronic equipment.

- MBRs are expensive devices and they can be bought, as a rule, at competitive bidding. Not all MBR functional units have the equal loading, equal thermal conditions and equal probability of failure, equal service life. Very loaded units like power supplies, input-output boards are most prone to failures such as electrical breakdown or thermal damage. Failures of central processor units are of absolutely different character and are usually connected not with physical damages, but with problems with the operating program. As a rule, units with input current and voltage transformers almost do not fail. In different models of MBRs of different producers, the reliability indices of certain units may differ considerably. However, nowadays it is absolutely impossible to use functional units of MBRs of one type instead of the units with the same purpose of an MBR of other type. Moreover, even the substitution of an MBR of one producer for the MBR of another producer at functional substation causes difficulties due to great variety of forms and dimensions of these devices (Fig. 7).

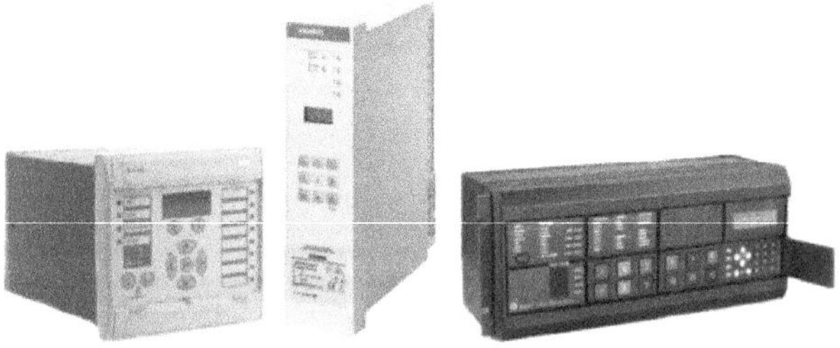

Figure. 7 – Modern MBR of world leading producers.

Digital (logic) inputs in MBRs of a very widely spread type REL, REC, RET 316 series are implemented as a set of absolutely identical cells which are functionally INHIBIT logic elements (Fig. 8). The input and output of this circuit through the insulation optrons Opt1 and Opt2 are cut into the MBR electrical circuits connected with microprocessor. A signal coming from optron Opt2 logically simulates (repeats) the presence or lack of input voltage.

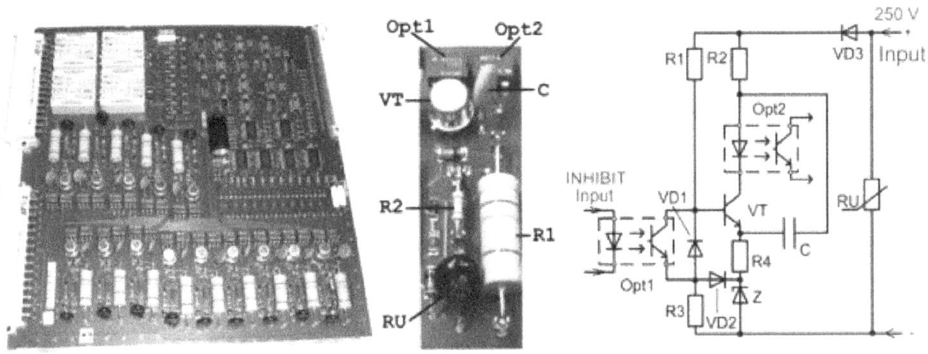

Figure. 8 – Digital (logic) inputs of MBR RE_*316 series: PCB, one channel and its circuit schematic.

The functioning of this circuit may be blocked on a software basis. With this, the blocking signal is sent to the inhibiting input of a cell (Opt1 input) from the microprocessor through corresponding electronic circuits. The problem of this logic cell consists in the high level input signal (220 – 250 V DC), which should be reduced to the 1.5 – 2 V at which the optron, Opt2, functions. The common current consumed by the circuit is divided between resistors R1 and R2 into approximately equal parts, this being the reason why they both, in fact, should have equal power. However two large resistors (like R1), set at a distance sufficient to ensure cooling, cannot be placed together on board. That is why the MBR designers decided to give the functions of an element dissipating surplus power to transistor VT which does not function in a key mode (as it is common in such circuits), but in a booster mode. Certainly, in such case, resistor R2 has low power and small dimensions. It is an original decision, but, in practice, this fall on miniaturization leads to serious problems: often resistor R2 burns out completely and this sometimes results in burning out of all parts of the board as well as the adjacent elements. The problem lies in transistor VT.

Constantly functioning in the booster mode and dissipating surplus power, transistor VT heats up to 70–80°C. Unlike a normal resistor, in which the resistance grows insignificantly with increasing temperature, the resistance of the direct transition of transistor decreases significantly, as a result of operating point shift at the characteristic and amplification factor increases. This leads to collector current increase, i.e., current through resistor R2. When the simultaneous heating of many R1 resistors and VT transistors in ten-fifteen input circuits occurs, the temperature may rise significantly in part of the MBR case (separated by partitions) in which the board of digital inputs is installed. This leads to a further shift of the transistor operating point and a current increase to its collector completes the burn out of resistor R1. In this situation, diode VD, which according to the scheme should stabilize the transistor operating point, appears to have little effect, as its temperature and the one of the transistor differ by 50–60 degrees. Thus, the producer's desire for reducing the MBR dimensions using this transistor instead of powerful resistor for dissipating surplus power has led to the deterioration in MBR reliability. The problem of insufficient reliability of this scheme connected with miniaturization lies not only in the difficulty described above. One more reason discovered in practice for burning the resistor R2 is poor level of insulation of miniature capacitor C. Partial breakdowns in its inner insulation and an increase of leakage current at a constant voltage 200V also leads to resistor R2 fusing. These breakdowns can also be due to the insufficient efficacy of varistors RU used in the MBR – the smallest (for saving room) and, therefore, the most low-powered of all existent types. Owing to their size they are not always capable of absorbing the energy of surge discharge. Additionally, it is not clear what protects what: does the varistor protect the electronic components or vice versa, as the clamping voltage of the varistors (650V) installed on the board significantly exceeds the voltage rating of electronic components, e.g. 350V for transistor VT. With such parameters ratio, the electronic components will fail until varistors are actuated.

- In order to increase the reliability of the MBR power supply, the direct operational current is used as the rule of a thumb. However, even using direct operational current does not exclude malfunctions in the accumulator circuits. For example, the supply of the direct current bus bars at the substation is provided only by a charger supplied from the direct current mains. In case of emergency conditions in high-voltage power mains protected by an MBR, the supply of the charger disappears and the MBR loses its power supply at the most crucial moment. Our research [6] has shown that under real operating conditions, MBRs of different types have time to generate a signal for switching the circuit breaker off only if they work in mode of momentary cutoff, i.e., without time delay.

OFFERED SOLUTIONS

In our opinion, the basic design principle for future MBRs should be similar to modern PCs: hundreds of varieties of containers, motherboards, storage units, and peripheral devices of many different producers perfectly combined with each other in such manner that they are interchangeable and allow upgrading certain units without affecting the other ones. Why not applying the same principle to new MBRs? We see the MBR of the future as certain functional modules implemented on the printed circuit boards in a single constructive style (standard) provided with plug-in connectors. We think that one should transfer this principle from the sphere of PCs also into the sphere of software: a basic "relay" prototype of WindowsTM and a set of applications realizing functions of concrete types of protection on a general basis is the most perspective line of software development.

MBRs should be placed in special metallic boxes (Fig. 9), produced according to a special technology ensuring reliable protection against external electromagnetic radiation over a wide frequency spectrum. Nowadays such boxes are produced by many companies, for example, R.F. Installations, Inc., Universal Shielding Corp., Eldon, Equipto Electronics Corp., European EMC Products Ltd., Amco Engineering, and many others. The boxes should have guides for inserting the printed circuit boards and corresponding connectors for connecting the boards. One such box may contain several different MBRs. Protection relays occupy a very small area of substation or station; therefore the problem of limitation of space occupied by these boxes should not be the prime consideration of new principles of protection devices design. On the contrary we think that the density of installation of PCBs implemented on usual elements should be reduced so as to allow a semiskilled technician to search for malfunctions and repair the device without difficulties. Trends of this kind, but not the above-mentioned FUF, should become the criterion of MBR technical level.

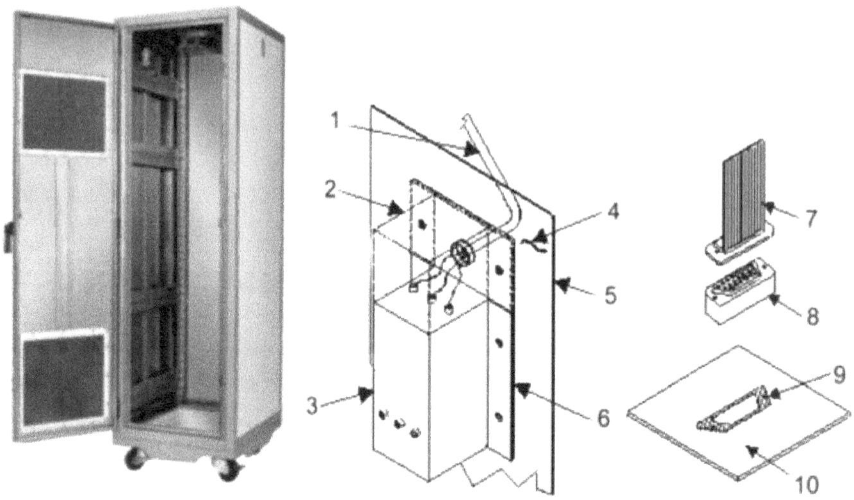

Figure. 9 – Special metallic container and elements of filters necessary for enhanced protection against external electromagnetic fields in a wide frequency spectrum.

1 – power shielded cable; 2 – section shielding power cable input and termination;

3 – power line filter with elements of protection against overvoltage;

4 and 5 – internal and external surfaces of the box;

6 and 9 – gaskets of special electroconductive rubber;

7 – signal cable; 8 – special high-frequency filter; 10 – container wall.

- The power of elements versus the dissipating power should be set (according to the author's experience) to exceed the dissipated power by 4 – 5 times. Only in this case one can ensure relatively low temperatures of forcesumming elements and their long service life. Limiting the values of current and voltage of applied electronic components should also be selected with a 4 – 5 fold reserve. One should select overvoltage protective elements (for example, varistors) with high enough dissipated power (disk diameter not less than 20 mm). Acceptable extended operating voltages of these protective elements and clamping voltages should be coordinated with real operating voltages and with a maximum acceptable voltage of applied elements.

Functional units of future digital protection devices, such as:

- Analog inputs (current and voltage transformers) provided with highly effective interference suppressive filters and electronic elements of primary conversion of signals;
- Logic (digital) inputs provided with high-frequency filters and effective overvoltage protection;
- Power-line filters with multistage overvoltage protection in supply circuit;
- Communication system and data transfer;
- Output relays unit including:
  a) Miniature relays with gold contacts for low-current low-voltage signal circuits;
  b) Industrial type electromagnetic relays with powerful contacts for controlling intermediate electromagnetic relays of automation systems;
  c) High-performance solid-state relays on basis of thyristors or IGBT-transistors with special drivers with optical isolation and with overvoltage protective elements – for switching on of trip coil;
- Power supplies;

Microprocessor recorders (meters) for recording of emergency modes and actuation of relays and other switching equipment, should be implemented as separate PCB-modules.

- The main processor unit with memory and all auxiliary elements should be placed in a separate shielding section and be galvanically insulated from all other units by means of optical coupling.
- One need not change the board of the main processor, connection and data transfer board, board of emergency situations meter implemented on microcomponents of surface mounting. The rest of the boards should be implemented using the usual components and the design of the PCB should be such that it can be searched for malfunctions and repaired. It is obvious

that this requirement will be urgent until every MBR board is extremely expensive but board repair remains profitable.

- Though there is a possibility of uniting functions of different types of protection devices placed in one box and to realize them on one general, powerful microprocessor, this possibility, in our opinion, should not be actualized because of the risk of excessive concentration of many protection functions in a single device. Even so, the registration of emergency modes and the activation of all protection devices installed in the box may be implemented on microprocesor and general flash-memory for the whole box. The connection system and data transfer can also be general for all protection devices installed in the box.
- Input transformers of current and voltage (analog input units) are highly reliable elements of MBR and one has not discovered cases of their damage in process of MBR service. This is why their power can be rated for share usage by all MBR devices installed in the box. Units of analog inputs should have functionally completed construction which allows removing the whole unit from general scheme of the MBR and replacing them with blocks of totally other constructions, for example, with optical inputs designed for receiving optical input analog signals from new generation optical transformers of current and voltage.
- Aluminum electrolytic capacitors of power supplies should be placed in a separate block (PCB). This block should be replaced with a new one every 5 years regardless of the capacitors condition.
- Power supplies should be general for the whole box. The power supply should contain two separate independent units: basic and reserve (which begins to function automatically as soon as basic unit fails). In addition, the power supplies should contain small hermetic non-serviced accumulators with charger as is done, for example, in fire alarm systems. It is rather instructive in this case of using this type of accumulator instead of such accumulator electrolytic capacitors of large capacity at voltages not less than 450 – 500V. In North Caroline University one developed capacitors with accumulated energy which exceeds energy of usual capacitors by 5 – 7 times [7]. It is also possible to use super capacitors which ensure power of relay protection box during some period of time sufficient for protection actuation in case of failures in centralized power system [6]. Power circuits of every separate module should be disconnected from power circuits of the next module such that damage in one of them cannot affect the functionality of the other.
- Relay protection boxes should have a board-module with highperformance, high-reliable electromechanical elements implemented, for example, on reed switches or hybrid reed switch-semi conductive elements. Such elements as these will ensure high-performance actuation of output protection relays around microprocessors at current big multiplication factors for protection device input, when the mode is definitely determined as emergency, and no other processing of input signal is not needed to determine mode as an emergency. As is shown in author's works [8, 9], simple modules on such elements have high performance (0.8 – 1.5 ms), which significantly exceeds the highest performance of MBRs (20 – 40 ms).
- As it is shown above, the analysis of MBR condition only by means of imbedded software, as it is currently done, is definitely not effective. On the other hand, real self-diagnostics, including the special sending of standard test signals automatically to analog and digital inputs of an MBR from an external source, and registration of changes in condition of relay and its outputs circuits is impossible without the complete disconnection of relay. In our opinion, the possibility of periodic automatic disconnection of an MBR with further automatic actuation of diagnostic system on active objects of electric-power industry is rather doubtful. Besides, to check certain functions of relay it is often needed changing (or canceling) some protection settings with further return on test completion. Complexity and costs of the system which provides checks such as these will be similar to cost of MBR itself. If we add the inevitable reduction of reliability to this whole complex because of its complication, and also the necessity of periodic calibration of diagnostics system, the hopelessness of this way of development is obvious, at least, in the nearest future. In our opinion, a much more attractive idea is the creation of a portable, mobile universal automatic diagnostic complex controlled by means of an external portable computer and a set of specialized programs for every type of protection, which operating in a half-automatic mode would take 15 – 30 minutes to effectively test the MBR. In general, a good basis for such diagnostic centers already exists: it is devices produced by companies such as Omicron, Doble and others. It is necessary to provide these devices with multi-channel interfaces for the synchronous connection of numerous inputs and outputs of the MBR, and also to develop a set of application programs, which ensure testing different MBR in an automatic mode. We think that with such diagnostic complexes, it makes sense to return to periodic check-ups of MBR with economically motivated intervals.

CONCLUSION

The principles of MBR construction that have been discussed can ensure, in our opinion, the increase in its reliability, flexibility and usability. This can provide opportunities for modernizing protection with minimum costs. If some MBR units function unsatisfactorily, one can easily replace them with the units of another producer. Having decided to use new generation microprocessor, one should not replace the whole device, including input circuits, power supplies, filters, etc., but only one or two PCBs that need to be changed. Introducing the use of optical transformers of current and voltage will involve changing only one PCB of analog inputs, but not the whole protection device, as it is required today. Likewise, one could change and modernize protection software. All this will lead to the rise in the number of new companies in the market specialized in production of separate functional blocks and separate application programs, as well as in the competitiveness, and, finally, to rash quality improvement of MBR with simultaneous cost reduction. The author understands that not all aspects of the problem have been addressed and not all principles of construction of future MBRs formulated. Therefore, he invites all interested specialists to discuss this topic.

REFERENCES

1. V. Gurevich: Electric Relays: Principles and Applications, Taylor & Francis Group, New York – London, 2005.
2. V. Gurevich: Nonconformance in Electromechanical Output Relays of MicroprocessorBased Protection Devices Under Actual Operating Conditions, Electrical Engineering and Electromechanics, No. 1, 2006, pp. 12 – 16.
3. V. Gurevich: Peculiarities of the Relays Intended for Operating Trip Coils of the High VoltageCircuit Breakers, Serbian Journal of Electrical Engineering, Vol. 4, No. 2, 2007, pp. 223 – 237.
4. V. Gurevich: Dealing with Problems in Output Relays Used in Microprocessor-Based Protection Devices Part I, Transmission and Distribution Magazine – Electricity Today, Vol.19, No. 1, 2007, pp. 40 – 50.
5. V. Gurevich: Dealing with Problems in Output Relays Used in Microprocessor-Based Protection Devices Part II, Transmission and Distribution Magazine – Electricity Today, Vol.19, No. 2, 2007, pp. 22 – 24.
6. V. Gurevich: Electromagnetic Terrorism: New Hazards, Electrical Engineering and Electromechanics, No. 4, 2005, pp. 84-86.
7. V. Gurevich: A Problem of Power Supply of Microprocessor-Based Protective Relays in Emergency Mode, Transmission and Distribution Magazine – Electricity Today, Vol. 18, No. 8, 2006, pp. 32 – 35.
8. Capacitors Hold Promise of Increase in Energy Storage Capability, RF Design, Jul 25, 2007.
9. V. Gurevich: Hybrid Reed – Solid-State Devices are a New Generation of Protective Relays, Serbian Journal of Electrical Engineering, Vol. 4, No. 1, 2007, pp. 85 – 94.
10. V. Gurevich: Simple Very High-Speed Overcurrent Protection Relay, Electrical Engineering and Electromechanics, No. 1, 2007, pp. 13 – 16.

CHAPTER 17 RELIABILITY OF MICROPROCESSOR-BASED RELAY PROTECTION DEVICES

Vladimir Gurevich

Keywords: Microprocessor-based protection devices, Relay protection, Reliability, AD converter, Central processor, Watchdog timer

INTRODUCTION

Malfunction of relay protection is one of the main causes of the heavy failures that periodically occur in power systems all over the world. According to the North American Electric Reliability Council [1] in 74 % cases the reason of heavy failures in power systems was the incorrect actions of relay protection in trying to avoid the failure. Thus the reliability of a power system depends on the reliability of relay protection in many respects. It is a fact that intensive research and development in the field of electromechanical protective relays (EMR) have been completely frozen for the past 30-35 years and all efforts of developers have been redirected to development of electronic, and then microprocessor-based protective devices (MPD). Meantime EMR completely provided and continue to provide until now reliable protection of all objects in electric power industry. The reason for the full disappearance of the EMR and transition to MPD is not the inability of EMR to carry out the functions, rather it is something completely other. Due to the large expenditures by leading of MPD-manufacturers in promoting the MPD, the development of new materials and technologies have not affected the EMR in any way. After tens of years in operation, today's EMP have worn out and become outdated and, consequently, are a cause of a fair amount of discontent amongst protective relays experts. On the other hand, demounting

EMR and transition to MPD in the electric power industry is connected with the necessity for investing significant amounts of money, not only for purchasing MPDs, but also for computers and special expensive test equipment, as well as for the replacement of expensive MPD failed units, which cannot be repaired. Significant capital investments are required as well for reconstruction of grounding systems of substations, for training of the relay personnel, etc. All this essentially disrupts the process of transition to MPD. According to [2], in 2002 in Russian power systems there were in operation 98.5% EMR and only 1.5% of various electronic devices of relay protection. According to [3] MPD constitutes about 0.12% of the total quantity of relay protection devices in Russia. In the West rates of replacement of relay protection in working power objects also is not so high (excluding new erected power objects, of course) According to [4] at existing rates it is required about 70 years for replacement of all old protective relays with microprocessor-based. Such low rates for updating protective relays in working power objects all over the world causes intensive advertising activity of MPD-manufacturers and their distributors. One of the main reasons usually presented in the vindication of MPD advantages is its considerably higher reliability, ostensibly, in comparison with electromechanical and electronic relays. This thesis is represented as being so obvious, that, usually, does not cause objections and frequently is repeated by managers and even by technical specialists of the power engineering companies. However, in a deeper analysis of a situation it appears that the basis of this thesis is made with a whole set of widespread myths about microprocessor protection [5].

RELIABILITY MYTHS

Myth 1: MPD Reliability is higher than EMR reliability because MPD does not contain moving internal elements [6].

EMR malfunction is usually associated with ageing and damage of wiring insulation (wear, drying), corrosion of the screws and terminal clips, deterioration of the mechanical

Citation: Vladimir Gurevich, Reliability of Microprocessor-Based Relay Protection Devices, doi:10.2298/SJEE0901167G

parts of the relay. However, the number of operation cycles (i.e., movement of mobile parts) over the entire EMR service life under real operating conditions in power systems does not exceed several hundreds. So to speak about mechanical deterioration of EMR mobile parts, for all practicality, it is only in the case of evident defects of the factorymanufacturer or use of improper materials for these purposes. As for corrosion of the metal elements or drying out of the insulation, this is a consequence of use poor-quality materials by the relay manufacturer. Such defects are typical for EMR of Russian manufacturer and do not come close to meeting the products of the leading Western companies which have been in operation for 30-40 years even in a tropical climate [7]. Thus, to speak of EMR as an insufficient mechanical resource, as a kind of the relay, is absolutely unreasonably. On the other hand, if the moving elements of the EMR are in movement only at the moment of operation (pick ups) of the relay, the thousands of electronic components in the MPD constantly are in work: signal generators, numerous transistor switches, amplifiers, comparators, timers, counters, logic elements, voltage stabilizers, constantly works; the microprocessor constantly exchanges signals with elements of memory, the analog-digital converter constantly conducts processing input signals, etc. Many elements constantly are under influence of a high working voltages (120 - 250 V) and voltage spikes which periodical arise in the input circuits and external power supply circuits; as well as constant power dissipation (that is, they are heated), etc. In especially heavy work in MPD conditions switching power supplies very frequently are the reason of MPD malfunction.

Myth 2: Reliability of semi-conductor (solid state) relays on discrete components is higher than reliability of electromechanical relays [8]. Reliability of electronic protective devices based on integral microcircuits (ICs) with a high degree of integration is higher, than reliability of devices on discrete electronic components [8]. Reliability of microprocessor-based relays is higher than reliability of electronic non-microprocessor devices.

The unconditional statement about the greater reliability of semi-conductor relays over electromechanical relays is a popular mistake [9]. Semi-conductor relays possess increased reliability only at very large number of switching cycles (hundreds of thousand, millions) or at high switching frequencies. In many other cases reliability of semi-conductor relays is essentially lower than electromechanical relays [10]. Discrete electronic elements have much higher capability for withstanding voltage spikes and other adverse influences than integrated microcircuits [11]. According to [12] 75 % of all damages to microprocessor devices are the result of voltage spike impact. Voltage spikes with amplitudes from tens of volts up to several kilovolts, arising from switching transients in circuits [13] or the impact of electrostatic discharges, are "fatal" for internal microcircuits and processor microcells. According to [12] normal transistors (discrete elements) can withstand a voltage of electrostatic discharges almost 70 times higher than, for example, a microchip of memory (EPROM) in a microprocessor system. The most calamitous of temporary failures caused by electromagnetic noise that occur in the microprocessor functioning can be time [14], such as spontaneous changes of the operative memory (RAM) and registers contents, and internal damages can have the latent character [15]. These kinds of damage do not come to light in any tests and can appear during at the most unexpected moments.

In [16] it is mentioned that in connection with low stability to transients and voltage spikes the MPD demand especially rigid requirements for the protection level against electromagnetic influences. Attempts of using an MPD without strengthened electromagnetic protection frequently lead to their malfunctioning [16,17]. Electronic devices with discrete components contain fewer components than similar devices on ICs, Fig. 1. This does not seem to promote higher reliability of ICs.

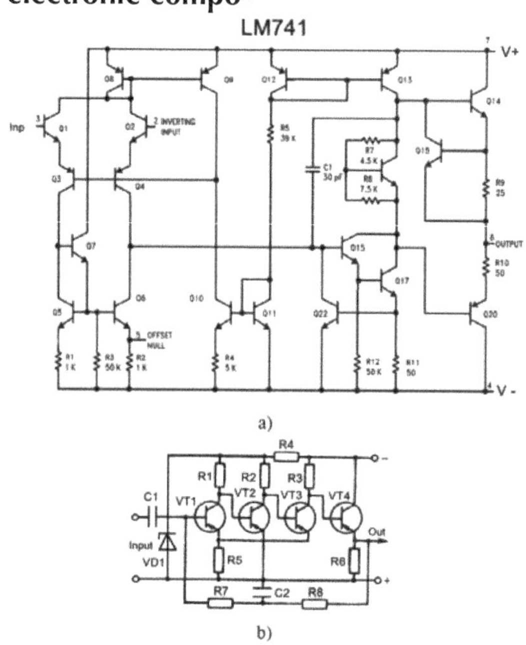

Figure. 1 – Circuit diagrams of two amplifiers: at the top of a widely used IC LM741 type containing 20 transistors; below - the amplifier on the discrete elements with same parameters, containing only 4 transistors.

The statistics on damages of MPD elements, collected by engineers of some MPD manufacturers, Fig. 2 [18], very persuasively denies the myth about higher reliability of ICs.

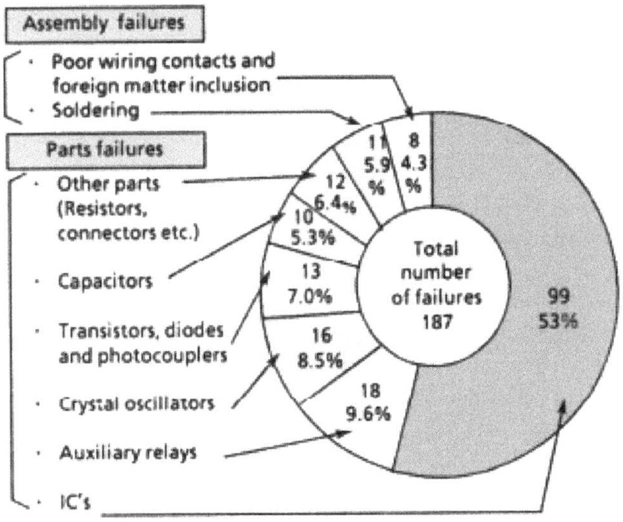

Figure. 2 – The statistical data on damages MPD based on investigations leading Japanese companies [18].

According to the statistics submitted in [8], it is extremely visible that the protective relays with electronic elements have three times the damageability than electromechanical relays, and microprocessor-based relays 50 times the damageability! Reliability of the microprocessors of such manufacturers as Intel and AMD can be actually very high, and in fact the microprocessor though not big is a very important part of the MPD which contains very many ICs. In [19] it is affirmed that the main processor unit (that is the printed-circuit-board with the microprocessor, memory, the analog-to-digital converter, library of programs and all auxiliary elements) part of MPD is the most subject to malfunctioning. Another strike against microcircuits lies not only in physical damage of the microprocessor, but also software failures - damages not known earlier for electromechanical and electronic relays. As it is pointed out in [19], program bugs are not always detected during MPD testing. An additional source of problems is the necessity for periodically upgrading the MPD program versions. Frequently during this process software-to- hardware incompatibilities [19] appear. Such problems can show up during the most unexpected moments and can lead to very heavy consequences for a power network. As is known, one of the reasons of the largest failures in a power supply system of the USA and

Myth 3: Reliability of MPD is much higher than reliability of all other types of the protective relays due to presence of the built-in selfdi-agnostics. With the self-diagnostics in MPD 70 - 80 % of all internal MPD's elements are covered [21, 30]

This thesis is very widespread and is met with in practically all publications devoted to MPD advantages. We shall consider features of this self-diagnostics more in detail.

Analog-digital converter (ADC)

This device transforms an input analog signal from CTs and VTs in binary code transmitted through special filters to processing in the microprocessor. All ADC work by sample of input values through the fixed intervals of time and thus will transform a sine wave signal to a set of the fixed amplitudes. As can be seen from Fig. 3, this is a rather complex device carrying out a complex algorithm and containing a sizable set of internal units.

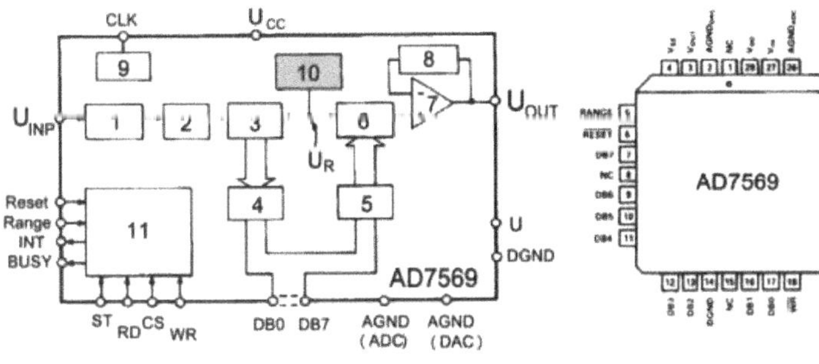

Figure. 3 – Structure of the analog digital converter AD7569 type. 1 – range network; 2 - unit of tracking/holding; 3 – analog digital converter (ADC); 4 – ADC latch; 5 – DAC register; 6 – digital analog converter (DAC); 7 - the amplifier; 8 – range network; 9 – synchronizing clock; 10 – reference power supply

Some modern ADCs are so complex that they include even a small microprocessor to manage their work. The ADC is actually the principal unit of the measuring device, and, as in any complex measuring device, there are various errors in the transformation of input values present in the ADC. These are errors in quantization, additive and multiplicative errors, differential and integrated nonlinearity of the transfer characteristic, an aperture error, aliasing, etc. How it is possible to supervise the proper functioning of such complex devices during the continuously change of input values when there is only a single element storing a constant level of a signal during ADC functioning as the reference voltage source 10 (see Fig. 3). The so-called "self-diagnostics" ADC is based on this reference voltage monitoring [21]. The efficiency and usefulness of such "self-diagnostics" the reader can estimate himself.

Memory

MPD employs two kinds of memory: ROM (Read Only Memory) intended for storing the managing program and setting, and RAM (Random Access Memory) intended for temporary storage of the results of input values measurement and intermediate calculations. The managing algorithm is a set of certain numerical codes. A certain control sum, which is remembered in a separate cell of memory, is made of these codes. During MPD functioning, the pre-recorded control sum is periodically compared with the actual sum. A mismatch of these sums should specify malfunction of a ROM [21]. Clearly, that process of calculation of the actual control sum and its comparison with pre-recorded sum is a discrete process, performed within certain intervals. And what happens if a malfunction occurs at a time between the intervals of comparison of the control sums? Won't there be a false operation of the protective relay and switching-off of a power network? The question is not completely hypothetical: such real cases which not were detected by of selfdiagnostics algorithm are described in [19]. The situation with the self-testing of the RAM is much more difficult as the contents of the RAM constantly change at a high frequency in the random mode during MPD operation. It is difficult to understand how, in a general, it is possible to test the memory cells which constantly re-record at the high frequency during operation, that is, to diagnose so-called «dynamic failures». MPD manufacturers have decided not to trouble themselves with the solving this problem and to test the RAM in an automatic mode by periodic recording the certain constant number in especially reserved memory cells and periodically reading this number with the subsequent comparison of these two numbers. If these numbers match, according to the manufacturers, this ostensibly confirm serviceability of the RAM [21], though it is absolutely not clear how it is possible to estimate serviceability of all RAM in the proper storing of the information in several cells of memory only. Besides it is wellknown that absence of the static errors in memory does not guarantee occurrence of dynamic errors [22, 23], that is, the errors arising directly during record and reading process. The problem with reliability of the MPD memory elements is actually much more complex. It appears that memory elements are subject to random unpredictable failures which have been not connected to physical damage of memory cells. Such random, temporary malfunctions caused by spontaneous changes of the contents in memory cells are referred to as "soft-failures" or "soft errors" (not to be confused with program failures - "software programming errors"). Such errors were not known earlier for the electronic devices based on discrete semiconductor elements or on usual microcircuits. Progress in the last years in the area of nanotechnology has led to a dramatic reduction in the sizes of semiconductor elements (along the order of microns and even parts of a micron), reduction of thickness of layers in semiconductor and insulation materials, reduction of working voltage, increase in the working speed (working frequency), reduction of electric capacity of separate memory cells, increase in density of placement of elementary logic cells in single chip. All this taken together has led to a sharp increase in the sensitivity of memory to ionizing radiation [24, 25]. This sensitivity became so high that usually (that is, completely normal) the radiating background at sea level became dangerous to memory cells. Streams of the high-energy elementary particles coming from space are especially dangerous. Even one such particle that hits a memory cell gives rise to secondary streams of electrons and ions causing spontaneous switching of the elementary transistor or discharging the capacity in charge storage memory elements. The problem is aggravated in modern microprocessor-based devices with the tendency toward the ever expanding use memory elements [25]. Many modern integrated microcircuits with high integration levels, included in the microprocessor-based device, contain the complex structures with embedded memory of such volume that serviceability, in general, is not supervised in any way. As shown in [26, 27], the problem of the sharp increase in sensitivity to ionizing radiations is actual not only for memory, but also for high-speed logic elements, comparators, etc., that is, practically, for all modern microelectronics.

CPU (Central Processor Unit)

As against the complexities described above with the monitoring of memory serviceability, the self-diagnostics of the CPU looks simple enough, Fig. 4.

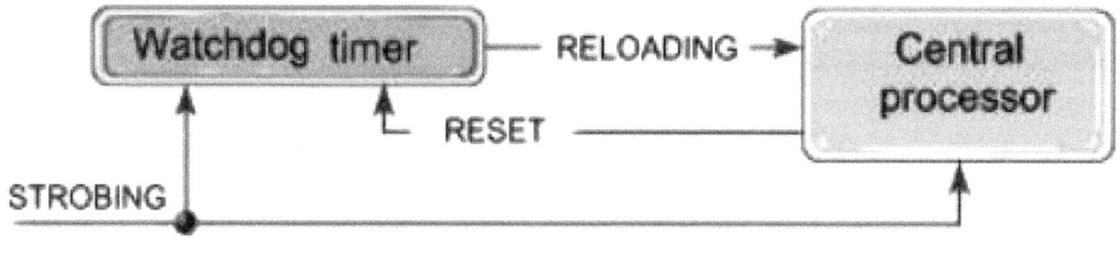

Figure. 4 – A principle of the monitoring of serviceability of the microprocessor with the help of the watchdog timer.

It simply sends control cycling pulses with a period set to the so-called «watchdog timer» - the watchdog timer is reset to an initial condition with arrival of each new control pulse, and then begins a new cycle of time reckoning. If at a certain moment the next control pulse from the CPU has not arrived, the timer starts the CPU reloads. A serious malfunction of the microprocessor and its "lag" during reloading process which is found out by the timer as the repeated absence of a control signal causes the locking of the CPU and transmitting a signal about CPU malfunction.

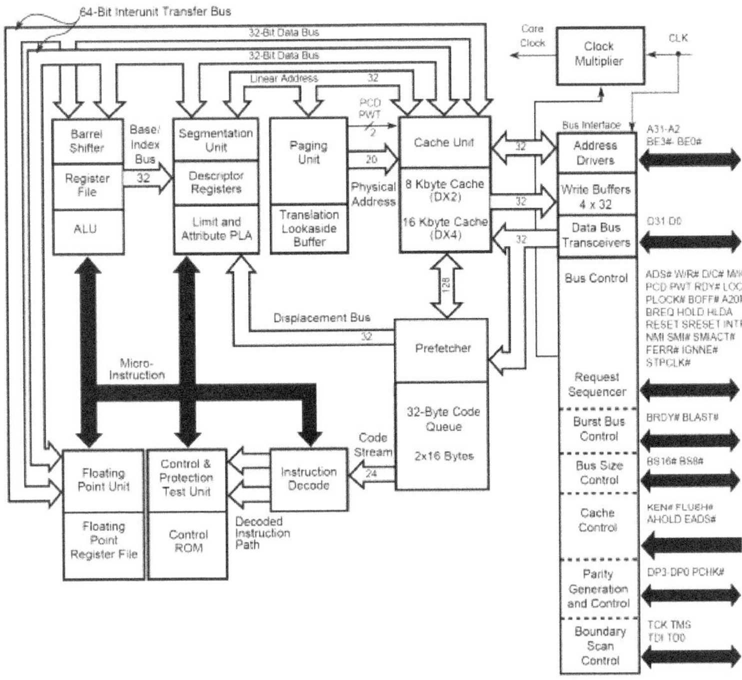

Figure. 5 – Intel 486 SX block diagram of microprocessor.

Process on tracking control pulses by the watchdog timer is synchronized with the help of external clock pulses (so-called "strobing"). Sometimes the watchdog timer is built directly in the microprocessor, and sometimes (which is more preferable) it is an external specialized integrated microcircuits (IC). As an example of such devices the IC series ADM690 – ADM695, manufactured by Analog Devices can serve. This small chip contains not only the watchdog timer, but also the monitor of a voltage level of a CPU feeding. The pause between control pulses of the watchdog timer of this series can be 0.1 or 1.6s. It is abundantly clear that to check the serviceability of hundreds of thousands of transistor nanostructures of which any microprocessor consists is absolutely impossible. At best all that can be said is that it monitors only the general working ability of the CPU, in other words, checking if it is alive or dead. Given the very complex internal structure of the CPU, Fig. 5, containing many, many units (microcells; registers for temporal storage of the commands, the data and addresses; the arithmetic logic device; system management and synchronization, etc.), the control signals from the CPU can continue to act on the watchdog timer even if part of the internal structure of the CPU is damaged.

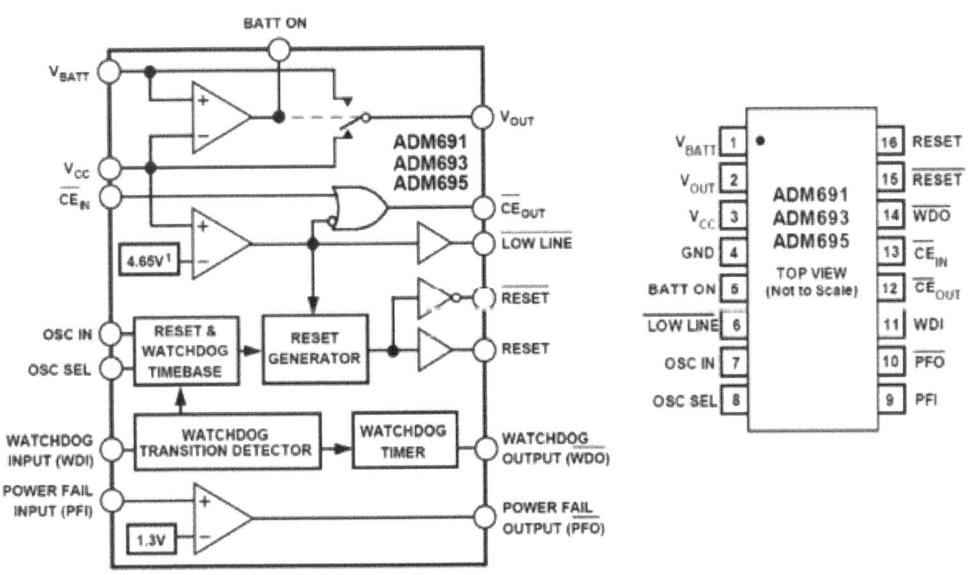

Figure. 6 – Series ADM691 - ADM695 watchdog timer, manufactured by Analog Devices Co.

It is obvious that damages of certain regions of the internal CPU structure (or parts of internal managing program) can be detected only in an operating mode (upon activization of these regions). If these regions of the CPU become active only at the input signals corresponding to emergency mode in an electric network, it means that the watchdog timer is a cold comfort. In itself, the watchdog timer is the device made with the same technology as all other microelectronics devices, Fig. 6, and is subject to malfunctioning and failures just as all other microelectronic devices. Owing to it working algorithm the watchdog malfunctioning during normal MPD operation can result in the locking of the CPU (i.e., locking of the whole MPD), or it will not notice "lag" of the CPU, in either case the relay protection will not work properly at emergency mode. Thus, the operational ability of the whole MPD Reliability of Microprocessor-Based Relay Protection Devices: Myths and Reality 177 appears to have a very strong dependence on the serviceability of one small chip named "watchdog".

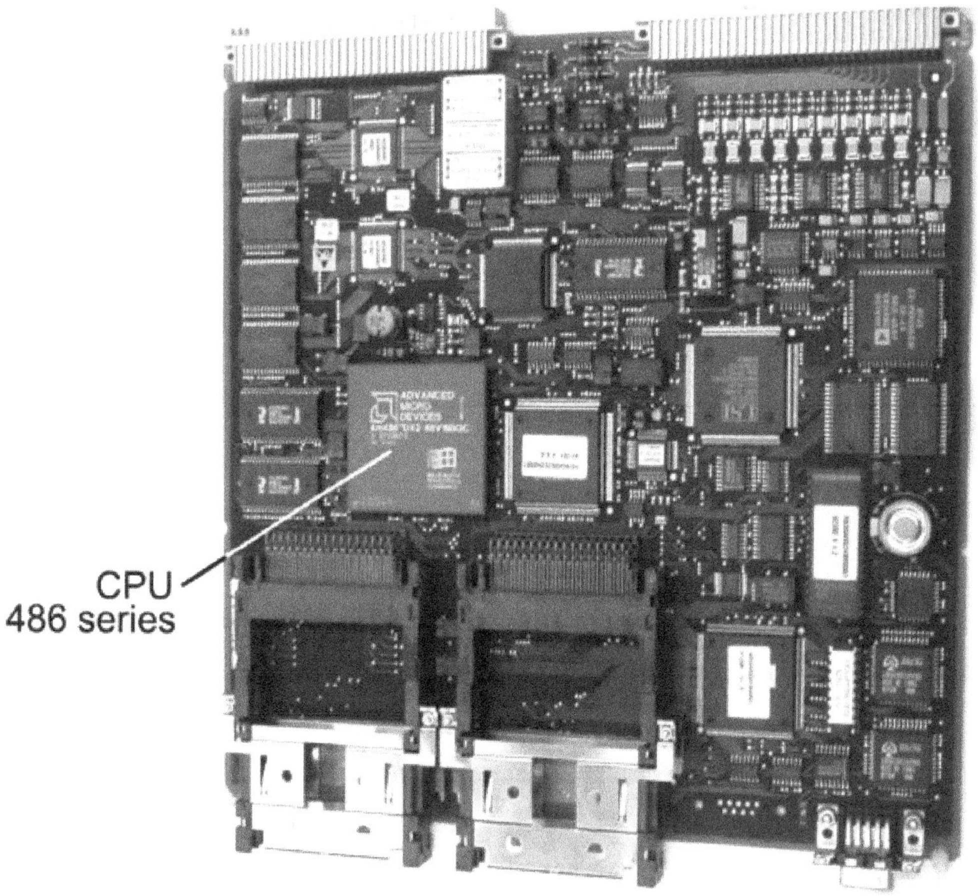

Figure. 7 – CPU of the of central processor unit of the MPD series RE *_316, manufactured by ABB.

One more important consideration is that the CPU is not the an isolated element upon which correct functioning in MPD structure depends, it also depends on the serviceability of tens of other complex integrated microcircuits to which CPU is connected, but for which self-diagnostics is not stipulated. It's enough to look at the printed-circuit-board (PCB) of the central processor unit, Fig. 7, to understand that serviceability of the CPU alone does not speak about serviceability of whole this PCB. Damage of any of numerous microelectronic (and not only!) components of this multilayered PCB will inevitably lead to infringement of correct MPD functioning and any watchdog here will not help, as proves to be true judging from the data results in [19].

Power supply

MPD of all types are supplied with what is called switching power supplies in which the input voltage (AC or DC) acts on the rectifier and the filter then interrupts with the high frequency (tens-hundred kilohertz) with the help of a powerful switching transistor, that turns in high-frequency AC. This highfrequency voltage is transformed by the high-frequency transformer to a low voltage (more often, 12 V), which is rectified, filtered and stabilized. Further lower voltages are formed from this DC voltage (5 V, for example), necessary for MPD functioning. Microprocessors, usually, are rather sensitive to a level of a feed voltage and can perform unpredictable operations at the voltage reduction below a certain value. In this connection, in MPD a constant monitoring of the level of a feed voltage of the CPU is carried out. As it was mentioned above, IC of family ADM 691-695 can be used for the continuous monitoring of the power supply voltage level. As well as in a case with the watchdog timer, the ADM 691-695 chip generates a signal for locking the CPU at an inadmissible voltage supply reductions. The locking signal remains until the voltage level is restored. Is it really possible to count such voltage level monitoring as selfdiagnostics of the power supply, raising it functioning reliability? Hardly, therefore it is the internal technological locking which only prevents failures in the CPU. Such monitoring has no relation to the reliability of the power supply. And meanwhile, power supplies are the most unreliable unit of the MPD. First, elements of the power supply work in a high load mode; they are constantly subjected to the influence of high values of a voltage and a current, voltage spikes, dissipated rather high power on the elements. Secondly, they contain a lot of aluminum electrolytic capacitors that rather badly carry the influence of high frequency currents on which switching power supplies work, and frequently are the reason of full breakdown of the power supply (and, consequently, of the whole MPD). We ask can monitoring of an output voltage level in this case help? Can it signal beforehand about the deterioration of a capacitor and thereby prevent the sudden MPD failure?

Output electromagnetic relays

As shown in previous researches [28,29] that the author conducted, contacts of the miniature electromechanical relays (usually used in all types modern MPDs as output elements) directly control the trip coils of high-voltage circuit breakers or the coils of auxiliary relays with a significant overload. Therefore reliability of these relays is essentially reduced to the comparison with the value which was normalized by the relays manufacturer. On the other hand, in promotional brochures of various manufacturers about MPD advantages it is usually emphasized that serviceability of such important elements as output relays is continuously supervised by means of a self- diagnostics system of the MPD. At first sight, it is rather difficult to represent that it is possible to check the serviceability of the electromechanical relay in working MPD if the contacts are inserted directly in circuit of the trip coil of the high-voltage circuit breaker. It is impossible to check the serviceability of contacts of the output relays in this case, so what check is possible? The coil only. This is what the MPD manufacturers have implemented: the supervision of the continuity of the coil of the relays. This is done by passing a constant weak current through the coil. But what if the most intense and unreliable part of the electromechanical relay is not the coil at all, but the contacts?! This is true, but it is not so important for the advertising campaign. It was necessary to declare loudly only to the MPD consumer about of self-diagnostics of output relays, but that these selfdiagnostics are completely inefficient, as a rule, almost nobody knows.

Units of digital and analog inputs

The digital inputs unit is a set of powerful resistors, opto-couplers, electronic filters, multiplexers, etc., mounted, usually, on a common PCB together with the output relays, Fig. 8. The analog inputs unit is a set of voltage and current transformers mounted on a separate PCB, Fig. 9. In [30] is noted that these units are only partly covered by self-diagnostics, without any explanation of how it is performed, but in [31] it is stated that they are not covered at all by self-diagnostics. PCBs of analog and digital inputs in the MPD have, as a usual, various configurations, Fig. 8. The given type of the PCB established in given MPD should be necessarily entered into its memory.

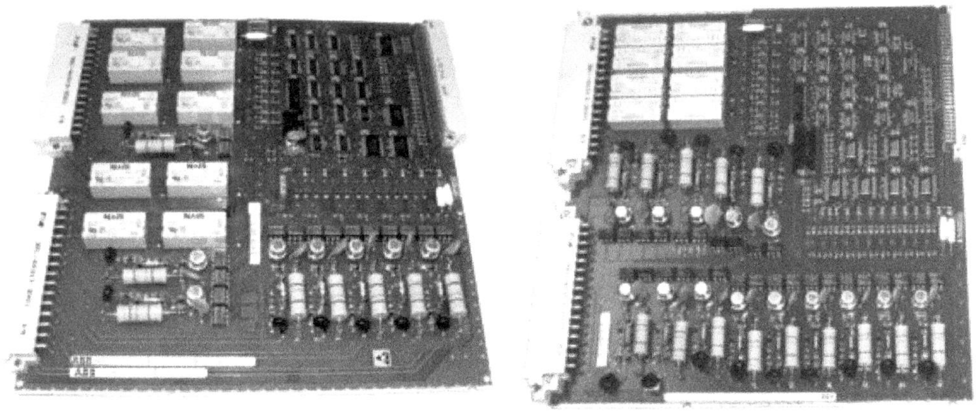

Figure. 8 – Units of digital inputs of various configurations from MPD series REL316.

To clear up the situation and to place points above i, we have replaced one such PCB (in MPD REL316 type) the type of which is stored in the MPD's memory, with a PCB of another type (Fig. 8) without changing the data in the MPD memory. Apparently the MPD loaded into a normal operating mode without noticing the substitution of the PCB. It is clear that it will not correctly function. What can be said about self-diagnostics of serviceability of internal components of these units in such situation? As they say, further comment is superfluous.

Figure. 9 – Units of analog inputs MPD containing voltage and current transformers.

In conclusion of this section it is necessary to note that, contrary to a popular belief, internal self-diagnostics actually is not the means intended for decreasing MPD failure rate, i.e., for increasing its reliability. The purpose of self-diagnostics is the locking of the MPD and delivering an alarm signal before the occurrence of emergency mode in a power network, instead of during a MPD failure.

Myth 4: MPD are essentially more reliable in comparison with relay protection devices of the previous generation as it contains considerably fewer number of components and these components are much less subject to physical ageing. MPD also contains smaller number of internal connections [32]

As for the claim that MPD contain a fewer number of components than relay protection devices of the previous generation, it turns out that actually the number of components making up some MPDs is more than the number of components of protective relays of previous generations. As for the claim of more intensive physical ageing of elements of the protective relays of the previous generation, this thesis also does not bear up. The author of this thesis compares the modern elements and technologies used in MPD with materials (impregnation and cover varnishes, plastics, insulation materials, contact and anticorrosion materials) that were in use in the USSR 50 years ago and employed in the protective relays for over several decades. As we already remarked above, old electromechanical relays of the western manufacture (BBC, Westinghouse, General Electric, etc.) in which high-quality materials and coverings were applied still work successfully and show no signs of ageing.

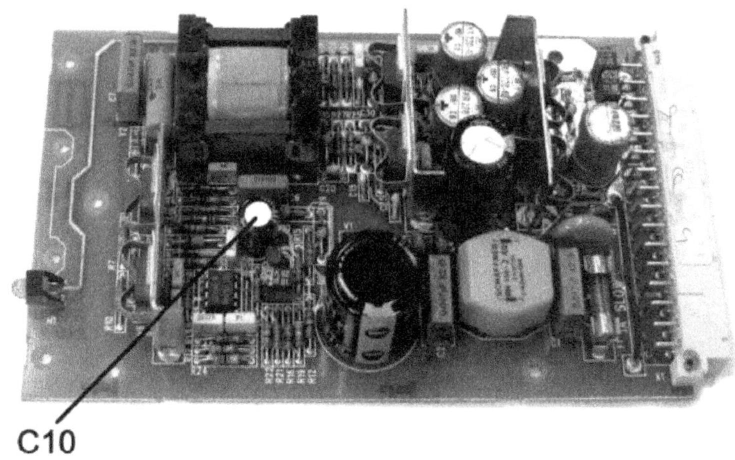

Figure. 10 – Switching power supply SPGU240A1 type used in MPDs various types. C10 - the capacitor, where a change of parameters results to full loss of working ability of the power supply.

Besides, the progress in the field of materials over the last decade has not been less than the progress in the field of microelectronics. On the other hand, not all is so bright with the ageing of the electronic components widely used in MPD. Even high-quality electrolytic capacitors of Japanese manufacture start to change the parameters after 7 - 10 years of operation under high-frequency used in MPD's switching power supplies. As a result a change of parameter of one such capacitor only, Fig. 10, completely stops power supplies functioning. For example, of power supplies such as SPGU240A1, used in MPD types SPAC, SPAD, SPAU, SPAJ have been shown to cause this. In other cases destruction not only of the electronic components takes place, but even dissolution copper streaks on PCB under action of the electrolyte which has leaked out from capacitors, Fig. 11.

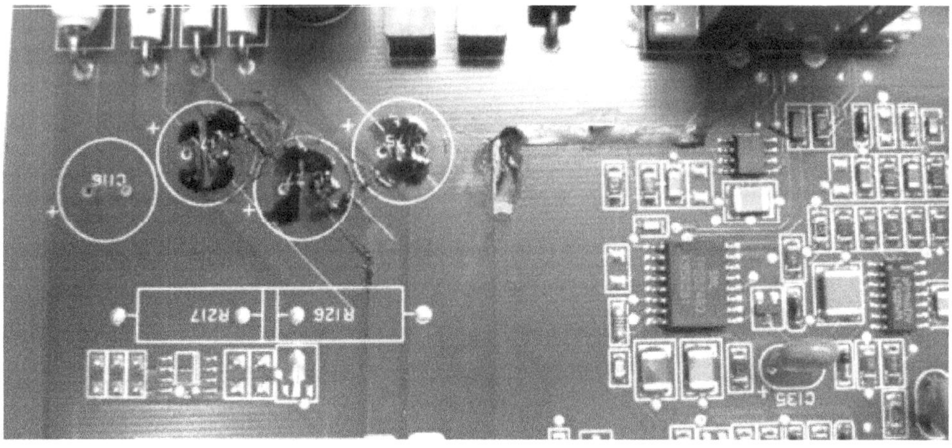

Figure. 11 – Destruction of copper streaks on the printed-circuit-board which is taking place under capacitors because electrolyte leakage.

One more problem is the aspiration of manufacturers for MPD miniaturization at any cost. This has led to using electronic elements in MPD working with an overload and dissipated increased value of heat that does not promote an increase in MPD reliability and reduction of elements ageing. This problem for circuits of digital inputs on which the voltage up to 250 V is applied [33] is especially prevalent. Multilayered PCB MPD involves a huge number of contact transitions (crosspieces) between layers. From the author's personal experience there have been cases of faulty MPD actions due to the increase of transitive resistance of these transitions. The design of many types of MPDs come with a motherboard with multicontact sockets and functional boards with the reciprocal sockets connected with a motherboard. Instead of a motherboard the flexible multicore trunks with the numerous contact sockets connecting among themselves, separate PCBs are sometimes used. These contact connections do not always provide a reliable transfer of low-voltage/low-current signals between boards. In any case, contrary to the widespread myth, MPD contains many more interconnections than the relay of the all previous generations.

ONE MORE CLASS OF PROBLEMS WHICH MPD MANUFACTURERS PREFER NOT TO MENTION

In view of the increased sensitivity of modern microelectronics to electromagnetic radiations, there is a problem for MPDs in connection with electromagnetic compatibilities (EMC). Many experts have noted the often incompatibility between real parameters of grounding systems in substations and the requirements showed by MPDs [34, 35] and, as result of it, on MPD failures. But little is known by the experts in the field of relay protection about a problem of "electromagnetic terrorism", the powerful electromagnetic radiations [36] that intentionally impacts on electronic devices, and also about a problem of hacker attacks (cyber security problem) [37]. These problems were unknown earlier in relay protection and became a reality only in connection with MPD applications, as their sensitivity to electromagnetic noises is 10,000 times higher than in electromechanical relays [34]. The built-in MPD software is also subject to external influences. And if, in addition to all the aforesaid, one takes into account that one MPD carries out the functions of 3 - 5 EMRs, the situation with MPD reliability is aggravated even more, as damage of one of the common MPD elements is equivalent, in consequence, to the simultaneous damaging of several kinds of protection at once. In this connection in [38] transition of microprocessor protection is offered to provide additional independent, simple, inexpensive, not microprocessor reserve protection for a cases of extreme situations.

Conclusion

1. Reliability MPD is lower than reliability of electromechanical relays and electronic relays on discrete elements.
2. Built-in self-diagnostics MPD is ineffective and is not a means at all for increasing of MPD reliability.
3. Nanotechnologies, used at manufacture of MPD's elements, leads to the occurrence of problems not known earlier for relay protection. Ignoring these problems can lead to catastrophic consequences. The managers making of the decision in the field of relay protection and the personnel of the power companies should be informed about these MPD features.
4. The recording function of emergency modes in power network and data transmission function on modern connection channels are not direct functions of relay protection and for their realization there are separate microprocessor devices which carry out these functions much better than MPDs. As against relay protection, failure of these devices does not lead to heavy failures in power systems. Therefore for relay protection devices the focus should be on other demands on reliability and, accordingly, to use other approaches at the designing, directed on increasing of reliability and decreasing in vulnerability.
5. The persons responsible for making the decisions on reconstruction of relay protection and ways of further developments should understand, precisely, the properties and features of the MPD, to take into account not only widely promoted MPD advantages, but also the serious lacks, one of which is lowered reliability.

REFERENCES

1. R.K. Hunt: Hidden Failure in Protective Relays: Supervision and Control, Thesis to Master of Science in Electrical Engineering, Virginia Polytechnic Institute, 1998.
2. E.V. Konovalova: Main Results of Relay Protection Devices Maintains in Power Systems in the Russian Federation, Relay Protection and Automatics of Power Systems Conference, Moscow, 2002.
3. A.K. Belotelov: Scientific and Technical Policy of the Russian Open Society EU of Russia in Development of Systems of Relay Protection and Automatics, Relay Protection and Automatics of Power Systems Conference, Moscow, 2002.
4. G. Johnson, M.Thomson: Reliability Consideration of Multifunction Protection, Protective Relaying Conference, 2001.
5. V. Gurevich: How to Equip the Relay Protection: Opinions of the Russian Experts and a View from the Side, News in Electric Power Industry, No. 2, 2007, pp. 52-59.
6. R. Projjalkumar: Is the Era of Electromechanical Relays Over?, Frost & Sullivan Market Insight, 2004.
7. V. Gurevich: Microprocessor-based Relay of Protection: An Alternative View, Electro-info, No. 4, 2006
8. C.R. Heising, R.C. Patterson: Reliability Expectations for Protective Relays, Developments in Power Protection-Fourth International Conference in Power Protection, 1989, pp. 23-26.
9. T.R. Mahaffey: Electromechanical Relays Versus Solid-State: Each Has Its Place, Electronic Design, 2002, pp. 44.
10. Electromechanical vs. Solid State Relay Characteristics Comparison, Application Note 13c3235, Tyco Electronics.

11. V. Gurevich: Electronic Devices on Discrete Components for Industrial and Power Engineering, Boca Raton-New York-London, CRC Press, 2008.
12. [12] O.M. Clark, R.E. Gavender: Lighting Protection for Microprocessor-based Electronic Systems, IEEE Transactions on Industry Applications, Vol. 26, No. 5, 1990, pp. 197-203.
13. K. Uchimura, J. Michida, S. Nozu, T. Aida: Multifunction of Digital Circuits by Noise Induced in Breaking Electric Contacts, Electronics and Communications in Japan, Vol. 72, No. 6, 2007
14. I.A. Henderson, J. McGhee, W. Szaniawski, P. Domaradzki: Incorporating High Reliability into the Design of Microprocessor-based Instrumentation, Science, Measurement and Technology, IEE Proceedings A, Vol. 138, No. 2, 1991, pp. 105-112.
15. A.G. Phadke: Hidden Failures in Electric Power Systems, International Journal of Critical Infrastructures, Vol. 1, No. 1, 2004, pp. 64-75.
16. [16] B.I. Kovalev, I.E. Naumkin: The Main Problems of Electromagnetic Compatibility of Secondary Circuits in High-voltage Substations, Relay Protection and Automatics of Power Systems Conference, Moscow, 2002.
17. Information Notice No. 94-20: Common-Cause Failures Due to Inadequate Design Control and Dedication, US Nuclear Regulatory Commission, Washington, 1994.
18. T. Matsuda, J. Kovayashi, H. Itah, T. Tanigushi, K. Seo, M. Hatata, F. Andow: Experience with Maintenance and Improvement in Reliability of Microprocessor-based Digital Protection Equipment for Power Transmission Systems, CIGRE, 1992.
19. S. He, L. Shen, J. Lui: Analyzing Protective Relay Misoperation Data and Enhancing Its Correct Operation Rate, IEEE/PES Transmission and Distribution Conference & Exhibition: Asia and Pacific, 2005, pp. 1-5.
20. Final Report on the August 14, 2003 Blackout in the United States and Canada: Causes and Recommendations, Tech. rep., U.S.-Canada Power System Outage Task Force, 2004.
21. V.Y. Shmuriev: Digital Protective Relays, Library of Electrical Engineering, Vol. 1, No. 4, Moscow, STF "Energoprogres" , 1999.
22. S. Hamdioui, Z. Al-Ars, A.J. Goor: Testing Static and Dynamic Faults in Random Access Memories, Proceedings of the 20th IEEE VLSI Test Symposium, 2002, pp.395-400.
23. S. Hamdioui, G.N. Gaudadjiev: Future Challenges in Memory Testing, Proceedings of PRORISC'03, 2003, pp. 78-83.
24. Soft Errors in Electronic Memory-A White Paper, Terrazon Semiconductor, 2004.
25. [25] R.C. Baumann: Soft Errors in Advanced Semiconductor Devices-Part I: The Three Radiation Sources, IEEE Transactions on Device and Material Reliability, Vol. 1, No. 1, 2001, pp. 17-22.
26. P.E. Dodd, M.R. Shaneyfelt., J.A. Felix, J.R. Schwank: Production and Propagation of Single-Event Transient in High-Speed Digital Logic ICs, IEEE Transactions on Nuclear Science, Vol. 51, No. 6, 2004, pp. 3278-3284.
27. A.H. Johnson, T.F. Miyahira, F. Irom, L.D. Edmonds: Single-Event Transients in HighSpeed Comparators, IEEE Transactions on Nuclear Science, Vol. 49, No. 6, Part 1, 2002, pp. 3082-3089.
28. V. Gurevich: Nonconformance in Electromechanical Output Relays of Microprocessorbased Protection Devices Under Actual Operation Conditions, Electrical Engineering & Electromechanics, No.1, 2006, pp. 12-16.
29. V. Gurevich: Peculiarities of the Relays Intended for Operating Trip Coil of the Highvoltage Circuit Breakers, Serbian Journal of Electrical Engineering, Vol. 4, No. 2, 2007, pp. 223-237.
30. J.J. Kumm, E.O. Schweitzer, D. Hou: Assessing the Effectiveness of Self-Test and Other Monitoring Means in Protective Relays, 21st Annual Western Protective Relay Conference, 1994.
31. Advanced Digital Relay Systems - Is testing still needed?, Omicron Electronics, Vol. 5, No. 1, 2000.
32. E. M. Shneerson: Digital Relay Protection. Energoatomizdat, 2007.
33. V. Gurevich: Microprocessor Protection Relays-the Present and the Future, Serbian Journal of Electrical Engineering, Vol. 5, No. 2, 2008, pp. 325-339.
34. R. Borisov: Insufficient attention to problem of EMC may turn as catastrophe, News in Electrical Engineering, 2001, v. 6(12), (Rus).
35. M. Matveev: Electromagnetic situation on objects determines EMC of the digital equipment, News in Electrical Engineering, Vol. 1 No. 13, 2002.
36. V. Gurevich: Electromagnetic Terrorism: New Hazards, Electrical Engineering & Electromechanics, No. 4, 2005.
37. IEEE Standard for Substation Intelligent Electronic Devices (IEDs) Cyber Security Capabilities, IEEE Std 1686-2007. [38] V.I. Puliaev: Results of Usage of Relay Protection and Automation in Open Society "United Electrical Systems", Relay protection and automatics of power systems Conference, Moscow, 2002.

CHAPTER 18 SMALL PRIVATE KEY MQPKS ON AN EMBEDDED MICROPROCESSOR

Hwajeong Seo, Jihyun Kim, Jongseok Choi, Taehwan Park, Zhe Liu and Howon Kim

Keywords: Public key cryptography, small private key, multivariate quadratic cryptography, embedded microprocessor, efficient software implementation, ATxmega128a1, AES accelerator, random number generator, signature generation

INTRODUCTION

The technology related to embedded systems has made significant progress, making many applications, such as home automation, surveillance systems and environment monitoring services, feasible. However, without secure and robust data protection from security threats, these services cannot be put into practice. To solve these problems, public key cryptography has been studied for several decades. The current main stream is Elliptic Curve Cryptography (ECC), which is an approach based on the algebraic structure of elliptic curves over finite fields. The use of elliptic curves in cryptography was suggested independently by Koblitz [1] and Miller [2] in 1985. The technology provides a short key-size and various applications, including Elliptic Curve Digital Signature Algorithm (ECDSA), Elliptic Curve Diffie-Hellman (ECDH).

The alternative multivariate quadratic ($\mathcal{MQ}$) cryptography provides encryption and digital signatures with modest computational resources [3]. There is no feasible attack to lattice-based cryptosystems that has been discovered yet under a quantum computing environment, while those to factoring for Rivest Shamir Adleman (RSA) and Digital Signature Algorithm (DSA) and Elliptic Curve Cryptography (ECC))-based systems already exist. However, the large size of the public and private keys required makes it difficult to fit such systems into low-cost devices like Radio-frequency identification (RFID) tags and smart-cards.

Citation: Hwajeong Seo, Jihyun Kim, Jongseok Choi, Taehwan Park, Zhe Liu and Howon Kim, Small Private Key MQPKS on an Embedded Microprocessor, doi:10.3390/s140305441.

In this paper, we study an efficient implementation of $\mathcal{MQ}$ cryptography in terms of shortening the private key and reducing the computational cost of signature generation and verification. We focus on $\mathcal{MQ}$ techniques on an embedded processor, because ECC has been studied for several decades and has reached its technological pinnacle [4]. In contrast, $\mathcal{MQ}$ cryptography has only recently begun to receive attention, and there is considerable room to improve its performance. Previously, small public key implementations have been actively studied, but the private key analogue has not. In this paper, we implement small private key $\mathcal{MQ}$ cryptography using a Pseudo-Random Number Generator (PRNG). To enhance its performance, we adopt an Advanced Encryption Standard (AES) module for the PRNG and hash function and use optimized techniques, including parallel computation, on-the-fly computation, vinegar monomials, optimized logarithmrepresentation and assembly programming.

The paper is organized as follows. In Section 2, we give an introduction to the basic structure of Multivariate Quadratic Public Key Scheme ($\mathcal{MQ}$ PKS)and related technologies. In Section 3, we present efficient implementation techniques for embedded microprocessors. In Section 4, we evaluate and analyze the performance of the proposed method. Finally, in Section 5, we conclude the paper with a brief summary of our contributions.

RELATED WORK

Unbalanced Oil and Vinegar

The idea of Unbalanced Oil and Vinegar (UOV)-signature schemes is to use a public multivariate quadratic map,

$\mathscr{P} : \mathbb{F}_q^n \to \mathbb{F}_q^m$ with:

$$\mathscr{P} = \begin{pmatrix} p^{(1)}(x_1, \ldots, x_n) \\ \vdots \\ p^{(m)}(x_1, \ldots, x_n) \end{pmatrix}$$

and:

$$p^{(k)}(x_1, \ldots, x_n) := \sum_{1 \le i \le j \le n} \alpha_{ij}^{(k)} x_i x_j = x^T \beta^{(k)} x$$

where $\beta^{(k)}$ is the $(n \times n)$ matrix describing the quadratic form of $p^{(k)}$ and $x^T = (x_1, \ldots, x_n)^T$ [5]. The trapdoor is given by a structured central map, F:Fnq→Fmq, with:

$$\mathscr{F} = \begin{pmatrix} f^{(1)}(u_1, \ldots, u_n) \\ \vdots \\ f^{(m)}(u_1, \ldots, u_n) \end{pmatrix}$$

and:

$$f^{(k)}(u_1, \ldots, u_n) := \sum_{1 \le i \le j \le n} \gamma_{ij}^{(k)} u_i u_j = u^T \delta^{(k)} u$$

In order to hide this trapdoor, secret linear transformation $\mathcal{S}$ is chosen, such that $\mathcal{P} := \mathcal{F} \circ \mathcal{S}$ [6]. For the UOV signature scheme, we define two variables called vinegar (u_i, $i \in V := \{1, \ldots, \upsilon\}$) and oil ($u_i$, $i \in O := \{\upsilon + 1, \ldots, n\}$). The central map, $\mathcal{F}$, is given by:

$$f^{(k)}(u_1, \ldots, u_n) := \sum_{i \in V, j \in V} \gamma_{ij}^{(k)} u_i u_j + \sum_{i \in V, j \in O} \gamma_{ij}^{(k)} u_i u_j$$

The number of vinegar variables is twice the number of oil variables to make the protocol secure. The transformation involves fully mixing the oil and vinegar variables, so that malicious users cannot obtain secret values by separating the oil and vinegar variables.

Signature Generation

To sign a document, d, a hash function, H:F∗q→Fmq, is used to compute the hash valueh=H(d)∈Fmq. Next, one computes $y = \mathcal{F}^{-1}(h)$ and then $z = S^{-1}(y)$. The signature of a document, d, is z∈Fnq.

Signature Verification

To verify the authenticity of a signature, the hash value, h, of the corresponding document and the value $h' = \mathcal{P}(z)$ must be computed. If $h = h'$ holds, the signature is accepted; otherwise, it is rejected.

Public Key Optimization

At CHES2011 (Cryptographic Hardware and Embedded Systems, 2011), the 0/1 UOV method for reducing the size of the public key was presented [7]. Choosing the special structure ($\mathcal{S}$, $\mathcal{P}$), it was reported that the key size and runtime of the verification algorithm could be reduced. The concept behind the reduction of the public key size is the use of a partially cyclic public key and GF(2) (Galois Field) elements for coefficients. The method proceeds by generating a partially circulant matrix and then computes the public key transformation matrix from a linear map to compute the corresponding secret key. If the specific requirements are met, we can generate the public key from small-size cyclic keys. Furthermore, coefficients of the GF(2) form are easily computable, reducing both the size and verification time.

Private Key Optimization

To reduce the private key size of $\mathcal{MQ}$ schemes, we can use a PRNG for key generation. This reduces the key size down to the size of the seed values. Recently, private key generation has been implemented using an RC4-based PRNG [8]. The basic idea is to generate a private key from symmetric cryptography, which can be used as the private coefficients. However, the method is implemented on PCs using JAVAso straight-forward implementation of this method on resource constrained device is infeasible, because PRNG has high overheads for embedded microprocessors. For a light-weight implementation, an embedded encryption module could be exploited. This approach was firstly introduced in INDOCRYPT2012 [9], where sub-operation of the RFSB-509 generating constant is concurrently computable by accumulating previous results. In terms of PRNG, the first implementation of AES-based PRNG was described in [10]. The method exploits AES counter mode of operation to generate high entropy random numbers with high throughput.

Previous Implementations on Embedded Microprocessors

A software implementation of enTTS (20, 28) on a MSP 430 microprocessor was reported in [11]. The signing and verification operations were executed within 71 ms and 726 ms, respectively. At CHES2004, Yang et al. reported signs of TTS(20, 28) in 144 ms, 170 ms and 60 ms and TTS(24, 32) in 191 ms, 227 ms and 85 ms for i8032AH, i8051AH and W77E59, respectively [12]. Recently, at CHES2012, implementations of UOV, Rainbow and enTTS on an eight-bit microprocessor were reported. The author implemented $\mathcal{MQ}$ signatures with security levels of 2^{64}, 2^{80}, and 2^{128} and demonstrated the feasibility of such protocols on resource-constrained devices. They also provided specific implementation techniques, such as self-invertible linear maps, LU decomposition and logarithm representation. First, self-invertible linear maps do not, by definition, require inversion, and their private key is smaller than a normal map. Second, LU decomposition factorizes a matrix into the product of lower triangular and upper triangular matrices. The decomposition representation reduced the straightforward implementation cost of Gaussian elimination. Finally, logarithm representation simply performs multiplication on a Galois field by computing addition. Until now, few implementations have been reported on embedded microprocessors, and even private key optimization has not been studied. For this reason, we have focused on a private key reduced model for embedded microprocessors.

Target Platform and Tools

We used ATxmega128a1on an Xplain board as our target platform. This microprocessor has a clock frequency of 32 MHz, 128 KB flash program memory and 8 KB SRAM. Furthermore, the device provides an AES crypto-accelerator that computes the encryption within 375 clock cycles. This is significant progress compared to the software implementation of AES on the ATmega128 processor, which requires 1,993 ~ 3,766 clock cycles [13,14] with pre-computations. In this paper, we provide a novel signature generation for modern microprocessors that uses an embedded AES accelerator for the PRNG and hash function. These approaches significantly reduce the size of the private key and optimize the computational performance, as well. To use AES module, we should trigger AES operation by following instructions as described in Algorithm 1. First, a status bit for AES operation is set. Second, a key and plain text are allocated to the AES accelerator, and then, we trigger the AES execution, which takes 375 clock cycles. During the execution, we can perform other operations using the microprocessor, because AES operations are independently executed on the AES accelerator. After the execution, we obtain the cipher-text generated from the AES accelerator by accessing the storage address. Our program is written in assembly for the main computations and partly C language for the interface. The development tool is the latest version of Atmel Studio 6.0.

Algorithm 1: AES encryption using AES accelerator	
Input: Secret key k, plain text p	
Output: Cipher-text c	
1.	AES accelerator setting
2.	Move k to key storage in AES accelerator
3.	Move p to plain text storage in AES accelerator
4.	Execute the AES accelerator
5.	Wait for completion (375 clock cycles)
6.	Get c from cipher-text storage in AES accelerator

Random Number Generator Based on a Block Cipher

Random numbers are widely used as seeds for cryptographic operations and secret key generation. Among various types of random number generators, a block cipher-based random number generator is considered in this paper to exploit the AES module in an embedded board. The following equation outlines the process of random number generation. The notations, enc, C, k and R, represent the encryption process, counter (secret seed), secret key and random number stream. First, the counter is encrypted with secret key, k. The output is then bitwise XORedwith counter value C_i, and the encryption can proceed. This process is iterated until we obtain a suitable size of random numbers.

$$\begin{aligned} R_1 &= enc_k(C_1) \\ R_{i+1} &= enc_k(C_i \oplus R_i) \\ where: C_i &= C_{i-1} + 1 \end{aligned} \tag{1}$$

The purpose of introducing the block cipher-based random number generator is that we will use it to generate the secret coefficients. This is possible using the AES accelerator on modern embedded boards, which can generate encrypted data conveniently. The AES accelerator is a peripheral device, so it operates independently of the microprocessor. We can order the encryption on the AES accelerator while simultaneously computing a signature on the microprocessor. This method can boost performance and reduce the required program memory. Furthermore, the PRNG process follows cipher-block chaining (CBC); this is efficiently executed using the CBC option on an embedded processor. In our implementation on ATXmega, timing and ROM cost around 40 (clock/bytes) and 204 (bytes), respectively. This result is reasonable for embedded processors. For randomness characteristic, we evaluated our random sequence on the National Institute of Standards and Technology (NIST)-test suite [15]. Firstly we collected a pseudo random number sequence from block cipher-based PRNG by two gigabytes. Then, we operated the NIST statistical test suite version 1.6 with the number of bit streams and the length of the bit set to 100 and 1,024, respectively. The results are described in Table 1, and reasonable proportion rates are achieved.

Table 1. Result of NIST random number generator (RNG) test.

Statistical Test	Proportion(%)
Frequency	99
Block-Frequency	99.4
Cumulative-sums	99.2
Runs	99.5
Longest-run	100
Rank	100
FFT	100
Serial	100
Lempel–Ziv	100
Linear-complexity	100

For the security concern, the strength is based on the bit-length of block-cipher. If the inner state of the generator is compromised once, the adversary could foresee future outputs. To resolve this problem, we should compute random numbers with refreshed seed values. One possible challenge to AES in counter mode would be a timing attack on the value of the counter. We can prevent these attacks by using a counter that always takes the same amount of time to increment its value, and AES-based PRNG could give a periodic generation. This drawback could be solved by reseeding the secret values on proper timing.

Hash Function Based on a Block Cipher

Hash functions compress an input of arbitrary length to a string of fixed length. Our main motivation for constructing a hash function based on a block cipher is to minimize the design and implementation effort. The hash function based on the block cipher is conducted according to Equation (2), where H, p_i, N and Edenote the hash code, plain text, nonce and encryption process, respectively. This structure follows the Davies–Meyer single-block-length compression function. The security level of the one-way hash function is determined by the minimum of the size of the key and the block length [16]. To ensure a sufficient security level, we used a 128-bit secret key for the AES-based hash function.

$$H_1 = enc_{p_1}(N) \oplus N$$
$$H_i = enc_{p_i}(H_{i-1}) \oplus H_{i-1} \tag{2}$$

In our hash function, we exploit the AES-accelerator for the block-cipher-based hash function. Previous$\mathcal{MQ}$PKS implementations on an embedded board were not concerned much with hash functions. However hash function should be included for practical purposes, because normally, a message is compressed in its own embedded board, not other places. In Table 2, we can find hash function implementations on the ATmega board. Compared with other results, our result improves speed and size altogether. In the case of speed, we use a dedicated hardware crypto module, so this is faster than other results that implement the functions in software. Furthermore, our result does not use much memory, because we only need to use hardware control code.

Table 2. Speed of compression functions.

Algorithm	Time (cycles/byte)	RAM (bytes)	ROM (bytes)
SHA-1 [17]	579	198	1.022
SHA-1 [18]	177	122	1.352
SHA-256 [17]	783	416	1.598
SHA-256 [18]	335	158	2.720
Blake-32 [17]	1.115	245	6.684
Blake-32 [19]	324	251	1.804
Blake-32 [18]	263	206	2.076
Skein-256 [18]	287	123	2.464
Ours (Davies *et al.*)	50	48	144

The $\mathcal{MQ}$PKS implementation consists of two parts, including signature generation and verification. The signature generation produces private coefficients and computes a signature message. In the case of signature verification, the signature is verified by checking the validity of the provided signature. By introducing AES-based PRNG, we significantly reduced the private key size, and with the optimized implementation, we show performance enhancements in both the signature generation and verification parts. The following subsections describe the optimization and implementation methods in detail.

Parallel Computation

Using the parallel feature of the AES accelerator, we can compute a signature while generating the private coefficients. Signature generation on the embedded processor is described in Figure 1. First, the message is hashed using the hash function. The output is 16 bytes each time and takes 375 clock cycles. Central map computation is then executed, while vinegar variables and private coefficients are generated. These operations are conducted in independent modules, so we can compute both operations together. For this reason, we do not need any additional computation costs to generate the private coefficients. After central map computation, we generate the coefficients of the linear map. As a result of this, the overheads of the key generation process are absorbed in the central and linear map computations. This parallel computation technique can be applied to the verification process for message hashing, which is described in Figure 2. We can the conduct hash function computing the verification process, so one operation is absorbed into other operations.

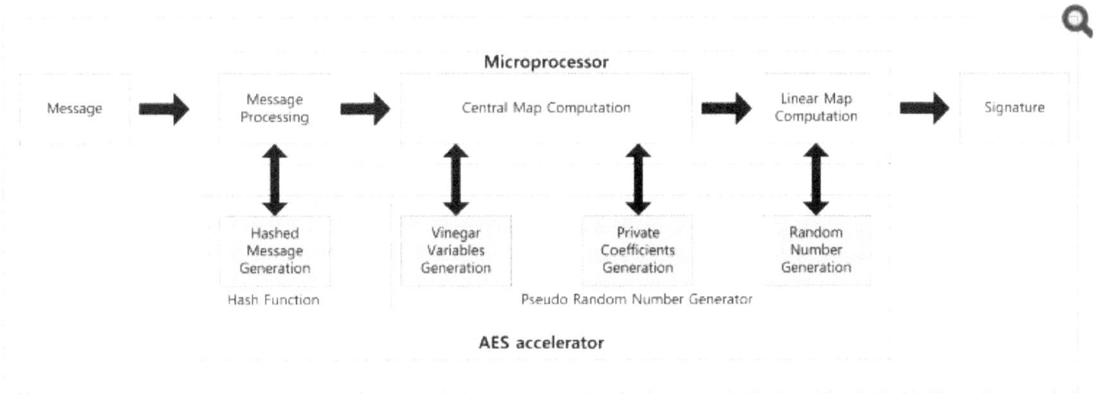

Figure 1. Signature generation process on a microprocessor.

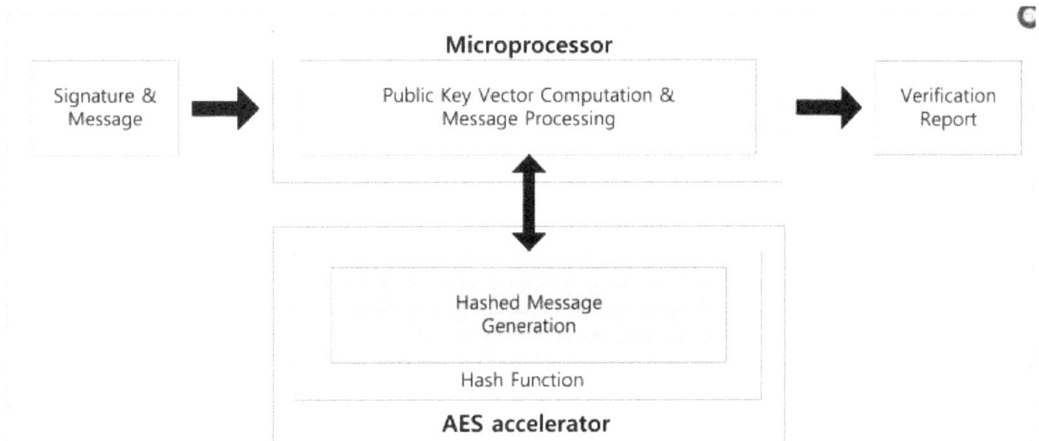

Figure 2. Signature verification process on a microprocessor.

IMPLEMENTATION OF SMALL PRIVATE KEY $\mathcal{MQ}$PKS

Logarithm Representation

Multiplication and inversion operations on GF(2^8) can be easily computed using logarithm representation in Algorithm 2, which transforms multiplication to a simple addition operation.

Algorithm 2: Implementation of Gaussian Elimination.

Input: Coefficients of Gaussian map g(i,j) , message mi ,

where $1 < i, j \leq o$, symbol o denotes the number of oil variables, the upper subscript describes the representation transition and the bottom subscript denotes the index. Steps from 1 to 18 describe forward elimination and steps from 19 to 26 describe backward elimination.

Output: Result r_i of Gaussian elimination.

for $i = 1$ to $(o - 1)$ do
for $t = i$ to o do
$temp\_g^l = inv^{n\to l}(g^n_{(t,i)})$
for $k = i$ to o do
$g^l_{(t,k)} = temp\_g^l + log^{n\to l}(g^n_{(t,k)})$
$g^n_{(t,k)} = exp^{l\to n}(g^l_{(t,k)})$
end for
$m^l_t = temp\_g^l + log^{n\to l}(m^n_t)$
$m^n_t = exp^{l\to n}(m^l_t)$
end for
for $k = i + 1$ to o do
for $t = i$ to o do
$g^n_{(k,t)} = g^n_{(i,t)} \oplus g^n_{(k,t)}$
end for
$m^n_k = m^n_i \oplus m^n_k$
end for
end for
$r^n_o = exp^{l\to n}(log^{n\to l}(m^n_o) + inv^{n\to l}(g^n_{(o,o)}))$
$count = 1$
for $i = o - 2$ to 0 do
for $j = o - 2$ to $o - 1 - count$ do
$r^n_i = r^n_i \oplus exp^{l\to n}\left(log^{n\to l}(r^n_{j+1}) + log^{n\to l}(g^n_{(i,j+1)})\right)$
end for
$r^l_i = r^n_i \oplus m^n_i$
$count = count + 1$
end for

To compute multiplication, values are first converted into their logarithm form by looking up the logarithmtable. Then, the relevant values are added, and the sum is returned to a normal representation by looking up the exponential table.

The representation setting is selected in an optimized way when we generate the private coefficients, vinegar values and coefficients of the linear map. The values are randomly generated and are considered to be inlogarithm form, because the private coefficients, linear map coefficient and vinegar values are directly multiplied from the first computation. Thus, storing values in logarithm representation is more efficient than leaving them in their normal representation when we consider the next operation. This method has one more advantage. In the logarithm representation, we can express the additional value of "0", which does not exist in the logarithm look-up table, so it cannot be used in the normal representation. However, the value exists in the exponential table, so we can use this representation for private coefficients.

To find the inverse of a value, we can use an inversion table, which transforms the value using the logarithmtable and then subtracts 0xff (255) before returning the resulting value to a normal representation. In our implementation, we use the modified inversion table described in Table A1, which outputs results inlogarithm representation with input variables in normal representation. These directly multiply the inverse value in the Gaussian elimination process described in Algorithm 3, in which the first column is inverted and then multiplied with the remaining values in the same row, thus setting the first column to one. After that, each row is bit-wise exclusive-ORed with other rows. This process, called forward elimination, continues to generate triangular form. In the backward elimination, from the last row, the equation is solved by each row.

Algorithm 3: Multiplication algorithm using logarithm representation written in assembly where ADD, ADC and CLR is addition, addition with carry and clear and Rd and Rr are destination and source registers, (ADD Rd, Rr : Rd←Rd+Rr, ADC Rd, Rr: Rd←Rd+Rr+C, CLR Rd: Rd←Rd ⊕Rd, Rd: destination register, Rr: source register, r_1 is cleared.)	
Input: Unsigned bytes $A^l(R_2)$, $B^l(R_3)$	
Output: Unsigned byte $A^l(R_2) = A^l(R_2) + B^l(R_3)$	
1.	ADD R_2, R_3
2.	ADC R_2, R_1

Assembly Programming

Assembly programming generally exhibits higher performance than high-level programming, such as C and JAVA. This is because we can optimize register allocation and use the status register, which provides a carry bit to determine a certain condition. In our implementation, we adopt assembly programming throughout the whole process to reach to highest performance. Multiplication in logarithm representation can be simplified in addition by using the assembly described in Algorithm 3. First, register r_1 is reset. Second, the operands are added. However, if the result is bigger than 0xff, an addition operation on the eight-bit microprocessor generates an output that is subtracted from 256, setting the carry bit. Therefore, this does not give the expected result. However, conducting an addition with the carry bit on the results, we can output the correct result, as we expected.

There is another case that exists. Algorithm 4 describes the exception condition. After central map computations, every parameter is stored into normal representation. The representation cannot map a zero variable into logarithm representation, so we should conduct exception handling for the zero variable. In this case, we directly output zero as a result without computation. In Algorithm 4, Step 1 clears destination register R_8. From Steps 2 to 5, input variables (R_2 and R_3) are compared with zero (R_1) to determine the zero variable. From Steps 6 to 13, logarithm mapping is conducted. In Steps 14 and 15, multiplication of R_2 andR_3 is conducted in logarithm representation. In Steps 16 to 19, mapping to normal representation is conducted.

Algorithm 4: Exception handling in multiplication for zero variables, where CP, BRCC and MOVW are compare, branch with carry and move register, (CP Rd, Rr: Rd-Rr, BRCC k: if (C=0) then PC ← PC+k+1, MOVW Rd, Rr: Rd+1:Rd ← Rr+1:Rr, ADD Rd, Rr: Rd←Rd+Rr, ADC Rd, Rr: Rd←Rd+Rr+C, CLR Rd: Rd←Rd ⊕Rd, Rd: destination register, Rr: source register, k: label, R_1 is cleared, R_4, R_5 indicate logarithm table, R_6, R_8 indicate exponential table, R_{28}, R_{29} indicate Ypointer, ZERO: label name.)	
Input: Unsigned bytes $A^n(R_2)$, $B^n(R_3)$	
Output: Unsigned byte $C^n(R_8) = A^n(R_2) \times B^n(R_3)$	
1.	CLR R_8
2.	CP R_1, R_2
3.	BRCC **ZERO**
4.	CP R_1, R_3
5.	BRCC **ZERO**
6.	MOVW R_{28}, R_4
7.	ADD R_{28}, R_2
8.	ADC R_{29}, R_1
9.	LD R_2, Y
10.	MOVW R_{28}, R_4
11.	ADD R_{28}, R_3
12.	ADC R_{29}, R_1
13.	LD R_3, Y
14.	ADD R_2, R_3
15.	ADC R_2, R_0
16.	MOVW R_{28}, R_6

17.	ADD R_{28}, R_2
18.	ADC R_{29}, R_1
19.	LD R_8, Y
20.	**ZERO:**

Vinegar Monomials

Central map computation is multiplying private coefficients with vinegar variables by the following equation:f(k)(u1,... ,un):=∑1≤i≤j≤nγ(k)ijuiuj=uTδ(k)u. These vinegar computations are executed in every central map computation, so vinegar monomials can reduce the vinegar variable computations throughout the whole processes by removing multiplication operations between each vinegar variable. Algorithm 5 describes the pre-computation of vinegar variables, in which we can generate vinegar monomials. The vinegar variables are generated in logarithm form. We left vinegar monomials as alogarithm form, because these variables are directly used for multiplication operations in $\sum_{i\in V, j\in V} \gamma_{ij} p_{(i,j)}$, where γ and p are the private coefficient and vinegar monomials, respectively.

Algorithm 5: Pre-computation of vinegar variables; symbols V and n denote the number of vinegar variables and the total number of vinegar and oil variables, respectively.	
Input: Vinegar variables u_i.	
Output: vinegar monomials $p_{(i,j)}$, $0 < i, j < V$.	
1.	for i = 1 to V do
2.	for j = i to V do
3.	$p_{(i,j)} = u_i \times u_j$
4.	end for
5.	end for

On-the-Fly Computation

$\mathcal{MQ}$PKS requires a large key size. If we firstly compute all private keys before we use them, these occupy a huge storage amount for retaining these values. To avoid this condition, we selected the on-the-fly method, which generates private keys, and then, these keys are used directly. The AES-based PRNG that we chose generates 16 bytes of secret information every computation. These variables are directly used for central and linear map computation. For efficient computation, we divide central map computation into vinegar and oil parts. Firstly, the vinegar part is computing ∑1≤i≤j≤υγ(k)ijuiuj with vinegar coefficients, vinegar monomials and message variables, and then, the oil part is executing ∑1≤i≤υ,1≤j≤oγ(k)ijui with vinegar and private coefficients.

Overview of the Computation Process

In Figure 3, briefly, we describe the representation of variables for each computation. Firstly, in Figure 3a, a message is hashed and outputted in normal representation. In the case of vinegar variables, logarithmrepresentation is selected. In Figure 3b, $\sum_{i\in V, j\in V} u_i u_j$ are the vinegar monomials for efficient computation and are stored in logarithm form. In Figure 3c, we firstly compute $\sum_{i\in V, j\in V} \gamma_{ij} p_{(i,j)}$, where γ and p are private coefficient and vinegar monomials, respectively. After that, we compute the remaining part, $\sum_{i\in V, j\in OV} \gamma_{ij} u_{(i,j)}$, to complete central map computation. In Figure 3d, Gaussian elimination is conducted with the results of the previous step. Finally, in Figure 3e, we generate linear map coefficients in logarithmrepresentation and then compute the linear map computations.

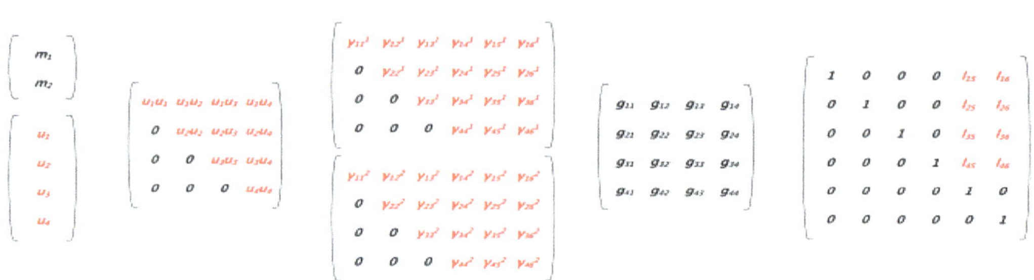

(a) (b) (c) (d) (e)

Figure 3. Overview of the representation of the variables for each computation: (**a**) message (normal), vinegar variables (logarithm); (**b**) vinegar monomials (logarithm); (**c**) private coefficients(logarithm); (**d**) Gaussian coefficient (normal); (**e**) linear map coefficients ...

RESULTS

In this section, we provide evaluation results on a UOV scheme implementation in terms of memory consumption and computational complexity. Memory consumption is mainly used for key storage. The private key size is o(oυ+υ(υ+1)2)+oυ for central map and linear map coefficients.

To compute the signature generation, previous implementations stored the private key in memory. However, our implementation stores only the seed values for the random number generator used for secret coefficient generation. For this reason, we can reduce the size of the private key by up to 99.9%. Furthermore, we show a performance enhancement by about 5.78% and 12.19% in signature generation and verification, respectively. This performance is achieved by adopting various optimization techniques that we explored before. The detailed evaluation results are available in Table 3 and 4,, respectively.

Table 3. Implementation results of the signature generation of the UOVscheme.

Scheme	Private Key (Byte)	Cycles $\times 10^3$	Time (ms) 32 MHz	ROM (Byte)	RAM (Byte)	Program Language
UOV (21,28) [20]	21,462	1,615	50.49	2,188	441	C + ASM
0/1 UOV (21,28) [20]	21,462	1,577	49.29	2,258	441	C + ASM
Our UOV (21,28)	**48**	**1,486**	**46.37**	**4,814**	**2,499**	**ASM**

Table 4. Implementation results of the signature verification of the UOV scheme.

Scheme	Private Key (Byte)	Cycles $\times 10^3$	Time (ms) 32 MHz	ROM (Byte)	RAM (Byte)	Program Language
UOV (21,28) [20]	25,725	1,690	52.83	466	n/a	C + ASM
0/1 UOV (21,28) [20]	4,851/20,874	1,395	43.60	578	n/a	C + ASM
Our UOV (21,28)	**25,725**	**1,225**	**38.30**	**2,069**	**562**	**ASM**

Computational Costs

Table 5 shows a detailed analysis of the computation costs for each operation. We separate whole computations into six categories. In central map computation, we divide into vinegar and oil parts. In Gaussian elimination, we divide into forward and backward eliminations. Our method exploits the AES operation for PRNG, so computations that need many private coefficients take many clock cycles. In our implementation, central map computation is the most expensive operation, because the private coefficient size is o(oυ), and it needs 882 AES operations. This result could be reduced more, because we compute two AES operations (32 outputs) in each column for 28 coefficients and did not use four remaining outputs, due to the difficult variable handling. If we fully use 32 outputs, the speed would be improved further.

Table 5. Computation costs for each operation in the case of UOV (21,28) on the ATxmega128a1.

Operation	Clock Cycle	Proportion (%)	Number of AES Operation
Vin-genand pre-com	9,396	0.63	2
Central-vinegar	514,070	34.59	546
Central-oil	**769,107**	**51.75**	**882**
Gaussian-forward	128,253	8.69	-
Gaussian-backward	1,460	0.09	-
Linear	62,896	4.23	42
Total	1,486,182	100	1,367

Source Code Storage (ROM) Requirements

Table 3 shows the reduction in size of the private key compared with traditional implementations. The size of the seed is computed with the following requirements (security level bit × 3, two for PRNG and one for the hash function). For all parameter variations, our implementation shows a 99.9% reduction in private key size. This is because our method generates private coefficients on the spot, so we store small seed values instead of whole private key values.

Variable Storage (RAM) Requirements

RAM stores persistent, counting or temporary variables for computation. The minimal RAM requirements are those for storing Gaussian elimination variables, which are generally of o^2 complexity. For better performance, we used more RAM to compute the vinegar variables and look-up tables. The computed vinegar variables, which multiply two vinegar variables, are used several times in the central map, so maintaining these values is more beneficial than computing them each time. These have a size of υ(υ+1)2. Second, the logarithm, exponential and inversion tables are used for converting the representations, and each table has a size of 256 bytes. By storing values in RAM, we can access data that is frequently required with lower overheads. The detailed information is available in Table 6.

Table 6. Minimal and our RAM requirements Minimal and our RAM requirements in bytes.

Scheme	UOV (21,28)	UOV (28,37)	UOV (44,59)	General
Minimal	441	784	1,936	o^2
Our	1,615	2,255	4,474	$o^2 + \frac{\upsilon(\upsilon+1)}{2} + 3(16 \times 16)$

Security Analysis

In this paper, we used representative block cipher AES as a core operation of random number generator and hash function to reduce the private key size. Therefore, the security levels highly rely on the strength of the block cipher. Recently, the vulnerability of a symmetric cryptosystem toward a quantum system has been proven by applying Grover's algorithm to break a symmetric algorithm by brute force, requiring $2^{n/2}$ of time, where n is the security bit [21,22]. For this reason, we should select a double-bit size to maintain the same security level of $\mathcal{MQ}$PK. This is the main strength of a quantum cryptosystem. To meet the n-bit security level of $\mathcal{MQ}$PK, we should select at least a 2n-security level. For this reason, we selected 128-bit AES encryption to meet the 64-bit security level of $\mathcal{MQ}$PK. To ensure the 96-, 128-bit security level, we could use 192-, 256-bit AES operations, which are also available in modern microprocessors. There is no specific conference and journal described.

Impacts on Other Protocols and Target Devices

We selected ATxmega128a1 as a target device to implement $\mathcal{MQ}$PKS. This does not mean that our method is limited to only the ATxmega128a1 board. There are only two requirements that exist for applying our methods. First, the microprocessor should support above an eight-bit word size, because our algorithm requires at least an eight-bit word size to use optimal GF(2^8). Recently, eight-, 16-, 32-bit machines have been most widely used in embedded environments. The representative target devices in eight-, 16- and 32-bit are the XMEGA, MSP and ARM series, respectively. This means the majority of the embedded system could be improved by our methods. Second, the AES accelerator should be embedded in target devices. This requirement is also commonly met in modern microprocessors. Previously, we mentioned the representative target devices, including XMEGA, MSP and ARM v8, provide the AES accelerator as a peripheral.

In the case of this scheme, our method would have huge impacts on other $\mathcal{MQ}$PKS. We implemented the UOV scheme in this paper, presenting private key reduction methods. This could be applied to other schemes, including UOV, Rainbow and enTTS, without difficulty. Because these are variant of the UOV scheme, the key generation process is similar to the UOV scheme. Furthermore, this is not limited to the size of the finite field, because our method is generally ideal; so it could be extended to other fields, as well.

CONCLUSIONS

The majority of previous results focused on small public key $\mathcal{MQ}$PKS implementations. However, no practical results on the reduction of private key size in embedded microprocessors have been reported. In this paper, we presented a novel parallel computing method using a block cipher-based random number generator and a hash function to reduce the size of the private key and to boost speed performance. The method generates private coefficients, computing the central and linear maps simultaneously, because the AES accelerator embedded in modern microprocessors, including the ATxmega, MSP430 and ARMv8 series, can compute AES operations independently with the microprocessor. The results showed a significant reduction in private key size and enhancement in computation costs for signature generation and verification. These results can be applied to other schemes, such as Rainbow and enTTS, to generate private coefficients. Future work involves implementing this scheme on recent platforms, including Compute Unified Device Architecture (CUDA), Open Computing Language (OpenCL), NEON, Streaming SIMD Extensions (SSE) and Advanced Vector Extensions (AVX).

ACKNOWLEDGMENTS

This work was supported by the Industrial Strategic Technology Development Program (no. 10043907, Development of high performance IoT device and Open Platform with Intelligent Software) funded by the Ministry of Science, ICT & Future Planning (MSIF, Korea).

APPENDIX

Inversion Table

Table A1 describes the inversion look-up table on GF(2^8). I = The input is the normal representation, but the output value is the inverse of the input in logarithm representation form.

Table A1. Inversion table (input: normal representation, output: logarithm representation).

	0	**1**	**2**	**3**	**4**	**5**	**6**	**7**	**8**	**9**	**a**	**b**	**c**	**d**	**e**	**f**
0	—	00	e6	fe	cd	fd	e5	39	b4	38	e4	97	cc	11	20	fc
1	9b	fb	1f	f1	cb	72	7e	10	b3	8e	f7	37	07	96	e3	3e
2	82	3d	e2	4a	06	46	d8	95	b2	1b	59	8d	65	36	f6	87
3	9a	d0	75	fa	de	f0	1e	db	ed	0f	7d	ba	ca	6c	25	71
4	69	70	24	42	c9	2f	31	6b	ec	a3	2d	0e	bf	b9	7c	c7
5	99	22	02	cf	40	f9	74	9d	4c	da	1d	67	dd	77	6e	ef
6	81	91	b7	3c	5c	49	e1	bd	c5	94	d7	ab	05	7a	c2	45
7	d4	86	f5	ea	64	60	a1	35	b1	2b	53	1a	0c	8c	58	a8
8	50	a7	57	af	0b	15	29	8b	b0	51	16	2a	18	19	52	17
9	d3	28	8a	85	14	e9	f4	0a	a6	34	a0	4f	63	56	ae	5f
a	80	f3	09	90	e8	3b	b6	13	27	bc	e0	d2	5b	89	84	48
b	33	44	c1	a5	04	9f	4e	79	c4	ad	5e	93	55	aa	d6	62
c	68	4d	78	6f	9e	41	23	03	43	6a	30	32	c8	c0	a4	2e
d	ac	c6	7b	c3	be	5d	92	b8	eb	d5	61	a2	a9	0d	2c	54
e	bb	ee	6d	26	dc	df	d1	76	4b	83	47	d9	88	66	1c	5a
f	98	b5	12	21	3a	ce	01	e7	f2	9c	73	7f	3f	08	8f	f8

Author Contributions

Jihyun Kim and Jongseok Choi contributed to algorithm analysis. Taehwan Park and Zhe Liu contributed to implementation of the cryptography. The professor, Howon Kim, gave fruitful advices to implement the cryptography.

Conflicts of Interests

The authors declare no conflict of interest.

REFERENCES

1. Koblitz N. Elliptic curve cryptosystems. Math. Comput. 1987;48:203–209.
2. Miller V.S. Advances in Cryptology—CRYPTO85 Proceedings. Springer; Berlin, Germany: 1986. Use of Ellipticin Curves in Cryptography; pp. 417–426.
3. Ding J., Schmidt D. Advances in Information Security. Citeseer; Forest Grove, OR, USA: 2006. Multivariate Public Key Cryptosystems.
4. Oliveira L.B., Aranha D.F., Gouvêa C.P., Scott M., Câmara D.F., López J., Dahab R. Tinypbc: Pairings for authenticated identity-based non-interactive key distribution in sensor networks. Comput. Commun.2011;34:485–493.
5. Braeken A., Wolf C., Preneel B. Topics in Cryptology–CT-RSA 2005. Springer; Berlin, Germany: 2005. A Study of the Security of Unbalanced Oil and Vinegar Signature Schemes; pp. 29–43.
6. Kipnis A., Patarin J., Goubin L. Advances in Cryptology—EUROCRYPT99. Springer; Berlin/Heidelberg, Germany: 1999. Unbalanced Oil and Vinegar Signature Schemes; pp. 206–222.
7. Petzoldt A., Thomae E., Bulygin S., Wolf C. Cryptographic Hardware and Embedded Systems–CHES 2011. Springer; Berlin, Germany: 2011. Small Public Keys and Fast Verification for Ultivariate Mathcal {Q} Uadratic Public Key Systems; pp. 475–490.
8. Borges F., Petzoldt A., Portugal R. Small Private Keys for Systems of Multivariate Quadratic Equations Using Symmetric Cryptography. [(accessed on 10 January 2014)]. Avaliable online:http://www.informatik.tu-darmstadt.de/fileadmin/user_upload/Group_TK/UOV_cnmac2012-final.pdf.
9. Von Maurich I., Güneysu T. Progress in Cryptology-INDOCRYPT 2012. Springer; Berlin, Germany: 2012. Embedded Syndrome-Based Hashing; pp. 339–357.
10. Prescott T. Random Number Generation Using Aes. Technical Report, Atmel. [(accessed on 10 January 2014)]. Availabel online: http://www.atmel.com/ja/jp/Images/article_random_number.pdf.
11. Yang B.-Y., Cheng C.-M., Chen B.-R., Chen J.-M. Security in Pervasive Computing. Springer; Berlin, Germany: 2006. Implementing Minimized Multivariate Pkc on Low-resource Embedded Systems; pp. 73–88.
12. Yang B.-Y., Chen J.-M., Chen Y.-H. Cryptographic Hardware and Embedded Systems-CHES 2004.Springer; Berlin, Germany: 2004. Tts: High-speed Signatures on a Low-cost Smart Card; pp. 371–385.
13. Eisenbarth T., Kumar S. A survey of lightweight-cryptography implementations. IEEE Des. Test Comput. 2007;24:522–533.
14. Osvik D.A., Bos J.W., Stefan D., Canright D. Fast Software Encryption. Springer; Berlin, Germany: 2010. Fast software aes encryption; pp. 75–93.
15. Rukhin A., Soto J., Nechvatal J., Smid M., Barker E. A Statistical Test Suite for Random and Pseudorandom Number Generators for Cryptographic Applications. National Institute of Standards and Technology; Gaithersburg, MD, USA: 2001.

16. Preneel B., Govaerts R., Vandewalle J. Advances in Cryptology—CRYPTO93. Springer; Berlin, Germany: 1994. Hash Functions Based on Block Ciphers: A Synthetic Approach; pp. 368–378.
17. Otte D. Avrcryptolib.2009. Technical Report. 2009. [(accessed on 10 January 2014)]. Availabel online:http://www.daslabor.org/wiki/AVRCryptoLib/en.
18. Osvik D.A. Fast embedded software hashing. IACR Cryptol. ePrint Arch. 2012;156:2012.
19. Eisenbarth T., Heyse S., Von Maurich I., Poeppelmann T., Rave J., Reuber C., Wild A. Evaluation of Sha-3 Candidates for 8-bit Embedded Processors. Proceedings of the Second SHA-3 Candidate Conference; Santa Barbara, CA, USA. 23–24 August 2010.
20. Czypek P., Heyse S., Thomae E. Cryptographic Hardware and Embedded Systems–CHES 2012.Springer; Berlin, Germany: 2012. Efficient Implementations of Mqpks on Constrained Devices; pp. 374–389.
21. Bennett C.H., Bernstein E., Brassard G., Vazirani U. Strengths and weaknesses of quantum computing.SIAM J. Comput. 1997;26:1510–1523.
22. Kobayashi H., Gall F.L. Dihedral hidden subgroup problem: A survey. IPSJ Digital Courier.2005;1:470–477.

CHAPTER 19

PERFORMANCE OF MULTI-CORE PROCESSORS

Nitin Chaturvedi and S. Gurunarayanan

Keywords: Chip Multiprocessor (CMP), Multiple-Chip Multiprocessor (M-CMP), Tiled Architecture

INTRODUCTION

Present sub-micron integrated circuit technologies have fueled microprocessor performance growth [3, 4]. Each new process technology increases the integration density allows for higher clock rates and offers new opportunities for micro-architectural innovation. Both of these are required to maintain microprocessor performance growth. Micro-architectural innovations employed by recent microprocessors include multiple instruction issue, dynamic scheduling, speculative execution, instruction level parallelism [6] and non-blocking caches. In the past, we have seen the trend towards CPUs with wider instruction issue and support for larger amounts of speculative execution but due to fundamental circuit limitations and limited amounts of instruction level parallelism, the superscalar execution model provides diminishing returns in performance for increasing issue width. Faced with this situation, building further a more complex wide issue superscalar processor was not at all the efficient use of silicon resources and a better utilization of silicon area. So researchers came up with a novel architecture which was constructed from simpler processors then super-scalar and multiple such processors are integrated on a single chip popularly known as chip multiprocessor or multi-core processors. To understand the performance trade-offs between wide-issue processors and single chip multiprocessors in a more quantitative way, researchers had compared performance of a six-issue dynamically scheduled superscalar processor with a 4 ´ two-issue multiprocessor. Comparison has a number of unique features. The results show that on applications that cannot be parallelized, the superscalar micro-architecture performs better than one processor of the multiprocessor architecture. For applications with fine grained thread-level parallelism the multiprocessor microarchitecture can exploit this parallelism so that the super-scalar micro-architecture is at most 10% better. For applications with large grained thread-level parallelism and multiprogramming workloads the multiprocessor micro-architecture performs 50–100% better than the wide superscalar micro-architecture. Today data centres powering web and transaction services face new requirements as they attempt to deliver higher levels of content-rich and high-bandwidth services to increasing numbers of users. These workloads share certain common characteristics. They exhibit high Thread-Level Parallelism (TLP) with multiple and independent processes running concurrently. Moreover, servers equipped with even more powerful and power-hungry microprocessors to meet higher computational demands are pushing the power and cooling capabilities of these servers to their limits, resulting in increased operating costs and decreased system reliability. Therefore for achieving high performance while maintaining existing power and thermal envelopes requires novel microprocessor designs that not only focuses on performance but rather on the aggregate performance per watt. The rest of the paper is organized as follows: In Section 2, the performance limits of superscalar design are discussed and it also presents the case for a single chip multiprocessor from an applications perspective and extended to M-CMP. Section 3 gives details of Simulation methodology. Section 4 presented the obtained results and finally Section 5 concludes.

Citation: Nitin Chaturvedi, S Gurunarayanan, Performance of Multi Core Processors, Doi 10.5121/ijdps.2013.4404.

LITERATURE REVIEW

Super-Scalar Processors

A trend in the microprocessor industry has been the design of CPUs with multiple instruction issue and the ability to execute instructions out of program order. This ability, called dynamic scheduling uses hardware to track register dependencies between instructions; an instruction is executed, possibly out of program order, as soon as all of its dependencies are satisfied. The register dependency checking was done with a hardware structure called the scoreboard. IBM used register renaming to improve the efficiency of dynamic scheduling using hardware structures called reservation stations [2]. It is possible to design a dynamically scheduled superscalar microprocessor using reservation stations, the most recent implementations of dynamic superscalar processors have used a structure similar to the one shown in Figure 1, which shows three major phases of instruction execution in a dynamic superscalar machine they are fetch, issue and execute.

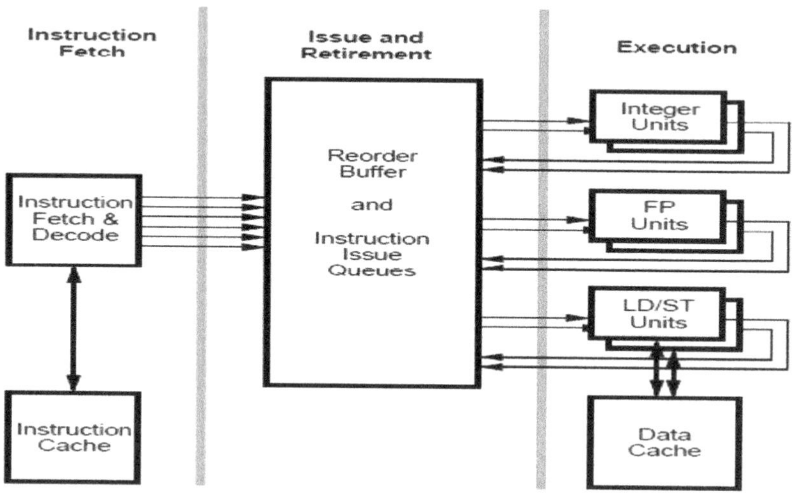

Figure1. Dynamic Super Scalar Microprocessor.

The goal of the fetch phase is to present the rest of the CPU with a large and accurate window of decoded instructions. Three factors constrain instruction fetch: miss predicted branches, instruction misalignment, and cache misses. The ability to predict branches correctly is crucial to establishing a large, accurate window of instructions. However, good branch prediction is not enough. As previous work has pointed out, it is also necessary to align a packet of instructions for the decoder. Even with good branch prediction and alignment a significant cache miss rate will limit the ability of the fetcher to maintain an adequate window of instructions. Previous research have shown that over 60% of the instruction cache miss latency can be hidden on a database benchmark with a 64KB two way set associative instruction cache. In the issue phase, a packet of renamed instructions is inserted into the instruction issue queue. An instruction is issued for execution once all of its operands are ready. The results have shown a quadratic increase in the size of the instruction issue queue and researchers believe that the instruction issue queue will fundamentally limit the performance of wide issue superscalar machines. In the execution phase, operand values are fetched from the register file or bypassed from earlier instructions to execute on the functional units. The wide superscalar execution model will encounter performance limits in the register file, in the bypass logic and in the functional units. Wider instruction issue requires a larger window of instructions, which implies more register renaming. Not only must the register file be larger to accommodate more renamed registers, but the number of ports required to satisfy the full instruction issue bandwidth also grows with issue width. Again, this causes a quadratic increase in the complexity of the register file with increases in issue width.

Single Chip Multiprocessor

In today's information era, commercial servers constitute the backbone of the global information and communication system infrastructure. Such servers run useful commercial applications that are essential to many aspects of everyday life such as banking, airline reservations, web searching and web browsing. As more people depend on these multi-threaded throughput-oriented applications, demand for more throughputs is likely to increase for the foreseeable future. Commercial servers must therefore improve their performance by providing more throughputs to keep up with the application demand.

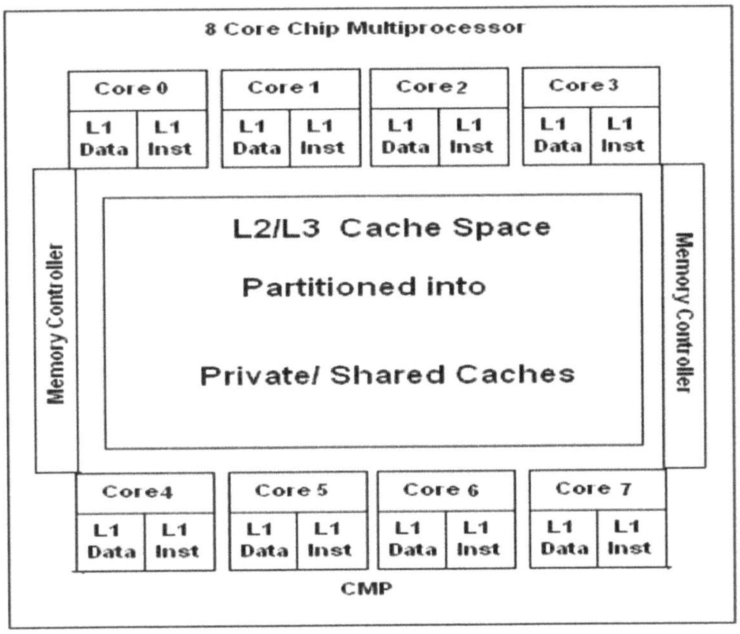

Figure 2. Single Chip Multiprocessor (CMP)

Since commercial applications have abundant thread-level parallelism, commercial servers were designed as multiprocessor systems—or clusters of multiprocessors—to provide sufficient throughput. While traditional symmetric multiprocessors (SMPs) can exploit thread-level parallelism, they also suffer from a performance penalty caused by memory stalls due to cache misses and cache-to-cache transfers, both of which require waiting for long off-chip delays. Several researchers have shown that the performance of commercial applications and database applications in particular, is often dominated by sharing misses that require cache-to-cache transfers. To avoid these overheads, architects proposed several schemes to integrate more resources on a single chip. Researchers have shown that chip-level integration of caches, memory controllers, cache coherence hardware and routers can improve performance of online transaction processing workloads by a factor of 1.5. Simultaneous multithreading designs allow the processor to execute several contexts (or threads) simultaneously by adding per-thread processor resources. This approach also improves the performance of database applications compared to a superscalar processor with comparable resources. The trend towards more integration of resources on a single chip is becoming more apparent in CMP designs where multiprocessor systems are built on a single chip. Chip multiprocessor (CMP) systems can provide the increased throughput required by multi-threaded applications while reducing the overhead incurred due to sharing misses in traditional shared-memory multiprocessors. A chip multiprocessor design is typically composed of two or more processor cores (with private level-one caches) sharing a second- level cache. CMPs in various forms are becoming popular building blocks for many current and future commercial servers. The increasing number of processor cores on a single chip increases the demand on two critical resources: the shared cache capacity and the off-chip pin bandwidth.

Multiple-Chip Multiprocessor (M-CMP)

The increasing number of transistors per chip now enables Chip Multiprocessors (CMPs), which implement multiple processor cores on a chip. CMP-based designs provide high-performance, cost-effective computing for workloads with abundant thread-level parallelism, such as commercial server workloads. Smaller-scale Single-CMP (S-CMP) systems, such as Stanford Hydra and Sun UltraSparc T1 [1], use a single CMP along with DRAM and support chips. Larger-scale Multiple-CMP (M-CMP) systems, such as Piranha [7] and IBM Power4, combine multiple CMPs to further increase performance.

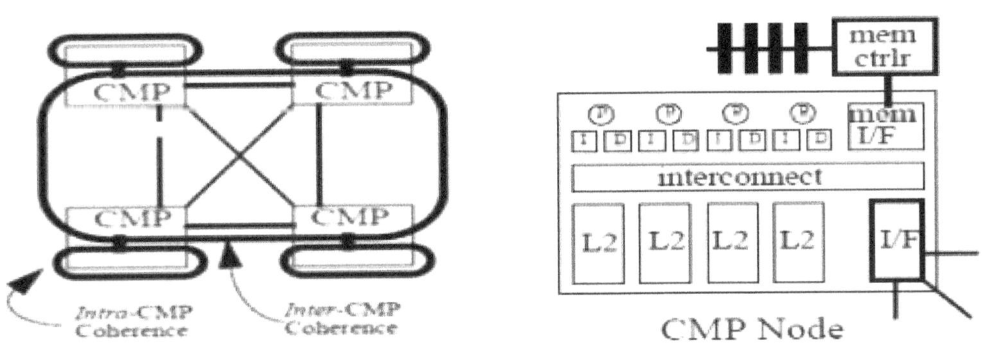

Figure 3. Multiple-CMP.

Because all of these systems use shared memory (to preserve operating system and application investment), a key challenge for M-CMP systems is implementing correct and high performance cache coherence protocols. These protocols keep caches transparent to software, usually by maintaining the coherence invariant that each block may have either one writer or multiple readers. S-CMP systems are conceptually straightforward and maintain coherence with traditional non-hierarchical snooping protocols (which rely on a logical bus) or directory protocols (which track cached copies at memory). M-CMPs present a greater challenge, because they must maintain both intra-CMP coherence and inter- CMP coherence

SIMULATION METHODOLOGY AND APPLICATION

For evaluating the performance of the novel CMP or M-CMP micro-architectures requires a way of simulating the environment in which we would expect these architectures to be used in real systems. We have used GEM5 [8] full system functional simulator extended with Multi-facet GEMS which is popularly used in the research community. The heart of GEM5 is the Ruby memory simulator.

- Execution time: The Ruby Cycle is our basic metric of simulated time used in the Ruby module of the GEMS simulator. The Ruby module is the basic module for the memory system configuration and interconnection network design The Ruby cycles are the recommended performance metric in GEMS.
- L1 cache misses: As the name says it represents the misses of L1 cache. It's calculated by dividing request missed by number of requests (Instruction + Data). It's an important metric for cache hierarchy.
- L2 miss/miss rate: This represents the total misses and miss rate of the L2 cache. It is calculated from the number of requests issued to the L2 and the misses of all banks of L2.
- L2/Dir replacement: Number of replacements of L2/ Directory entries. It's caused by capacity misses and conflict misses.
- Miss latency average: Average of the L1 miss latency in Ruby cycles. It is measured from the moment a memory request is issued to the moment when the date is retrieved.
- Memory requests: Number of reads and writes issued to main memory.
- Applications: The benchmark used is a multithreaded implementation of a simulation of 3D Lattice-Boltzmann magneto-hydrodynamics (in other words, plasma turbulence). Pthreads are used to implement threading.

RESULTS

In order to analyze the impact of various parameters on the performance of Multi-core Architectures, we have varied the number of cores, which in turn changes L2 cache size further we have varied directory size from 64 to 2048 entries on a four node, 8 node 16 node and 64 node Chip multiprocessor.

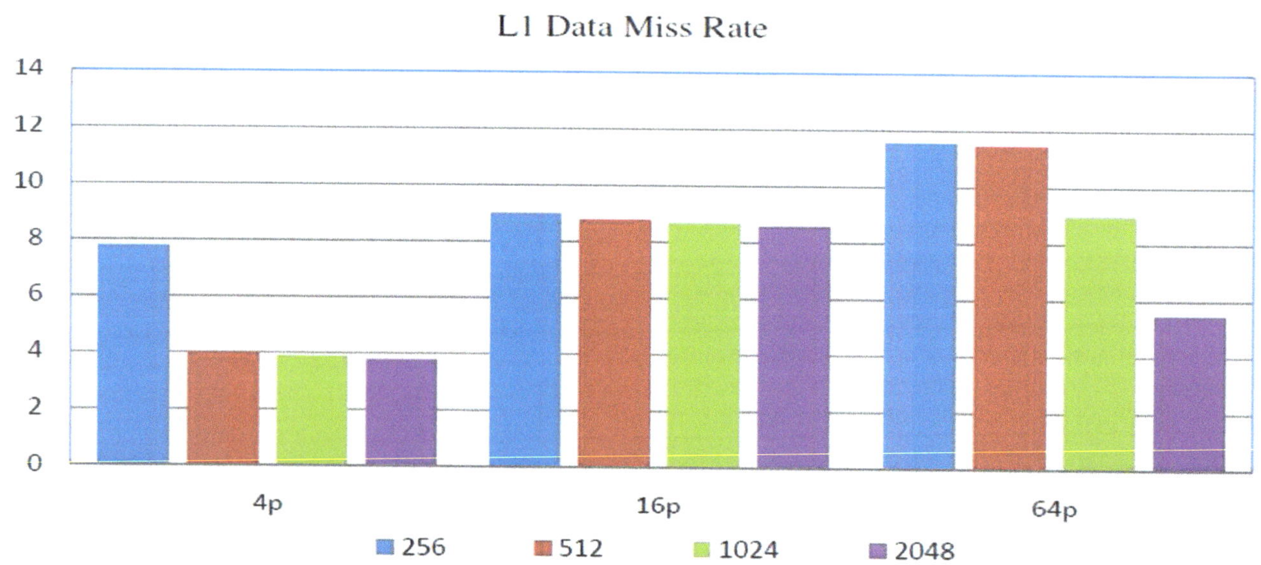

Figure 4. L1 Data Miss Rate.

All simulations in this section are configured with same cache size: 32KB L1I + 32KB L1D and 512KB L2 cache per core. The L1 data cache miss rate was calculated as dividing number of L1D misses by data requests. It was observed that L1 miss was greatly affected by directory si ze. It again influences L2 requests and on-chip traffics. Therefore, it's an important factor that determines CMP performance.

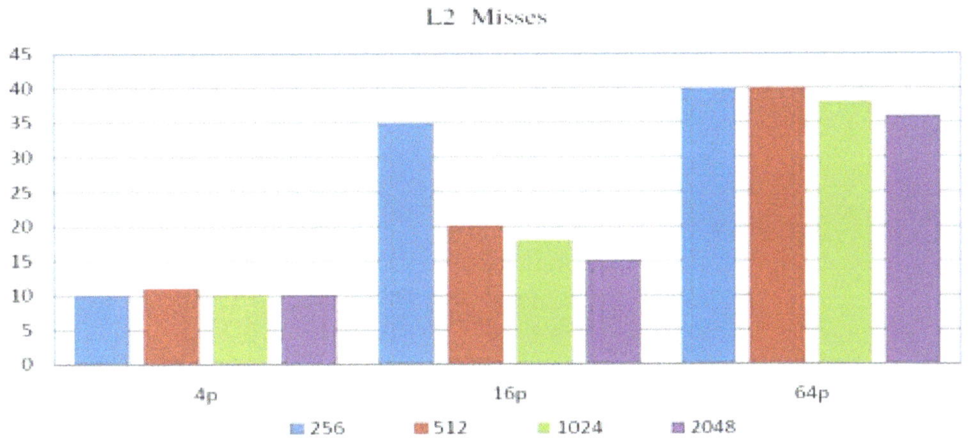

The number of L2 misses increases as no of cores within the system increases, nevertheless the miss rate decrease to 40%-80% because of the larger total L2 cache on chip

Figure6. L2 Miss Rate

It was observed that L2 miss rate decreases with increase in director size. L2 misses is mainly determined by L2 cache size and more importantly application working set.

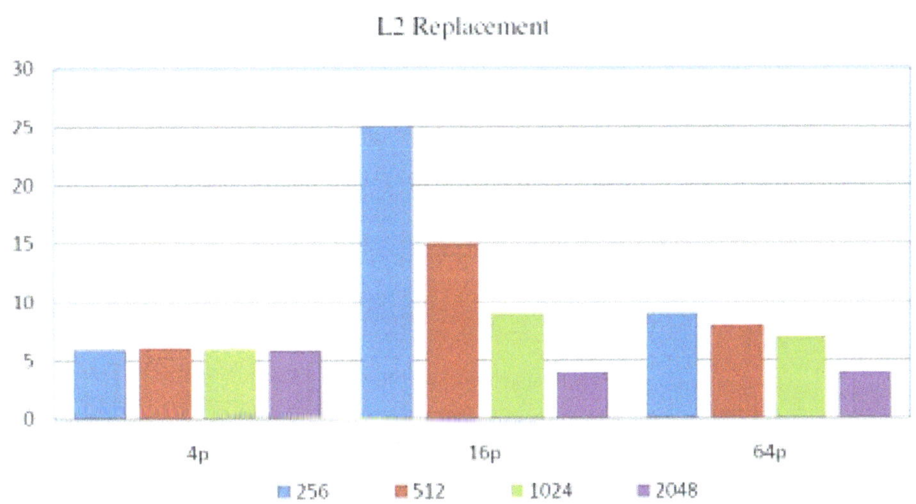

Figure 7. L2 Replacement

L2 Replacement occurs when the L2 cache is full and another allocation is required. According to LRU policy, a block will be chosen and replaced by a new one. If this block is clean just ignore it and process allocation without pause. Otherwise the data block needed to be written back to main memory. It was observed that the replacements are influence by directory and L2 size and on the applications.

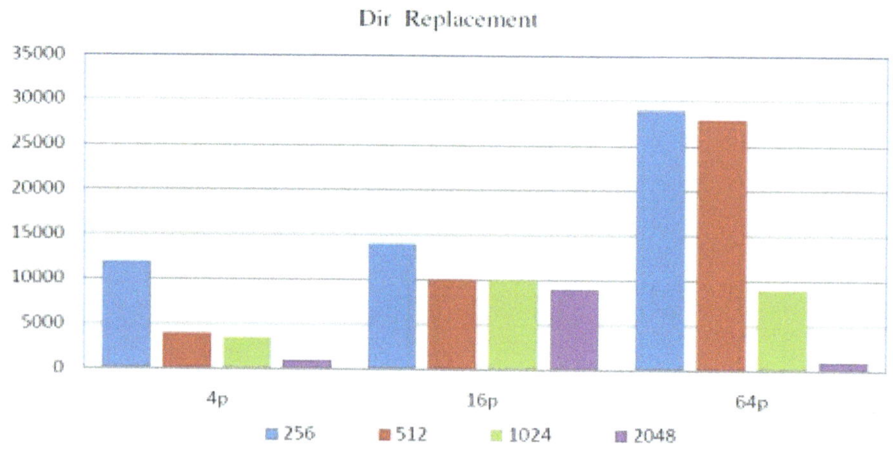

Figure8. Directory Replacements.

On 16 core node we have observed that larger directory size does not improve the directory replacements. The reason is that so many data are mapped to the same location resulting in many conflicts. One possible solution is increase set associativity of directory to avoid the conflicts. The trend seems normal in other two cases.

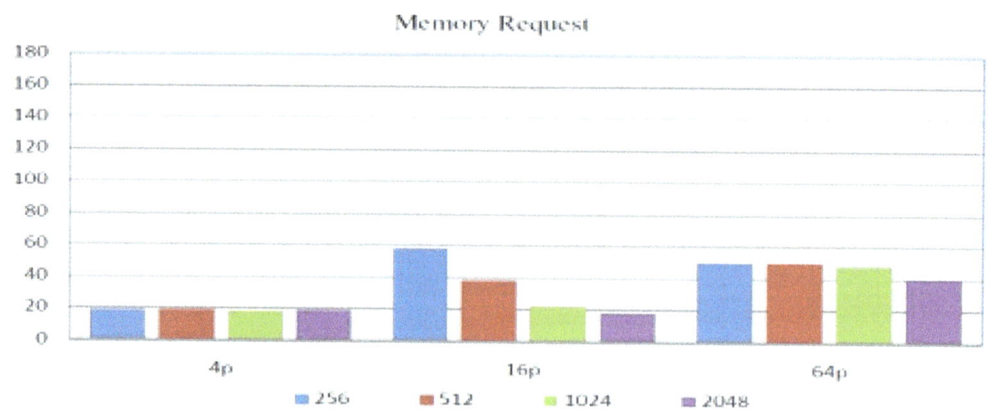

Figure 9. Memory reques.

Memory request to some extent represents the requirement of memory bandwidth of applications. Comparing with L2 misses in previous figure it is observed that the memory read requests caused by L2 miss is the dominant faction of total memory requests.

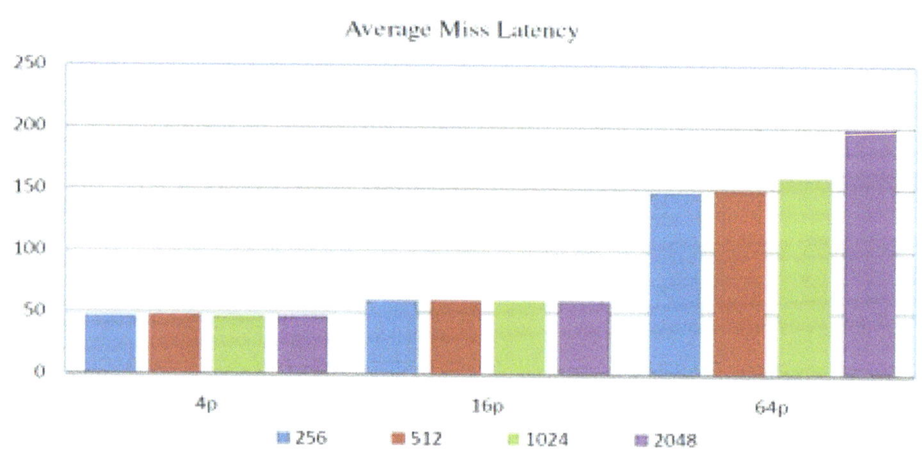

Figure10. Average Miss Latency

The miss latency increases 50% from 4-core to 16- core and 150%-230% from 16-core to 64- core. On a L1 miss, there are up to 3 nodes involved to fulfill the miss: local node, home node and remote node. The 3- way communication aggravates the latency of on-chip communication.

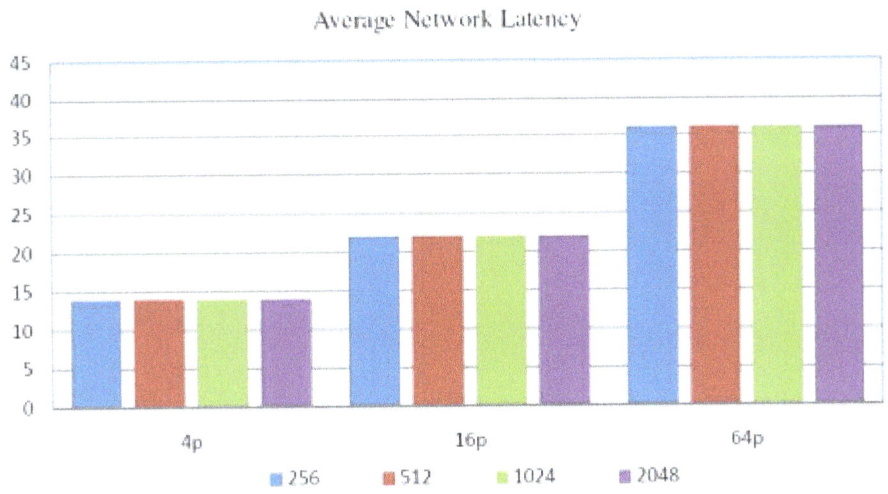

Figure11. Average Network Latency

With Hierarchical network model, we can measure the average network latency in detail. It goes from 14, 22 to 36 in there configurations. For all structure, the latency looks almost the same, which depends on network topology and on-chip link latency.

CONCLUSIONS

The current advanced submicron integrated circuit technologies require us to look for new ways for designing novel microprocessors architectures to utilize large numbers of gates and mitigate the effects of high interconnect delays. In this paper we have discussed the details of implementing both a wide, dynamically scheduled superscalar processor and a single chip multiprocessor. The alternative S-CMP is composed of simpler processors and can be implemented in approximately the same area. We believe that the S-CMP will be easier to implement and will reach a higher clock rate. Results show that on applications that cannot be parallelized the superscalar processor performs marginally better than one processor of the multiprocessor architecture. On applications with large grained thread-level parallelism and multiprogramming workloads the single chip multiprocessor performs 50–100% better than the wide superscalar processor. Novel architectures like M-CMP that exploits large bandwidth instead of extremely high frequency to achieve the target throughput performance by executing multiple threads concurrently in a shallow with simple pipeline effectively hides instruction and memory latency while utilizing resources. Results shows that the performance of these S-CMP and M-CMP depends on L2 cache size, no of directory entry and no of processors integrated on the single chip. Therefore, paper presents an open area of research on CMP to achieving high performance while maintaining existing power and thermal envelopes requirements and the designs must focus not only on performance but rather on the aggregate performance per watt, optimized interconnect topology, novel intra and inter CMP cache coherence protocol. Further performance gap between processors and memory require adaptive novel techniques to manage on chip cache memory judiciously

REFERENCES

1. S. Leon et al., "The UltraSPARC T1 processor: CMT reliability," in Proc. IEEE Custom Integrated Circuits Conf., Sep. 2006, pp. 555–562.
2. D. W. Anderson, F. J. Sparacio, and R. M. Tomasulo, "The IBM System/360 model 91: Machine philosophy and instruction-handling," IBM Journal of Research and Development, vol. 11, pp. 8–24, 1967.
3. T. C. Chen, "Where CMOS is going: trendy hype vs. real technology," presented at the IEEE Int. Solid-State Circuits Conf. (ISSCC), Sanrancisco, CA, Feb. 2006, Plenary Session 1.1.
4. M. Horowitz and W. Dally, "How scaling will change processor architecture," in IEEE ISSCC Dig. Tech. Papers, Feb. 2004, pp. 132–133.
5. K. Farkas, N. Jouppi, and P. Chow, "Register file considerations in dynamically scheduled processors," Proceedings of the 2nd Int. Symp. on High-Performance Computer Architecture, pp. 40–51, San Jose, CA, Feb 96.
6. D. W. Wall, "Limits of Instruction-Level Parallelism," Digital Western Research Laboratory, WRL Research Report 93/6, November 1993.
7. L. A. Barroso, K. Gharachorloo, R. McNamara, A. Nowatzyk, S. Qadeer, B. Sano, S. Smith,R. Stets, and B. Verghese. Piranha: A Scalable Architecture Based on Single-Chip Multiprocessing.In Proceedings of the 27th Annual International Symposium on ComputerArchitecture, pages 282–293, June 2000.
8. Milo M.K. Martin, Daniel J. Sorin, Bradford M. Beckmann, Michael R. Marty, Min Xu, Alaa R. Alameldeen, Kevin E. Moore, Mark D. Hill, and David A. Wood, "Multifacet's General Execution-driven Multiprocessor Simulator (GEMS) Toolset" Computer Architecture News (CAN), pages 92-99, November 2005.

INDEX

A

advanced configuration and power interface (ACPI) 64
Advanced Encryption Standard (AES) 203
Advanced Vector Extensions (AVX). 212
All-Digital Phase-Locked Loop (ADPLL) 74
application-specific integrated circuits (ASICs) 176
automatic tuning system 91

B

best-case execution time (BCET) 16
Boundary-Scan (BS) 137

C

cache memory 29, 32, 33, 34, 35, 36
central processing unit (CPU) 50
Chip-Multiprocessor 99
Chip multiprocessor (CMP) 217
chip multiprocessors (CMP) 59
chip multiprocessors (CMPs) 11
Clock Domain Crossings (CDC) 112
complex instruction set computing (CISC) 30
computed tomography (CT) 163
computer-aided design (CAD) 73
condensed simultaneity graph (CSG) 51
condensed simultaneity graph (DSG) 52
Conditional Partial Order Graph (CPOG) 176

D

dependency graph (DG) 52
design space exploration (DSE) 1
Digital Signature Algorithm (DSA) 203
dynamic thermal management (DTM) 59, 60
dynamic voltage and frequency scaling (DVFS) 60

E

electromechanical protective relays 185
electronic systems level (ESL) 99
Elliptic Curve Cryptography (ECC) 203
Elliptic Curve Diffie-Hellman (ECDH) 203
Elliptic Curve Digital Signature Algorithm (ECDSA) 203
Energy-efficiency 175

F

Factor of Unit Functionality\» (FUF) 185
field programmable gate array (FPGA) 49
field-programmable gate array (FPGA) 161, 176
Figure Of Merit (FOM) 47
finite state machine (FSM) 49
Finite State Machines (FSMs) 176
flash memory 29, 30, 31, 32, 34, 35, 36, 37, 38
Frequency Detector (FD) 44
Front-Side-Bus (FSB) 100, 104
functional-level power analysis (FLPA) 8

G

general-purpose processors (GPPs) 161, 162

H

hardware description languages (HDL) 49
high-performance reconfigurable computing (HPRC) 161

I

instruction level parallelism (ILP) 13
instruction set architecture (ISA) 12
instruction-set simulator (ISS) 1
integrated microcircuits (IC) 197
Intelligent Sensor InterFace (ISIF) 74

J

Joint Test Action Group (JTAG) 137

K

Kahn Process Network (KPN) 2

L

least-recently-used (LRU) 32
light-dependent sensor (LDR) 90
long timing effect (LTE) 13
Loop Filter (LPF) 42

M

maximum scheduling temperature (MST) 64
microprocessor-based protective devices (MPD) 193
microprocessor-based protective relays (MBR) 185
Module Test and Maintenance Bus (MTM-Bus) 137, 146
Motion Accurate Picture Processing (MAPP2) 123
Multiplying Delay Lock Loop (MDLL) 42

N

National Institute of Standards and Technology (NIST) 205
Networks on Chip (NoC) 111, 132
Noise Transfer Functions (NTFs) 42

P

Phase Frequency Detector (PFD) 42
Phase-Locked Loop (PLL) 74
Power Spectral Densities (PSDs) 42
printed circuit board (PCB) 186
printed-circuit-board (PCB) 198
programmable logic controller (PLC) 87
Programmable logic controllers (PLCs) 49
Pseudo-Random Number Generator (PRNG) 203
pulse width modulation (PWM) 75, 88

Q

Quality-of-Service (QoS) 175

R

RAM (Random Access Memory) 196
reconfigurable function unit (RFU) 13
register update unit (RUU) 3
Relay protection 193, 198, 200, 201
revolutions per minute (rpm) 89

S

Serial Vector Format (SVF) 138
simultaneity graph (SG) 52
simultaneous multithreading (SMT) 15
sinusoidal PWM (SPWM) 78
space vector pulse width modulation (SVPWM) 75
special-purpose processor (SPP) 161
stage, and extracts data memory (SRAM) 30
static random-access memory (SRAM) 29
symmetric multiprocessors (SMPs) 217
synergistic processors (SPs) 15
synthetic aperture radar (SAR) 161, 162, 163
system request queue (SRQ) 100
Systems on Chip (SoC) 111, 131

T

test access port (TAP) 138
Thermal-Aware Scheduling (TAS) 63
Thread-Level Parallelism (TLP) 215
time-to-digital converter (TDC) 74
total harmonic distortion (THD) 75

U

uninterruptible power supplies (UPS's) 75

V

variable speed drives (VSDs) 75
variable-voltage variable-frequency (VVVF) 75
very high-speed integrated circuit hardware description language (VHDL) 49
very long instruction word (VLIW) 14
Voltage Controlled Oscillator (VCO) 41
Voltage source inverter (VSI) 75

W

worst-case execution time (WCET) 11, 16